Practical **MEMS**

VILLE KAAJAKARI
Louisiana Tech University

small
gear
publishing

Las Vegas, Nevada

Published by Small Gear Publishing
848 N. Rainbow Blvd. #2753
Las Vegas, NV 89107

All brands, company names, trademarks, trade names, and product names used in this book are for reference purpose only and they belong to their respective owners. The information presented is for educational purpose only and does constitute endorsement, advertisement, comparison, or merchantiability of any product.

Authorized Use
No part of this book may be reproduced, stored in a retrieval system, or transmitted in any form or by any means, electronic, mechanical, photocopying, recording, scanning, or otherwise without the prior permission from the author.

Limit of Liability/Disclaimer of Warranty
While the publisher and author have used their best efforts in preparing this book, they make no representations of warranties with respect to the accuracy of this book and specifically disclaim any implied warranties or merchantiability of fitness for a particular purpose. Neither the publisher nor the author shall be liable for any loss of profit or any other commercial damages, including but not limited to special, incidental, consequential, or other damages.

ISBN: 978-0-9822991-0-4

Library of Congress Control Number: 2009900294

Library of Congress subject headings:
1. Microelectromechanical systems. 2. Transducers. 3. Accelerometers. 4. Gyroscopes. 5. Pressure transducers. 6. Actuators. 7. Microfluidics. 8. Integrated circuits.

∞The paper used in this book meets the requirements of the American National Standards Institute/National Information Standards Organization Permanence of Paper for Publications and Documents in Libraries and Archives, ANSI/NISO Z39.48-1992.

Version 1.03

Contents

Preface

This book is primarily for learning the operational principles of microelectrome-chanical devices. The book is written for senior level undergraduates or graduate students as well as practicing microsystem engineers. Although some under-standing of microdevice fabrication is assumed, the book is self-contained. It is intended for engineers from all disciplines, as the book focuses on physical principles of device operation.

The book is fashioned after a textbook I was hoping to read when I first encountered micromechanics. While several textbooks cover the fabrication and the device applications, the analysis of merits and limitations of the microde-vices is lacking in the literature. For example, many books describe how ac-celerometers work and the steps to fabricate them, but quantitative performance analysis is not carried out. The analysis of the expected performance level is es-pecially important in micromechanical systems, as the fabrication tools can be used to make almost any structure. Yet, only a handful of devices (accelerom-eters, pressure sensors, gyroscopes, microphones, and optical mirror displays) have been proven commercially successful.

To address this gap in the existing literature, this text focuses on systematic analysis of large volume or high potential MEMS applications. This book's mission is show why some microdevices have been successful. In addition to the commercially proven applications, "failed" applications are also covered to understand why the large research effort has not translated into profits.

The theory in this book is supported by over 100 analysis examples. By working through the examples, the reader will learn how to do the "back-of-the-envelope" calculations that are invaluable in the complex task of optimizing microsystem designs. In addition, each chapter has a number of carefully se-lected problems. Most problems have been classroom tested and lead the reader to further investigate the exciting microsystem technology.

This book was written over three years while the author was teaching the microsystem design class at the Louisiana Tech University. A special thank you (and apology) to all the students who were forced to participate in debugging this book.

Best efforts have been taken in checking and double checking the book. If you do come across a typo and an error, please let the author know by sending him an e-mail at `ville@kaajakari.net`. Errata and additional supporting material for this book can be found at author's web-site `www.kaajakari.net/PracticalMEMS`.

Ville Kaajakari
Assistant Professor, Louisiana Tech University
Ruston, Louisiana, March 2009

Symbols and units

Any MEMS book faces the challenge of using a consistent system of units. Same symbols are well established in different fields. For example, to electrical engineers R represents the resistance but to chemical engineers R usually represents the gas constant. Inventing a new system of units for the book is not meaningful as the old conventions are well established and familiar to readers.

This book attempts to alleviate the problem by choosing the symbols with a minimal ambiguity and overlap. For example, the Young's modulus is commonly represented with the symbol Y or with the symbol E. In this book, the symbol Y is used to indicate position along the Y-axis so we have chosen to adopt E for the Young's modulus. Similarly, velocity can be represented with either v or $\dot{x}$. Here, we have chosen to use $\dot{x}$ for the velocity and the symbol v represents voltage. Some overlap is unavoidable. For example, symbol ρ may represent either density or electrical resistivity.

This book uses SI units. For example, distances are measured in meters (m) and masses are given in kilograms (kg) which is the base unit in the SI system. Prefixes are added before the base unit. For example nm means 10^{-9} m and μkg means 10^{-6} kg.

List of symbols

A	amplifier gain or transfer function
A	area, m^2
BW	bandwidth, Hz
C	capacitance, F
C_a	analyte concentration, moles/m^3
C_f	fringe capacitance, F
C_{ox}	transistor gate oxide capacitance density, F/m^2
C_m	motional capacitance, F
c_P	heat capacity, J/kg^3
D	diffusion coefficient, m^2/s

d electrode gap, m

d_0 electrode gap at dc-operation point, m

d_g effective gas molecule diameter, m

E Young's modulus, Pa

e piezoelectric coefficient, C/m^2

$\mathcal{E}$ electric field, V/m

F force, N

F_0 amplitude or magnitude of force, N

F_E external force, N

F_e electrical force, N

F_m mechanical force, N

f_0 resonant frequency, Hz

f_L low frequency limit, Hz

f_H high frequency limit, Hz

G shear modulus, Pa

$\mathbf{G}$ acceleration in standard gravity, 9.81 m/s^2

GF gauge factor

g_m transistor small signal transconductance, A/V

$H(s)$ transfer function

h height,m

I second moment of inertia, m^4

I electrical current, A

$I_{\mathbf{dc}}$ dc current, A

i ac current, A

i small signal current, A

$i_{\mathbf{mot}}$ motional current, A

j imaginary number

k_B Boltzmann's constant, $1.380 \cdot 10^{-23}$ J/K

k spring constant, N/m

L inductance, H

L length, m

L transistor gate length, m

L_m motional inductance, H

l length, m

m mass, kg

N_A Avogadro's number, $6.022 \cdot 10^{23}$ 1/mol

n carrier density, m^{-3}

n_V number of gas molecules in unit volume, m^{-3}

P power, W

p pressure, Pa

Q quality factor

q charge, C

q_e electron charge, $1.6 \cdot 10^{-19}$ C

$q_{\mathbf{rms}}$ root mean square charge noise, C

R gas constant, 8.3145 J/mol·K

R resistance, Ω

R_m motional resistance, Ω

$\mathbf{Re}$ Reynold's number

r_i ion radius, m

S strain

s Laplace domain $s = j\omega$

T stress, Pa

T temperature, K

ΔT temperature change, K

t thickness, m

$u(X, Y, Z)$ displacement at position (X, Y, Z)

V voltage, V

$V_{\mathbf{dc}}$ dc voltage, V

V_P pull-in voltage, V

V_T transistor threshold voltage, V

v small signal voltage, V

v ac voltage, V

v_0 amplitude of ac voltage, V

v_n instantaneous noise voltage, V

$\bar{v}_n$ voltage noise spectral density, V/$\sqrt{\text{Hz}}$

$v_{\mathbf{rms}}$ root mean square noise voltage, V

W energy, J

W transistor gate width, m

W_e electrical energy, J

W_k kinetic energy, J

W_p potential energy, J

w width, m

x displacement of lumped element, m

x_0 amplitude of displacement, m

x_0 equilibrium displacement, m

$\dot{x}$ velocity, m/s

$\ddot{x}$ acceleration, m/s^2

x_n instantaneous noise displacement, m

$\bar{x}_n$ displacement noise spectral density, m/$\sqrt{\text{Hz}}$

x_P pull-in point, m

$x_{\mathbf{rms}}$ root mean square noise displacement, m

X, Y, Z location in 3D space, m

X_1, X_2, X_3 location in 3D space, m

Z electrical impedance, Ω

Z_0	characteristic impedance (typically 50 Ω), Ω
z_i	ion charge
α	thermal expansion coefficient, 1/K
β	sensor efficiency
γ	damping coefficient, Ns/m
γ	surface tension, N/m
γ_c	transistor channel noise parameter
ϵ	permittivity ($\epsilon = \epsilon_R \epsilon_0$), F/m
ϵ_0	permittivity of free space, $8.85 \cdot 10^{-12}$ F/m
ϵ_R	relative permittivity
η	electromechanical transduction factor, F/V
κ	thermal conductance, W/m·K
κ	nonlinearity factor, $1/m^2$
λ	mean free path for molecules, m
λ	wavelength, m
μ	viscosity, Pa·s
μ_γ	material loss viscosity, Pa·s
μ_0	permeability free space, $4\pi \cdot 10^{-4}$ NA^{-2}
μ_n	electron mobility in transistors, m^2/Vs
ν	Poisson's ratio
π	piezoresistive coefficient, Pa^{-1}
π_t	transverse piezoresistive coefficient, Pa^{-1}
π_l	longitudinal piezoresistive coefficient, Pa^{-1}
ρ	density, kg/m^3
ρ	electrical resistivity, Ωm
σ	electrical conductance, $1/\Omega$m
τ	time constant, s
ω	angular frequency, Hz
ω_0	resonance frequency ($\omega_0 = 2\pi f_0$), Hz
ω_{0d}	drive-mode resonance frequency, Hz
ω_{0s}	sense-mode resonance frequency, Hz
ω_d	damped oscillation frequency

1

Introduction

Microelectromechanical systems (MEMS) is a loosely defined term for man-made mechanical components that are characterized by small size. Translated literally, MEMS should have dimensions in the micron-scale and have both electrical and mechanical components that form a system. Many MEMS devices do not meet these requirements. For example, microfluidic channels may not have any electrical components. In Europe, MEMS is often called *microsystems*. This term may be more accurate but MEMS is more catchy and is used in United States, Asia, and increasingly in Europe.

While the exact definition of MEMS is difficult to formalize, most people agree on the idea of MEMS. Typically, the MEMS device introduces a paradigm shift in manufacturing and/or application. For example, miniature silicon accelerometers have largely replaced the costly macroscopic piezoelectric accelerometers. The MEMS accelerometers are smaller, but their real advantage is the manufacturing process that utilizes the batch fabrication processes originally developed for the integrated circuit technology. Batch fabrication enables simultaneous processing of thousands of identical devices on a single wafer. This is in contrast to the traditional series manufacturing one device at the time. Batch fabrication has made the accelerometers economical, and with the lower cost of silicon accelerometers, the use of inertial sensors has widened first in the automotive industry and more recently in the consumer market.

In addition to providing a cheaper and/or better alternative to the existing technology, MEMS has enabled completely new devices: Inkjet print heads have made low-cost color printing a reality. Micromirror arrays containing more than one million individual mirrors were developed for high definition television, and are used in data projectors in offices, classrooms, auditoriums, and in homes for video games and home theaters.

This chapter gives an overview of the MEMS industry, the history, the type of devices on market, and the fabrication methods used to make them. The overview will pave the way for the detailed device studies in the subsequent chapters.

1.1 History of MEMS

The history of micromachining is tied to the development of integrated circuit (IC) technology. Starting in the 1960s, researchers experimented with using IC fabrication technologies (for example lithography, silicon etching, and thin film growth) to make mechanical structures. Some of the early devices such as the resonant gate transistor [1] were not commercially successful but the work lead to a commercial adoption of pressure sensors and accelerometers in the 1970s. Many of the early processes and applications are documented in the classical paper by Petersen [2].

The significant research and development effort in the 1980s and 1990s have lead to new fabrication technologies, devices, and markets for MEMS. Notably, surface micromachining has enabled integration of mechanical components with the integrated circuits leading to low cost accelerometers and micromirror arrays. Commercialization of the MEMS technologies has finally started to impact society on a larger scale. Today, most consumers have, knowingly or unknowingly, encountered MEMS based products.

Building on the technological advancements, the MEMS market is currently growing by all measures. The number of MEMS devices sold is increasing and new products are coming out every year. Devices such as microphones that once looked too expensive to implement with microfabrication technologies are sold by millions [3]. While the future for the MEMS is bright, it remains to be seen whether the technology will saturate or if new manufacturing processes and applications will be invented to further drive the development.

1.2 MEMS applications are diverse

The MEMS market size is currently over \$6B per year, but as shown in Figure 1.1, only a few devices are genuinely mass produced. The oldest application, pressure sensors, commands revenue of over \$500M per year. As the price and size of pressure sensors keep decreasing, new applications emerge. For example, integration of pressure sensors inside hypothermic needles [4] or sport watches is possible due to minute size of MEMS pressure sensors.

The other major sensor market, the inertial sensors, has historically been dominated by the automotive industry. Recently, the reduction in price has enabled adoption of MEMS accelerometers in consumer devices such as orientation sensors in digital cameras and game console user interfaces. Gyroscopes have

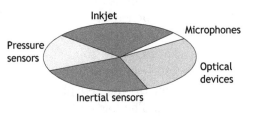

Emerging applications:

- RF resonators
- RF switches
- Lab-on-a-chip
- Drug delivery systems
- Optical switches
- Microspectrometers

Figure 1.1: The MEMS market is dominated by few applications all generating revenues over $500M/year. Emerging device fields currently generate revenue less than $100M/year but show huge growth potentials.

also entered into mass production and show double-digit growth in revenue. The main markets for the gyroscopes are the automotive industry, inertial navigation to aid GPS, and image stabilization for digital cameras. It is expected that the gyroscopes will also be utilized to enhance the human-computer user interface.

For average consumers, the inkjet print heads may be the most familiar microdevice. Each replacement inkjet cartridge has a micromachined inkjet nozzle head. The inkjet print heads are frequently regarded as the largest MEMS market in terms of revenue. However, as the inkjet manufacturers sell the MEMS component as a part of the printer system, it is difficult to attach an accurate dollar value to the MEMS component portion of the inkjet market. Regardless, the inkjet revenue is measured in hundreds of millions.

The lucrative digital microdisplay (DMD) market is dominated by a single manufacturer, Texas Instruments, who holds the key patents in the field. In projection displays, the high contrast ratio of mechanically actuated mirrors enables the micromirrors to compete against the more common LCD technology. The MEMS displays are a unique MEMS product in that they contain millions of moving structures. Fabrication on this scale would not be possible without batch fabrication methods.

Another more recent display product is the reflection based display for portable devices introduced by Qualcomm. The device is based on interferometry, and unlike LCD displays, it does not require any back light. Other optical MEMS devices, such as switches for fiber optical communications, hold promise but may take several years to gain acceptance.

Silicon microphones are the latest entry to the mass market. The growth is driven by cell phone industry that demands solderability – a characteristic not met by otherwise excellent traditional electret microphones. The microphones are an encouraging example of a MEMS product that only a few years ago was deemed too expensive but is now rapidly gaining market share.

Beyond these established markets, a number of MEMS devices hold promise.

The optical switches were regarded the next killer application but the collapse of several network companies in 2001-2002 has dampened the interest in optical networking. Radio frequency (RF) switches is an interesting application as micromechanical switches offer performance advantages over solid-state devices. Currently, the MEMS switches are considered for radars and test equipment where the high performance is needed. Adoption by the cell phone industry appears likely but lifetime and cost issues remain. The low cost and the small size of RF microresonators have also raised interest. This market is attractive as the revenue is high, but in terms of power handling and signal-to-noise ratio, the miniature resonators are not as robust as the established macroscopic resonators. Biomedical devices are appealing, as the small size naturally interfaces well with biological systems. It remains to be seen whether these or some other application will break the $100M barrier.

1.3 MEMS fabrication is based on batch processing

Microfabrication has historically been tied to integrated circuit (IC) fabrication and most of the processing tools and terminology have been adopted directly from IC manufacturing. The parallels are so deep that sometimes a retiring IC manufacturing plant is converted to MEMS fabrication that does not require the latest fabrication technology. But the MEMS fabrication technology is not easy to master. The MEMS specific challenges include packaging of movable mechanical structures, manufacturing of thick structures, and obtaining good absolute dimensional control.

The focus of this book is on device design but some exposure to fabrication technologies is necessary in order to understand the limitations of the technology. In other words, the understanding of the fabrication is not required to explain how a particular MEMS device operates but it explains why the device looks the way it does. The overview given here is enough to explain the general fabrication steps for the devices covered in this book. To supplement the fabrication overview in this book, there are a number of good introductory [5–8] and advanced [9,10] books about microfabrication. In addition, microfabrication has been reviewed in several journal papers that provide concise introduction to the field [2, 3, 11, 12]. These books and review papers combined with the large choice of IC fabrication textbooks [7, 13, 14], provide a solid foundation for micromanufacturing and process integration.

The batch fabrication is a radical departure from the traditional series manufacturing, and is well suited for making relatively simple, mechanical components on a large scale. As illustrated in Figure 1.2, individual devices are photolithographically defined onto a wafer using a photomask and ultraviolet light. The batch fabrication process, specifically the use of photolithography, allows the defining of any shape on the surface of the wafer but it is difficult to fab-

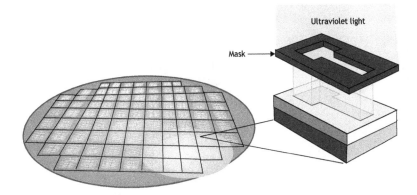

Figure 1.2: An illustration of the batch fabrication. Thousands of components can simultaneously be defined on a single wafer using photolithography.

ricate truly three-dimensional shapes. The process can be compared to carving shapes out of cardboard. Due to fabrication limitations, the MEMS components often look flat or two-dimensional.

The cost for processing the MEMS wafer does not depend on the number of devices on it and the batch fabrication is an economical way to make a large number of devices. On a typical wafer, there can be thousands of devices and even a small MEMS foundry can fabricate components by millions. Once all the processing steps have been completed, the wafer is diced into individual pieces or dies. Finally, the dies are packaged, often together with an IC.

Numerous MEMS fabrication processes have been developed. Traditionally, the MEMS processes have been divided into surface micromachining and bulk micromachining. We will review these two technologies in the following sections.

1.3.1   Surface micromachining makes thin structures

Surface micromachining is based on patterning thin films on top of a substrate wafer [11]. The surface micromachined structures are relatively flat which simplifies the subsequent wafer processing. A typical fabrication process is illustrated in Figure 1.3 where steps of thin film deposition followed by selective etching are repeated to form semi-3D structures. The thickness of each layer may vary but it is typically less than 5 μm. The simplest structures, such as accelerometers, have two structural layers and one sacrificial layer as shown in Figure 1.3. The record in complexity is the five structural layer process by Sandia National Laboratories that was developed for complex moving devices (microengines and gears) [15].

The surface micromachining resembles the traditional IC manufacturing that is also based on processing thin films on a silicon wafer. The compatibility with the IC processing is one of the main advantages of surface micromachining,

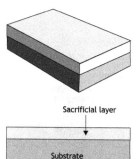
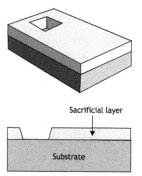

(a) Surface micromachining starts with a substrate wafer, typically silicon with diameter of 100-200 mm and thickness of 500-700 μm.

(b) A sacrificial layer is deposited or grown on the wafer. Silicon dioxide with thickness of 1-2 μm is commonly used.

(c) A hole is made by lithography and etching of the sacrificial layer. The smallest dimensions are usually 2-3 μm.

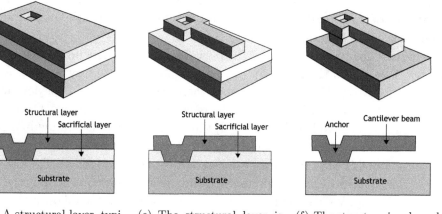

(d) A structural layer, typically 1-5 μm polycrystalline silicon is deposited.

(e) The structural layer is defined using lithography and etching.

(f) The structure is released by removing the sacrificial layer by etching.

Figure 1.3: Typical surface micromachining process involves combinations of layer depositions, optical lithography, and etches to fabricate thin microstructures.

as it is relatively easy to integrate surface micromachining with IC:s to combine mechanical and electrical components on the same chip. The single-chip integration may lead to better performance and reduced packaging cost especially if a large number of connections between the mechanical and electrical parts are needed. For example, the realization of micromirror arrays with more than a million individually controlled mirrors would not be possible without on-chip control for the individual mirrors.

1.3.2   Bulk micromachining makes thick structures

Unlike surface micromachining, which is based on thin film deposition on top of a substrate wafer, bulk micromachining defines structures by selectively etching the substrate [12]. This can result in relatively thick structures; the typical wafer thickness of 500-700 μm is about 100 times the typical thickness for surface micromachines. The large thickness is useful, for example, in inertial sensors that benefit from a large mass. In addition to the thickness, the bulk micromachined structures can be made of single-crystal silicon as opposed to amorphous or polycrystalline thin films. The predictable and stable material parameters of crystalline silicon are desirable for mechanical sensors.

Bulk micromachined accelerometers and pressure sensors were the first commercialized MEMS products. These devices have been hugely successful, for example, MEMS pressure sensors represent over 90% of all sold pressure sensor units [3]. Early bulk micromachined devices were made by wet etching [16] and several exotic processes have been developed. For example, thin membranes for pressure sensors have been defined using epitaxial growth of silicon combined with electrochemical etch stop [2]. The wet processes are still used but advances in plasma processing have made dry etching the mainstream. Especially the combination of deep reactive ion etching (DRIE) and silicon on insulator (SOI) technology has simplified the bulk manufacturing and reduced the device size [17–19].

DRIE enables etching narrow channels through the entire wafer and results in almost vertical sidewalls. It is possible to make channels with aspect ratio of 50 to 1 or better, meaning that a 500 μm deep trench can be only 10 μm wide. In contrast to the wet etching where the depth and width of a trench are typically equal, the reduction of device size can be significant and more devices are obtained from a single wafer. As the cost of processing one wafer is approximately constant, the smaller device size directly reduces the device cost.

Figure 1.4 shows a possible process to make accelerometers using SOI wafers, DRIE etching, and wafer bonding. The final structure is hermetically sealed at the wafer level, which greatly reduces the cost of final assembly and packaging. The manufacturing process is quite straightforward and results in a compact structure with well-defined features.

The SOI wafers used in MEMS are manufactured by bonding two silicon wafers together with a 1-2 μm layer of silicon dioxide in between them. The silicon dioxide acts as a natural etch stop and all etched structures have the desired thickness determined by the SOI thickness. The silicon dioxide can also be used as a sacrificial layer for making free structures of single-crystal silicon. As two silicon wafers are used for making one SOI wafer, the SOI wafers are more expensive than bare silicon wafers. However, the material cost increase is compensated by the processing costs savings.

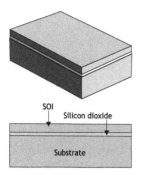
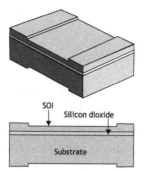
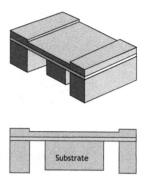

(a) The process is started with a SOI wafer with substrate thickness of 550-650 μm and SOI thickness of 10-20 μm.

(b) A 2 μm recess is etched on front and back to define where movable structure is.

(c) Trenches are etched using DRIE on the back of the wafer to define the sensing mass. The oxide stops the etch.

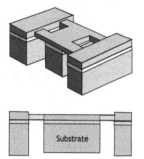
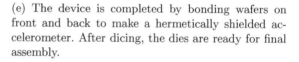

(d) Final DRIE on the front of the wafer defines the supporting beams that hold the mass.

(e) The device is completed by bonding wafers on front and back to make a hermetically shielded accelerometer. After dicing, the dies are ready for final assembly.

Figure 1.4: Bulk micromachining process for making a silicon accelerometer.

1.4   Introduction to the *Practical MEMS* book

This book is focused on in-depth analysis of microdevice operation. The emphasis sets the *Practical MEMS* book apart from other MEMS books that cover both the fabrication and device operation. The integrated approach of including both fabrication and analysis has merits and an integrated textbook is a good first introduction to the microsystems. However, the depth of analysis in textbooks that cover both fabrication and applications is limited to describing the device operation.

This book goes further into exploring why certain devices are successful and others have failed. The first part of this book covers the traditional microsensors (accelerometers and pressure sensors) that are a major and growing commercial microsystem application. The emphasis is on measuring small signals that is a fundamental challenge when making small sensor systems. Since ability to do simplified analytical design analysis is invaluable in the early stage of any sensor design, the physical principles behind the sensor operation are illustrated by numerous calculated examples. These examples are carefully chosen to both illuminate the device operation and to quantify the performance trade-offs in the microsensors.

The second half of the book introduces actuators. The merits of different actuation schemes are illustrated by developing scaling laws for different actuation schemes. Capacitive, thermal, piezoelectric actuation theory is developed and illustrated with examples. Applications ranging from optical, RF, and sensing (gyroscopes) are explored with emphasis on critical evaluation of whether MEMS has a competitive advantage to replace the current technologies. Again, the physical challenges of miniaturization are illustrated with several calculated examples. For example, the effect of mirror size is studied in optical MEMS applications such as optical displays and microscanners.

Finally, the book is concluded with an introduction to MEMS fabrication economics. The cost, yield, and profits in batch fabrication are investigated. Several case studies are used to illustrate the challenges of making a profit with microfabrication.

Key concepts

- MEMS stands for microelectromechanical system. Europeans prefer the shorter and often more accurate word *microsystem*.

- Microdevices have been developed since the 1960s. Since the 1990s, the field has been growing rapidly.

- Pressure sensors, accelerometers, gyroscopes, microphones, optical displays, and inkjet printers are established commercial MEMS applications.

- Optical networking and RF MEMS hold commercial promise but have not yet become significant industry.

- Batch fabrication enables fabrication of thousands or even millions of identical mechanical components on a single wafer.

- Optical lithography is used to define the shape of the structures. Large number of devices can be made using the same optical mask.

- Surface micromachining is based on processing and patterning thin films on a wafer.

- Bulk micromachining is based on etching the wafer to make relatively thick structures.

Exercises

Exercise 1.1
Obtain the classical review paper from 1982 by Petersen [2] and recent review by Bryzek *et al.* [3] and answer to the following questions: (1) How does the MEMS fabrication processes in the papers differ? (2) What applications are highlighted in papers? (3) How does applications in the two papers differ and is this difference reflected in manufacturing processes?

Exercise 1.2
Using your favorite search engine, find at least three estimates for the world wide MEMS market size. Comment on how reliable you feel the sources are and whether there is discrepancy between the estimates.

Exercise 1.3
Using your favorite search engine, find an estimate of the world wide market size for integrated circuits (ICs) and compare it to the MEMS market size. Noting that silicon microcircuits and silicon microsensors were invented around the same time in the 1960s, think of reasons that could explain the difference in the market size.

Exercise 1.4
Why is surface micromachining more compatible with integrated circuit fabrication than bulk micromachining?

Exercise 1.5
List MEMS applications that you have personally encountered.

Exercise 1.6
Why is optical lithography important in microfacrication?

Exercise 1.7
The number of citations is a relatively objective way to judge the importance of academic publications. Most publications are cited less than ten times, the papers that resonate well with the academic community will be cited more than 30 times, and seminal papers receive over a hundred citations.

Google Scholar (`scholar.google.com`) is a free tool to search scholarly literature. Google Scholar will also give an estimate of the number of citations for each search result has received. The more cited articles will have a higher ranking and will appear first. For example, search "RF MEMS" will display the article by C. Goldsmith, *et al.* titled "Performance of low-loss RF MEMS capacitive switches," that has been cited more than 200 times.

Try out Google Scholar to find a highly cited journal paper on: i. MEMS accelerometer, ii. MEMS pressure sensor, iii. MEMS gyroscope, iv. RF MEMS switch, v. optical MEMS, vi. MEMS microphone, and vii. MEMS inkjet print head. Note the number of citations manuscripts in different applications have received and compare the search results. How does the importance to the academic community correlate with the commercial interest? Note that in old applications such as accelerometers the word MEMS may not appear in the article. Other relevant keywords include silicon and solid-state.

2

Noise in micromechanical systems

Miniaturization of mechanical sensors is not limited by our ability to fabricate small structures. Researchers have demonstrated nanomechanical devices that look like micromechanical devices but are smaller. The challenge is in obtaining small structures that perform a useful function. In sensors, the noise sets the limit for the smallest acceptable size. Hence, it is natural to start this book by reviewing the noise in electrical and mechanical systems.

It is easy to understand why noise is so important in micro- and nanodevices when we note that the thermal noise energy $k_B T$ is constant irrespective of the system size [20,21]. However, the signal power is usually lowered when the sensor size is reduced. For small enough devices, the signal level falls below the thermal noise floor. The small size, however, may be desirable for variety of reasons. For example, it may be of interest to obtain higher resonance frequency or simply to lower the cost by incorporating more devices on a silicon wafer. Hence, the understanding of the noise limitations is crucial for optimal sensor design.

In this chapter, the noise in electrical and mechanical systems is covered. The focus is on the thermal noise as it is present and sets the fundamental limit for the smallest measurable signal. In addition, the low-frequency $1/f$-noise is also covered due to its prevalence in practical measurements. The theory is illustrated with the analysis of noise in electrical and mechanical systems.

2.1 Noise as a statistical quantity

Noise is random fluctuation of electrons, atoms, or molecules. The random motion of the atomistic particles results in measurable noise on a micro- and macro-scale. For example, the random motion of electrons in conductors results in voltage noise. In mechanical domain, the thermal vibrations of atoms in a mass results in random fluctuations of the mass position. As the electrical noise is the most familiar form of noise, we will use the noise voltage to illustrate the noise statistic. The results of this section can be applied to other noise variables such as noise induced displacement or noise current.

Because of noise, it is not possible to predict the instantaneous value for the measured signal. However, the probability distribution for the measurement is usually well characterized. If we know the characteristics of the noise, we can calculate the probability that the voltage falls within a certain range. For most physical noise sources, the probability follows the Gaussian distribution

$$f(v_n) = \frac{1}{v_{\rm rms}\sqrt{2\pi}} e^{-\frac{v_n^2}{2v_{\rm rms}^2}}, \tag{2.1}$$

where $v_{\rm rms}$ is the standard deviation from the average voltage. The probability that the measured noise voltage v_n is between v_1 ans v_2 is obtained by integrating Equation (2.1)

$$P = \int_{v_1}^{v_2} f(v_n) {\rm d}v_n. \tag{2.2}$$

Figure 2.1 shows an example of thousand voltage measurements of noise with $v_{\rm rms} = 2$ nV. The probability of noise being within $v_{\rm rms}$ of the mean voltage is 68% and the probability of measuring noise voltages over $3v_{\rm rms}$ is less than 0.3%.

The standard deviation $v_{\rm rms}$ is also called the rms-voltage where the "rms" stands for "root mean square". Statistically, if a large number of measurements are made, the square root of the mean of the squared voltages approaches $v_{\rm rms}$. Mathematically, the definition is

$$v_{\rm rms} = \lim_{N\to\infty} \sqrt{\frac{1}{N}\sum_{n=1}^{N} v_n^2}, \tag{2.3}$$

where v_n are the individual noise voltage measurements.

The noise sources in this book are assumed uncorrelated unless otherwise stated. For a system with several noise sources, the powers of the uncorrelated sources are summed. For example, if two noise sources produce uncorrelated noise voltages $v_{\rm rms,1}$ and $v_{\rm rms,2}$, the total noise is

$$v_{\rm rms,tot} = \sqrt{v_{\rm rms,1}^2 + v_{\rm rms,2}^2}. \quad \text{(uncorrelated)} \tag{2.4}$$

For correlated sources, we sum the noise voltages

$$v_{\rm rms,tot} = v_{\rm rms,1} + v_{\rm rms,2}. \quad \text{(correlated)} \tag{2.5}$$

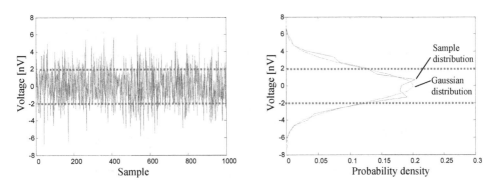

Figure 2.1: Example of Gaussian noise with $v_{\mathrm{rms}} = 2$ nV. The cumulative distribution of 1,000 samples approaches the true Gaussian and 68% of the measurements are within v_{rms} of the mean.

2.2 Noise in frequency domain

The rms-noise characterizes the noise amplitude in time domain but gives no insight into what frequency the noise appears. The power spectral density $\overline{v_n^2}$ and spectral density $\overline{v}_n \equiv \sqrt{\overline{v_n^2}}$ tell how much, on average, noise per unit bandwidth there is on a certain frequency. Knowing the noise frequency spectrum allows calculating the effect of system bandwidth on total noise. The rms-noise is obtained from integrating the power spectral density

$$v_{\mathrm{rms}} = \sqrt{\int \overline{v_n^2}(f)\mathrm{d}f}, \tag{2.6}$$

where the integration is carried out over all the frequencies. The two most important frequency distributions for the spectral density are the white noise and the $1/f$-noise. We will cover these two noises in the next two sections.

2.2.1 White noise

White noise is independent of frequency and the noise spectral density is constant $\overline{v}_n(f) = \overline{v}_n$. Due to finite bandwidth of any real system, the measured noise is not white but is always shaped by the system transfer function $H(f)$ as illustrated in Figure 2.2. Mathematically, the measured noise is

$$\overline{v_{n,\mathrm{out}}^2}(f) = |H(f)|^2 \overline{v_n^2}, \tag{2.7}$$

where

$$H(f) \equiv \frac{v_{\mathrm{out}}}{v_{\mathrm{in}}} \tag{2.8}$$

is the system transfer function that relates the input voltage to the output voltage.

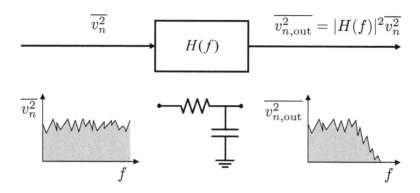

Figure 2.2: Noise is shaped by the system transfer function $H(f)$. All practical measurements have limited noise bandwidth due to capacitances, inductances, and resistances that shape the frequency response.

The output rms-noise is obtained by integrating the noise power spectral density given by Equation (2.7) over all frequencies

$$v_{\mathrm{rms}} = \sqrt{\int |H(f)|^2 \overline{v_n^2}\,\mathrm{d}f} = \overline{v_n}\sqrt{\int |H(f)|^2\,\mathrm{d}f}, \qquad (2.9)$$

where we have taken the constant $\sqrt{\overline{v_n^2}} = \overline{v_n}$ in front of the integral. Defining the noise bandwidth as

$$BW = \frac{1}{|H_{pk}|^2} \int |H^2(f)|\,\mathrm{d}f, \qquad (2.10)$$

where $|H_{pk}|$ is the peak value of the transfer function allows us to write Equation (2.9) as

$$v_{\mathrm{rms}} = |H_{pk}|\sqrt{BW}\,\overline{v_n}. \qquad (2.11)$$

Equation (2.11) is useful for quick evaluation of rms-noise due to noise density $\overline{v_n}$, noise bandwidth $\sqrt{BW}$, and gain $|H_{pk}|$.

The most common white noise mechanism is the thermal noise. Strictly speaking, the thermal noise is not constant at very high frequencies ($f > 80$ THz). Equation (2.7) explains why the white noise approximation holds so well in practice: the measurement systems always have a finite bandwidth and even the measurement wires effectively filter the noise at very high frequencies ($f > 100$ Ghz). Thus, the thermal noise is white at all frequencies of interest.

Example 2.1: Noise spectrum and rms-noise
Problem: An amplifier input has noise density $\overline{v_n} = 4 \text{ nV}/\sqrt{\text{Hz}}$. What is the rms-noise voltage at the amplifier output if the amplifier gain is $|H_{pk}| = 10$ and the noise bandwidth is $BW = 10$ kHz?
Solution: The rms-noise from Equation (2.11) is

$$v_{\text{rms}} = |H_{pk}|\sqrt{BW}\,\overline{v_n} = 4 \ \mu\text{V}.$$

Example 2.2: The rms-noise in RC filter
Problem: Consider a single-pole RC low-pass filter in Figure 2.2 with the transfer function $H(f) = 1/(1 + jf/f_{-3 \text{ dB}})$ where the -3-dB bandwidth is $f_{-3 \text{ dB}} = 1/2\pi RC$. Derive expressions for the noise bandwidth and evaluate it for $f_{-3 \text{ dB}} = 10$ kHz. What is the noise spectral density and rms-noise voltage at the output for a white noise source with $\overline{v_n} = 10 \text{ nV}/\sqrt{\text{Hz}}$ at the filter input if the resistor noise is ignored?
Solution: The peak of the transfer function is $|H_{pk}| = 1$. From Equation (2.10), the noise bandwidth is

$$BW = \frac{1}{|H_{pk}|^2} \int_0^\infty |H(f)|^2 \mathrm{d}f = \frac{1}{1^2} \int_0^\infty \frac{1}{1 + \left(\frac{f}{f_{-3 \text{ dB}}}\right)^2} \mathrm{d}f = \frac{\pi}{2} f_{-3 \text{ dB}}.$$

The noise band-width is $BW = \pi f_{-3 \text{ dB}}/2$ or about 1.57 times the -3-dB band-width $f_{-3 \text{ dB}}$ due to the slow roll-off of single-pole filter. For higher order filters with steeper roll-off, the noise bandwidth is closer to the -3-dB bandwidth. The noise bandwidth $f_{-3 \text{ dB}} = 10$ kHz is $BW = \pi f_{-3 \text{ dB}}/2 \approx 15.7$ kHz. The rms-noise at the output is

$$v_{\text{rms}} = \overline{v_n}|H_{pk}|\sqrt{BW} = 1.25 \ \mu\text{V}.$$

The noise spectral density at the output is

$$\overline{v_n^2}(f) = |H(f)|^2 \overline{v_n^2} = \frac{\overline{v_n^2}}{1 + (f/f_{-3 \text{ dB}})^2} = \frac{10^2 \text{ nV}^2/\text{Hz}}{1 + (f/10^4 \text{ Hz})^2}.$$

Thus, the noise spectral density at low frequency ($f \ll f_{-3 \text{ dB}}$) is $\overline{v_n} = \sqrt{\overline{v_n^2}} = 10 \text{ nV}/\sqrt{\text{Hz}}$.

2.2.2 $1/f$-noise

The $1/f$-noise, also known as the flicker noise, is inversely proportional to the frequency. As the noise increases with decreasing frequency, the $1/f$-noise is especially troublesome at low frequencies. The low-frequency $1/f$-characteristic is surprisingly universal but the underling noise mechanism varies. In electrical components, the $1/f$-noise is typically caused by random trapping and release of electrical carriers. In the very high quality factor mechanical quartz resonators, the $1/f$-noise is attributed to nonlinear forces in quartz crystals and surface related effects [22, 23]. For the reported MEMS devices, the electrical $1/f$-noise dominates but advancements in measurement techniques and scaling to nano-scale could make the mechanical $1/f$-noise significant.

The $1/f$-noise can be written as

$$\overline{v_{1/f}} = \sqrt{\overline{v_n^2}\frac{f_c}{f}}, \tag{2.12}$$

where f_c is the $1/f$-corner frequency defined as the frequency where the $1/f$-noise and the white noise $\overline{v_n^2}$ have equal amplitude as illustrated in Figure 2.3. The corner frequency varies from device to device and may depend on operating conditions. A model for the $1/f$-noise in resistors is given in Section 5.4 and the transistor noise model is covered in Section 9.1.

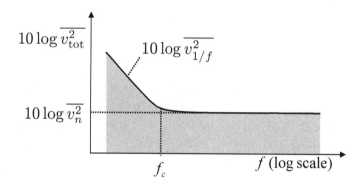

Figure 2.3: Below the corner frequency f_c, the $1/f$-noise is greater than the thermal noise.

The rms-noise due to the $1/f$-noise is obtain by integrating Equation (2.12) over the frequency band of interest

$$v_{\text{rms}}^2 = \int_{f_L}^{f_H} \overline{v_{1/f}^2}\,\mathrm{d}f = \int_{f_L}^{f_H} \overline{v_n^2}\frac{f_c}{f}\,\mathrm{d}f = \overline{v_n^2}f_c \ln f_H/f_L, \tag{2.13}$$

where f_L and f_H are the lower and upper frequency limit, respectively. The lower frequency of integration f_L raises questions: If the lower frequency limit is taken as zero, Equation (2.13) yields infinite value for the rms-noise. However, the assumption $f_L = 0$ corresponds to an infinitely long measurement time and is not realistic. Taking the lower limit to be 0.01 Hz would correspond to measurement time of $t \sim 1/f_L = 100$ s. Due to the logarithmic dependency, the total noise is not a strong function of f_L. As shown in Table 2.1, increasing the f_L beyond one week has only a small effect in the total noise. Clearly, the v_{rms} stabilizes for practical measurements as the measurement time increases.

Table 2.1: The integrated $1/f$-noise with $f_H = 1$ kHz.

f_L[Hz]	Time span	$\ln f_H/f_L$
1	1 s	6.9
10^{-2}	1 min 40 s	12
10^{-4}	3 hours	16
10^{-6}	12 days	21
10^{-8}	3 years	25

Although the $1/f$-noise is prevalent in low frequency electrical measurements, it is not a fundamental noise limit: The $1/f$-noise can be reduced with a proper design or the measurements can be carried out at frequencies above the corner frequency where the thermal noise dominates. However, the effort of reducing the $1/f$-noise can be costly and often the $1/f$-noise sets the practical noise limits.

Example 2.3: The rms-noise and $1/f$-noise
Problem: Piezoresistive sensor has the thermal noise density $\overline{v_n} = 2$ nV/$\sqrt{\text{Hz}}$ and the $1/f$-corner frequency $f_c = 4$ kHz. Calculate the total sensor noise for the 0.1 Hz to 10 kHz frequency band.
Solution: The rms-noise due to thermal noise is

$$v_{\text{rms},1} = \overline{v_n}\sqrt{BW} = 2 \text{ nV}/\sqrt{\text{Hz}} \cdot \sqrt{10^4 \text{ Hz}} = 0.2 \ \mu\text{V}.$$

The rms-noise due to $1/f$-noise is

$$v_{\text{rms},2} = \sqrt{\overline{v_n^2} f_c \ln f_H/f_L} = \sqrt{4 \text{ nV}^2/\text{Hz} \cdot 4 \text{ kHz} \ln 10^4/10^{-1}} = 0.43 \ \mu\text{V}.$$

The $1/f$-noise and thermal noise are uncorrelated and the noise powers add. The total noise is

$$v_{\text{rms}} = \sqrt{v_{\text{rms},1}^2 + v_{\text{rms},2}^2} = 0.47 \ \mu\text{V},$$

which is dominated by the $1/f$-noise.

2.3   Equipartition theorem and noise

The equipartition theorem of thermodynamics states that the average thermal energy for each degree of freedom is $W = \frac{1}{2}k_B T$, where $k_B = 1.38 \cdot 10^{-23}$ J/K is the Boltzmann's constant. A degree of freedom is defined as capable of storing energy independently of other variables. For example, ideal gas molecules have three degrees of freedom, velocities in X-, Y-, and Z-directions, and harmonic resonator has two degrees of freedom, velocity of the mass and position of the spring. The mechanical motion due to thermal fluctuations was first observed by Mr. Brown, a botanist who studied pollen under a microscope. Consequently, the mechanical noise is sometimes referred to as the *Brownian motion*. In electrical components, the thermal noise was first explained by Johnson and Nyquist in 1928 and the thermal noise often called Johnson-Nyquist noise [24].

Here the analysis of thermal noise is covered in electrical and mechanical systems. In both systems, the underlying physics are the same. Consequently, our analysis follows the same steps.

2.3.1   Thermal noise in electrical systems

In an electrical network, the number of degrees of freedom is the number of independent inductors plus the number of independent capacitors. By independent, we mean inductors that are not connected directly in series with other inductors and capacitors that are not connected directly in parallel with other capacitors. In this case, each component can store energy independently of other components.

From the equipartition theorem, we can calculate the rms-voltage over each capacitor and the rms-current through each inductor. The average thermal energy stored in a capacitor is

$$W_C = \frac{1}{2}k_B T. \tag{2.14}$$

The energy due to a voltage v_{rms} stored in the capacitor is

$$W_C = \frac{1}{2}C v_{\text{rms}}^2. \tag{2.15}$$

Equations (2.14) and (2.15) give the rms-noise voltage on the capacitor

$$v_{\text{rms}} = \sqrt{\frac{k_B T}{C}}. \tag{2.16}$$

As Equation (2.16) shows, the rms-noise over each independent capacitor depends only on capacitance value and temperature.

The average energy stored in an inductor is

$$W_L = \frac{1}{2}L i_{\text{rms}}^2 = \frac{1}{2}k_B T \tag{2.17}$$

and the rms-current is therefore

$$i_{\text{rms}} = \sqrt{\frac{k_B T}{L}}. \tag{2.18}$$

Again, the noise current depends only on the inductance value and temperature. Figure 2.4 illustrates how Equations (2.16) and (2.18) can be used to calculate the rms-voltages over capacitors and the rms-currents through inductors.

The frequency spectrum of the noise depends on the resistances (damping) in the system. The noise spectral density is obtained by associating each resistor with a white noise voltage generator

$$\overline{v_n^2} = 4k_B T R \tag{2.19}$$

as illustrated in Figure 2.4. The noise voltage generator is just another manifestation of the thermal noise. As is shown in Appendix D, the noise voltage generator given by Equation (2.19) follows directly from Equations (2.16) and (2.18). Alternatively, the rms-noise voltages and currents can be obtained by calculating the currents and voltages due to the noise generator(s) and integrating over the frequency. This is illustrated in Example 2.4.

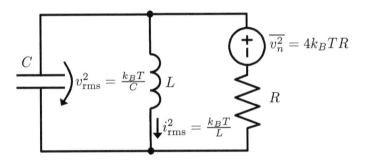

Figure 2.4: The rms-noise voltages and currents are obtained directly from the equipartition theorem. The noise spectral densities are obtained by associating each resistor with noise voltage generator $\overline{v_n^2} = 4k_B T R$.

Example 2.4: Noise in RC circuit
Problem: Consider the RC circuit shown in Figure 2.5. Calculate the noise spectrum and rms-noise over the capacitor.

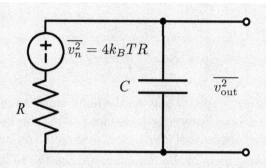

Figure 2.5: The noise in RC circuit.

Solution: The voltage over capacitor due to the resistor noise generator $\overline{v_n}$ is

$$\overline{v_{n,\text{out}}} = \frac{1}{|1 + RCs|}\overline{v_n} \equiv |H(s)|\overline{v_n},$$

where $s = j\omega$. The power spectral density is

$$\overline{v_{n,\text{out}}^2} = |H(\omega)|^2\overline{v_n^2} = \frac{1}{1 + R^2C^2\omega^2}\overline{v_n^2} = \frac{4k_BTR}{1 + R^2C^2\omega^2}.$$

The rms-noise voltage is obtained by integrating the power spectral density to give

$$v_{\text{rms}}^2 = \int_0^\infty \overline{v_{n,\text{out}}^2}\,\mathrm{d}f = \int_0^\infty \frac{4k_BTR}{1 + R^2C^2\omega^2}\,\mathrm{d}f = \frac{k_BT}{C}.$$

This result could also have been obtained directly from Equation (2.16).

The total noise at the output is independent of the value of R and depends only on C. For example, with a 1-pF capacitor, the rms-noise voltage is 64 μV at $T = 300$ K. The 1-pF capacitor corresponds to a typical input capacitance of a low-noise CMOS amplifier. The fact that the total noise can only be decreased by increasing the capacitance introduces difficulties for amplifier design as large capacitance requires both a large silicon area and reduces the amplifier bandwidth.

An alternative representation for the noise voltage generator is shown in Figure 2.6. According to Norton's theorem, the voltage generator $\overline{v_n}$ can be represented with a current generator

$$\overline{i_n} = \frac{\overline{v_n}}{R} = \sqrt{\frac{4k_BT}{R}}. \tag{2.20}$$

The current source generator given by Equation (2.20) is often more convenient for circuit analysis but the results from Equations (2.19) and (2.20) are identical.

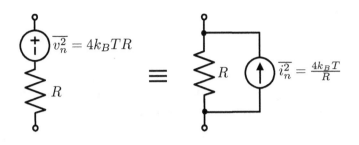

Figure 2.6: Equivalent representations for the thermal noise generator.

2.3.2 Thermal noise in mechanical systems

In mechanical devices, the number of degrees of freedom is the number of ways energy can be stored. For example, gas atoms can have kinetic energy in three directions (X, Y, and Z) so there are three degrees of freedom. Each degree of freedom has, on average, thermal energy $\frac{1}{2}k_BT$ and the total thermal energy for a gas atom is $\frac{3}{2}k_BT$. Here, we are mainly concerned about mechanical resonators. The one dimensional resonator shown in Figure 2.7 can have kinetic energy stored in the mass and potential energy stored in the spring. As there are two degrees of freedom, the total thermal energy is k_BT. The equipartition theorem states that, on average, the thermal energy is equally divided between the potential and kinetic energies.

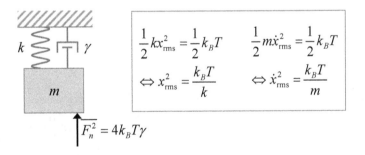

Figure 2.7: The rms-noise velocities and displacements are obtained directly from the equipartition theorem. The noise spectral densities are obtained by associating damper with noise force generator $\overline{F_n^2} = 4k_BT\gamma$.

From the thermal energy, we can calculated the rms-noise velocity $\dot{x}_{\mathrm{rms}}$ and the rms-noise displacement x_{rms}. The average kinetic energy is

$$W_{\mathrm{kin}} = \frac{1}{2}m\dot{x}_{\mathrm{rms}}^2 = \frac{1}{2}k_BT \tag{2.21}$$

and the rms-velocity is therefore

$$\dot{x}_{\text{rms}} = \sqrt{\frac{k_B T}{m}}. \tag{2.22}$$

This rms-noise velocity depends only on the mass and the temperature. The average potential energy stored in the spring is

$$W_{\text{pot}} = \frac{1}{2} k x_{\text{rms}}^2 = \frac{1}{2} k_B T \tag{2.23}$$

and the rms-displacement is therefore

$$x_{\text{rms}} = \sqrt{\frac{k_B T}{k}}. \tag{2.24}$$

The noise displacement depends only on the value of spring constant and temperature.

The frequency content of the noise depends on the mechanical damping in the system. The noise spectral density is obtained by associating the damper with a noise force generator

$$\overline{F_n^2} = 4 k_B T \gamma. \tag{2.25}$$

Notice the similarity between noise generators for resistors and mechanical dampers given by Equations (2.19) and (2.25), respectively. As is shown in Appendix D, Equation (2.25) follows directly from the equipartition theorem. Hence, the rms-noise displacements and velocities can also be obtained from Equation (2.25) as is illustrated in Example 2.5.

Example 2.5: Velocity of floating silicon beads
Problem: Silicon beads are suspended in a liquid. The frictional force affecting the beads is proportional to the bead velocity. Calculate the rms-velocity for the beads due to Brownian noise as a function of beam diameter and evaluate it for bead diameters ranging from 1 cm to 10 nm.
Solution: The equation of motion for the beads is

$$m\ddot{x} + \gamma\dot{x} = F,$$

where m is the bead mass, γ is the damping coefficient, and F is the force acting on the beads. The transfer function between the velocity $\dot{x}$ and the force F is

$$\dot{x}/F = \frac{1}{ms + \gamma} \equiv H(s).$$

The power spectral density for bead velocity due to the thermal noise generator is

$$\overline{\dot{x}_n^2} = |H(s)^2| \overline{F_n^2} = \frac{\overline{F_n^2}}{\omega^2 m^2 + \gamma^2} = \frac{4 k_B T \gamma}{(2\pi f)^2 m^2 + \gamma^2},$$

where we have used $\overline{F_n^2} = 4 k_B T \gamma$ and $\omega = 2\pi f$. The rms-noise velocity is obtained by integrating the power spectral density to give

$$\dot{x}_{\text{rms}}^2 = \int_0^\infty \overline{\dot{x}_n^2} \, \mathrm{d}f = \int_0^\infty \frac{4 k_B T \gamma}{\omega^2 m^2 + \gamma^2} \, \mathrm{d}f = \frac{k_B T}{m}.$$

This result could also have been obtained directly from Equation (2.22).

The rms-velocity depends only on the mass and temperature. The velocities are tabulated in Table 2.2. For macroscopic parts, the noise velocities are insignificant, but as the size of the spheres decreases to micro- and nano-scale, the velocities became substantial.

Table 2.2: Noise velocities for silicon ($\rho = 2330$ kg/m^3) spheres with a diameter D suspended in a liquid.

D[m]	10^{-2}	10^{-4}	10^{-6}	10^{-8}
m[kg]	$12 \cdot 10^{-4}$	$12 \cdot 10^{-10}$	$12 \cdot 10^{-16}$	$12 \cdot 10^{-22}$
$\dot{x}_{\text{rms}}$[m/s]	$2 \cdot 10^{-9}$	$2 \cdot 10^{-6}$	$2 \cdot 10^{-3}$	2

Example 2.6: Vibration amplitude of anchored beads

Problem: The silicon beads of Example 2.5 are anchored with a spring k that scales with the mass so that the resonant frequency $\omega_0 = 1$ kHz is constant. What is the rms-vibration amplitude for the beads?

Solution: The rms-noise displacement obtained from Equation (2.24) is

$$x_{\text{rms}}^2 = \frac{k_B T}{k} = \frac{k_B T}{m \omega_0^2}.$$

The bead displacements are tabulated in Table 2.3. For macroscopic parts, the noise displacements are small but become substantial as the size decreases to micron-scale. The rms-velocities remain unchanged from Example 2.5.

Table 2.3: Noise displacements for silicon ($\rho = 2330$ kg/m^3) spheres with a diameter D anchored with a spring $k = m\omega_0^2$.

D[m]	10^{-2}	10^{-4}	10^{-6}	10^{-8}
m[kg]	$9 \cdot 10^{-4}$	$9 \cdot 10^{-10}$	$9 \cdot 10^{-16}$	$9 \cdot 10^{-22}$
x_{rms}[m]	$2 \cdot 10^{-12}$	$2 \cdot 10^{-9}$	$2 \cdot 10^{-6}$	$2 \cdot 10^{-3}$

2.4   Signal-to-noise ratio

Intuitively, the measured signal should be larger than the noise level in the system. To quantify the sensor noise performance, we need to look at the sensor signal-to-noise ratio (SNR) that characterizes the signal quality. The SNR is defined as

$$SNR = \frac{\text{signal power, } P_s}{\text{noise power, } P_n}. \tag{2.26}$$

The use of power ratio makes the definition of SNR applicable in all situations. In electrical systems, the signal power is $P_s = v_{s,\text{rms}}^2/R$ and the noise power is $P_n = v_{n,\text{rms}}^2/R$ where $v_{s,\text{rms}}$ is the rms-signal voltage, $v_{n,\text{rms}}$ is the rms-noise voltage, and R is the system resistance. As we are concerned with the power ratios, the resistance R cancels out and Equation (2.26) becomes

$$SNR = \frac{P_s}{P_n} = \frac{v_{s,\text{rms}}^2}{v_{n,\text{rms}}^2}. \tag{2.27}$$

The SNR in dB units is (see Appendix H)

$$SNR_{\text{dB}} = 10\log_{10}\frac{P_s}{P_n} = 20\log_{10}\frac{v_{s,\text{rms}}}{v_{n,\text{rms}}}. \tag{2.28}$$

Figure 2.8 illustrates the usefulness of the signal-to-noise ratio for comparing two pressure sensors. Both sense the same pressure differentials but the performance of the sensors differ. The sensor #1 has lower noise than sensor #2 but comparing the sensor performance on noise magnitude alone is not meaningful as it is always possible attenuate the noise and signal by an equal factor. Calculating the signal-to-noise ratios show that sensor #2 has 25 higher SNR.

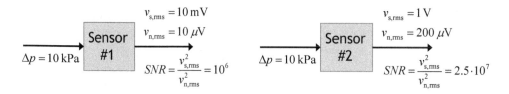

Figure 2.8: Illustration of sensor noise performance: pressure sensor with lowest absolute noise may not have the best signal-to-noise ratio.

To help visualize the SNR, Figure 2.9 shows sine-waves at three different signal-to-noise ratios. Figure 2.9 leads to a question of how large the signal-to-noise ratio should be? The answer depends on how large the probability for erroneous signal can be. For example, in a safety critical application such as car air bags, a wrong positive or negative signal could have disastrous consequences. In this application, $SNR \gg 10$ is desired to reduce the probability of wrong air bag deployment to an insignificant number. Consumer applications such as game controllers are more forgiving and the sensor with a signal-to-noise ratio of one may well be usable.

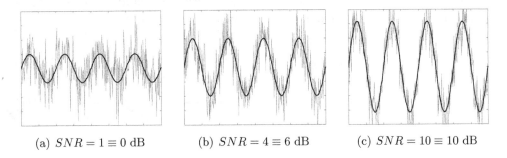

(a) $SNR = 1 \equiv 0$ dB   (b) $SNR = 4 \equiv 6$ dB   (c) $SNR = 10 \equiv 10$ dB

Figure 2.9: Plots of sine-waves in the presence of noise.

2.5   Input referred noise

In addition to amplifying the signal, the amplifier will introduce noise at the amplifier output. To quantify the effect of the noise, we need to compare noise to the signal level at the output which requires accounting for the amplifier gain. Alternatively, we can calculate the input referred noise

$$v_{n,\text{in}} = \frac{v_{n,\text{out}}}{A} \tag{2.29}$$

to compare the signal and noise at the amplifier input as is illustrated in Figure 2.10. As the input referred noise already accounts for the amplifier gain it allows fair comparison of different sensor circuits; the system with the smallest input referred noise can measure the smallest signals.

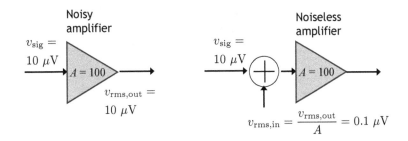

Figure 2.10: Calculation of input referred noise. Both circuits in the figure represent the same amplifier, have the same $SNR = 100$, and have equal noise at the amplifier output.

Example 2.7: Input referred noise of amplifier chain

Problem: Figure 2.11 shows two amplifiers with different gain and noise characteristics. If the amplifiers are connected in a series, which amplifier should be connected first for the best noise performance?

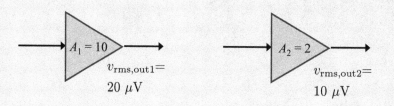

Figure 2.11: Two amplifiers with different gain and noise at the output.

Solution: Figure 2.12 shows the amplifiers connected with amplifier #1 first. The input referred noise due to the first amplifier is

$$v_{rms,in1} = \frac{v_{rms,out1}}{A_1}.$$

The input referred noise due to the second amplifier is

$$v_{rms,in2} = \frac{v_{rms,out2}}{A_1 A_2}.$$

Summing the noise powers, the total input referred noise is

$$v_{rms,in} = \sqrt{\frac{v_{rms,out1}^2}{A_1^2} + \frac{v_{rms,out2}^2}{A_1^2 A_2^2}} = \sqrt{4 \ \mu V^2 + 0.25 \ \mu V^2} = 2.1 \ \mu V.$$

As the second amplifier noise is divided by the first and second amplifier, its contribution to the total noise is small. From the input referred noise, the total noise at the output is $v_{rms,out} = A_1 A_2 v_{rms,in} = 41 \ \mu V$.

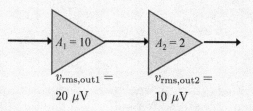

Figure 2.12: Amplifiers connected in series with amplifier #1 first.

Figure 2.13 shows the amplifiers connected with amplifier #2 first. Summing the noise powers, the total input referred noise is

$$v_{\mathrm{rms,in}} = \sqrt{\frac{v_{\mathrm{rms,out2}}^2}{A_2^2} + \frac{v_{\mathrm{rms,out1}}^2}{A_1^2 A_2^2}} = \sqrt{25 \ \mu\mathrm{V}^2 + 1 \ \mu\mathrm{V}^2} = 5.1 \ \mu\mathrm{V},$$

which is more than two times higher than for the previous configuration. The total noise at the output is $v_{\mathrm{rms,out}} = A_1 A_2 v_{\mathrm{rms,in}} = 102 \ \mu\mathrm{V}$.

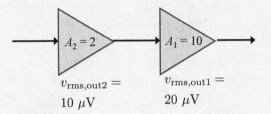

Figure 2.13: Amplifiers connected in series with amplifier #2 first.

This example shows that, as the noise of the first state is amplified by subsequent stages, the first stage noise performance is critical. Conversely, if the first stage already provides amplification, the relative importance of subsequent stages is reduced. A low noise pre-amplifier is therefore critical for good sensor performance.

2.6 Averaging signals

A common technique of reducing the noise is to average multiple measurements. For white noise, the measurements are uncorrelated and using Equation (2.4) the average noise from N measurements is

$$< v_{\mathrm{rms}} > = \frac{\sqrt{v_{\mathrm{rms},1}^2 + v_{\mathrm{rms},2}^2 + \cdots + v_{\mathrm{rms,N}}^2}}{N} = \frac{v_{\mathrm{rms}}}{\sqrt{N}}. \qquad (2.30)$$

Assuming that the signal stays constant during the measurement, the average signal is

$$< v_{\mathrm{s}} > = \frac{v_{\mathrm{s},1} + v_{\mathrm{s},2} + \cdots + v_{\mathrm{s,N}}}{N} = v_{\mathrm{s}}, \qquad (2.31)$$

which is not affected by averaging. The signal averaging in the presense of white noise therefore increases the signal-to-noise ratio. The drawback of averaging is that the signal needs to stay constant during the averaging time and the averaging therefore effectively reduces the measurement bandwidth. From a signal processing point of view, averaging can be considered as a simple filtering operation. For excellent discussion on sample filtering, noise, and frequency response, the reader is referred to reference [25].

The signal averaging in the presence of $1/f$-noise does not improve the measurement accuracy. The $1/f$-noise is less random than other noise types and there is noise correlation between the current and prior measurements. In comparison, white noise has no memory of the past; current values of the process are independent of the past values but $1/f$-noise processes somehow posses memory of the past. This memory is long: The $1/f$-noise in MOSFET's has been measured down to $10^{-6.3}$ Hz or one cycle in three weeks and the $1/f$-noise in weather data has been computed down to 10^{-10} Hz or 1 cycle in 300 years [26]!

As the $1/f$-noise is correlated, the noise measurements sum. From Equation (2.5), the average noise from N measurements is

$$< v_{\mathrm{rms}} > = \frac{v_{\mathrm{rms},1} + v_{\mathrm{rms},2} + \cdots + v_{\mathrm{rms,N}}}{N} = v_{\mathrm{rms}} \qquad (2.32)$$

and averaging does not reduce the $1/f$-noise. This is a significant result as it means that there is no way to filter the $1/f$-noise no matter how long the measurement. Thus, the emphasis should be on minimizing the $1/f$-noise or moving the measurement to a higher frequency where the $1/f$-noise is not as dominant.

Key concepts

- The noise sets the fundamental limit to the smallest measurable signals.

- Thermal noise is white meaning that it has flat frequency spectrum.

- The average thermal noise energy per degree of freedom is $\frac{1}{2}k_B T$.

- The rms-noise can be obtained by equating the kinetic or potential energy with the thermal noise energy. For example, rms-noise displacement is obtained from $\frac{1}{2}kx_{\mathrm{rms}}^2 = \frac{1}{2}k_B T$.

- The frequency spectrum for the noise depends on the system damping. It can be calculated by associating each damping element with a noise force generator ($\bar{v} = \sqrt{4k_B T R}$ in electrical systems and $\bar{F} = \sqrt{4k_B T \gamma}$ in mechanical systems).

- The signal-to-noise ratio and the input-referred noise are used quantify the system noise.

- Averaging can be used to reduce the white noise but not the $1/f$-noise.

Exercises

Exercise 2.1
Resistors $R_1 = 1$ kΩ and $R_1 = 3$ kΩ are connected in series. What are the total noise voltage spectral density and current spectral density for the resistors combined.

Exercise 2.2
Resistors $R_1 = 1$ kΩ and $R_1 = 3$ kΩ are connected in parallel. What are the total noise spectral density and current spectral density for the resistors combined.

Exercise 2.3
For data conversion, the analog signal is sampled and stored in a 1-pF capacitor. To reduce the $k_B T/C$-noise, multiple samples are taken and averaged. How many samples are needed to obtain rms-noise voltage $v_{\mathrm{rms}} = 1$ μV?

Exercise 2.4
An CMOS amplifier has an input impedance of $C_{\mathrm{in}} = 100$ fF and it is connected to a sensor with resistance $R = 1$ kΩ. Calculate the rms-noise due to thermal noise and the noise spectral density at low frequencies at the amplifier input.

Exercise 2.5
What is the thermal noise induced rms-displacement for an atomic force microscope cantilever with $m = 25$ pkg and $f_0 = 20$ kHz? How does this correspond to typical atomic spacing and can individual atoms be discerned?

Exercise 2.6
What is the thermal noise induced rms-velocity for the atomic force microscope cantilever in Exercise 2.5?

Exercise 2.7
Figure 2.14 shows two RC circuits. Assume $R = 1$ kΩ and $C = 10$ pF.
(1) Which circuit is high pass and which is low pass filter?
(2) What is the general expression for $|v_{\mathrm{out}}/v_{\mathrm{in}}|$ for the two circuits?
(3) What is $|v_{\mathrm{out}}/v_{\mathrm{in}}|$ for the circuits at $f = 10$ MHz?
(4) What is the expression for the noise spectral density at filter output due to the resistor noise. Assume that the inputs are grounded?
(5) What is the rms-noise at the outputs with the inputs grounded?

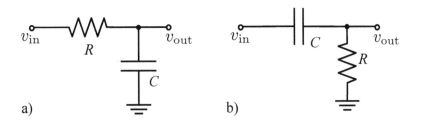

Figure 2.14: RC-filter circuits

Exercise 2.8

An amplifier has output noise density $\overline{v_n} = 4$ nV/$\sqrt{\text{Hz}}$. What is the rms-noise voltage if the noise bandwidth is 10 kHz?

Exercise 2.9

Piezoresistive sensor has the thermal noise density $\overline{v_n} = 10$ nV/$\sqrt{\text{Hz}}$ and the $1/f$-corner frequency $f_c = 1$ kHz. Calculate the total sensor noise for the 0.1 Hz to 10 kHz frequency band.

Exercise 2.10

Silicon accelerometer consist of a mass attached to a spring with a spring constant $k = 30$ N/m. What is the noise induced rms-displacement for the mass?

Exercise 2.11

An accelerometer has sensitivity $A = 100$ mV/G and output noise is $v_{\text{rms}} = 2$ mV. What is the input referred noise equivalent acceleration?

Exercise 2.12

Calculate the SNR after each amplifier in Figure 2.11 if signal $v_{\text{sig}} = 1$ mV is fed to amplifier chain input.

Exercise 2.13

Calculate the input referred pressure noise for the two pressure sensors in Figure 2.8.

3

Accelerometers

Accelerometers are one of the highest volume MEMS products: The annual worldwide sales are more than 100 million units and growing steadily. Historically, the automotive industry has been the growth driver. Today, all cars employ at least high-G crash sensors for air bag deployment. In addition, low-G sensors are used for active suspensions and vehicle stabilization controls. More recently, as the accelerometer prices have dropped to a dollar range, consumer applications have become economically feasible. For example, the latest generation of game consoles contains accelerometers for measuring the game controller movement to enable motion based user interface. Recently, cell phones enhanced with a motion based user interface have also become available. Accelerometers are also used by runners to determine the running speed and in digital cameras to determine the picture orientation. Some laptop hard drives utilize accelerometer based "free fall" detection to protect the hard drive from impacts. With the decreasing price, the number of accelerometer applications is going to increase in the coming years.

In this book, the accelerometers will be used to illustrate the microsensing techniques presented in later chapters. Chapter 4 covers the design of micromechanical springs used for the construction of microsensors. The piezoresistive and capacitive sensing principles are covered in detail in Chapters 5 and 6, respectively, and piezoelectric sensing is analyzed in Chapter 7. The interface electronics and the associated noise is covered in Chapter 8 and Chapter 9, respectively. Switched capacitor circuits suitable for low power capacitive accelerometers are introduced in Chapter 10.

This chapter focuses on the fundamental principles of acceleration sensing. First, we will analyze the mechanical response of accelerometers. The response function is derived and studied in the frequency and time domain. After analyzing the effect of damping and resonant frequency on the accelerometer response, we will cover the fundamental mechanical noise limitations. The chapter is concluded with case studies on surface and bulk micromachined accelerometers.

3.1   Operation principle

The accelerometer structure is illustrated in Figure 3.1. A proof mass m is connected to the frame by a flexible spring k. Due to the mass inertia, the proof mass motion will lag the frame motion. To prevent excessive ringing, the vibrations are damped by introducing gas (or liquid such as oil in macroscopic sensors) inside the package. This damping is represented with a dashpot γ.

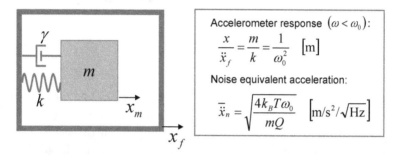

Figure 3.1: The basic structure of an accelerometer consists of a proof mass m that is suspended with a spring k to a frame. Due to inertia of the proof mass, the motion of the mass does not follow the frame motion and the difference in displacement $x = x_f - x_m$ can be used to measure the acceleration. Also provided is the noise-equivalent acceleration (due to mechanical Brownian noise).

The accelerometers can either be single axis or they can measure acceleration in multiple directions. In principle, a three axis accelerometer can be based on a single proof mass that can move in X-, Y-, and Z-directions. By measuring the mass displacement in all three directions, the acceleration can be deduced. Most practical sensors, however, measure the mass displacement along just one or two directions and multiple independent masses are used for three axis accelerometers [27].

The following sensing principles are used for micromechanical accelerometers [28]:

Piezoresistive sensing is based on piezoresistors integrated onto the spring. The piezoresistor resistance changes when subjected to the acceleration induced stress. Thus, by measuring the resistance change, the acceleration is deduced. The first micromachined silicon accelerometers developed in the 70's were based on piezoresistive sensing [29]. The piezoresistive sensing is robust and simple to implement but has poor noise and power performance. The piezoresistive sensing is covered in Chapter 5.

Capacitive sensing is based on detecting small change in capacitance due to relative movement of the proof mass and the frame. The capacitive accelerometers are currently the most widely used accelerometers as they are inexpensive, have good noise performance, and low power consumption. The capacitive sensing is covered in Chapter 6.

Piezoelectric sensing is based on a charge polarization of piezoelectric materials due to the strain caused by the inertial force. In the simplest configuration, the proof mass is attached to a piezoelectric plate that acts as a spring. The piezoelectric plate generates current that is proportional to the change in acceleration. As the sensor generates the current, the sensor is called self-generating. The drawback of piezoelectric sensors is that the sensor only measures the changes in acceleration and cannot be used to measure dc-acceleration. The piezoelectric sensing is commonly used for macroscopic sensors but is rarely used for microscopic accelerometers. The piezoelectric sensing is covered in Chapter 7.

In addition to the above sensing principles, optical and magnetic position detection have been used for macroscopic sensors [28], but these methods are not practical for microsensors.

3.2   Accelerometer equation

To analyze the accelerometer in Figure 3.1, we start with the equation of motion for the proof mass given by

$$m\frac{\partial^2 x_m}{\partial t^2} + \gamma\frac{\partial(x_m - x_f)}{\partial t} + k(x_m - x_f) = F_E, \qquad (3.1)$$

where x_m and x_f are the positions of the mass and frame, respectively, and F_E is the external force acting on the mass for example due to actuation or Brownian noise. Equation (3.1) can be simplified by subtracting $m\frac{\partial^2 x_f}{\partial t^2}$ from both sides leading to

$$m\frac{\partial^2(x_m - x_f)}{\partial t^2} + \gamma\frac{\partial(x_m - x_f)}{\partial t} + k(x_m - x_f) = -m\frac{\partial^2 x_f}{\partial t^2} + F_E. \qquad (3.2)$$

Recognizing that $x = x_f - x_m$ is the difference of the frame and mass positions leads to the familiar one degree-of-freedom damped resonator governed by

$$m\frac{\partial^2 x}{\partial t^2} + \gamma\frac{\partial x}{\partial t} + kx = F, \qquad (3.3)$$

where F is the sum of inertial and external forces $F = m\frac{\partial^2 x_f}{\partial t^2} - F_E = m\ddot{x}_f - F_E$.

Solving Equation (3.3) using Laplace transformation (see Appendixes A and B) and defining the quality factor as $Q = \omega_0 m/\gamma$ gives

$$x = \frac{F/m}{s^2 + s\omega_0/Q + \omega_0^2} = \frac{\ddot{x}_f}{s^2 + s\omega_0/Q + \omega_0^2} \equiv H(s)\ddot{x}, \qquad (3.4)$$

where we have assumed that there are no external forces ($F_E = 0$) and $\ddot{x} = \frac{\partial^2 x_f}{\partial t^2}$ is the frame acceleration.

The frequency responses $H(s)$ for a critically damped accelerometer ($Q = 0.5$) as a function of the resonance frequency ω_0 are plotted in Figure 3.2. The low frequency response increases with the decreasing resonance frequency ω_0 and the high frequency displacement is seen as independent of the resonance frequency. Thus, reducing the resonant frequency increases the sensitivity but decreases -3 dB bandwidth.

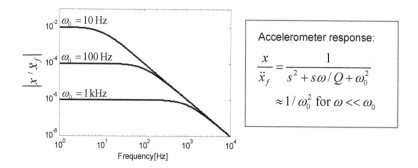

Figure 3.2: The frequency responses $x/\ddot{x}_f \equiv H(s)$ for a critically damped accelerometer ($Q = 0.5$).

3.2.1   Low-frequency response

The low-frequency response describes the accelerometer operation below its mechanical resonant frequency. Most MEMS sensors operate in this region with the typical resonant frequencies in the 10 Hz to 10 kHz range. From Equation (3.4), the low frequency response to acceleration $\ddot{x}_f$ is

$$x \approx \frac{m\ddot{x}_f}{k} = \frac{\ddot{x}_f}{\omega_0^2} \quad \text{for} \quad \omega \ll \omega_0. \qquad (3.5)$$

Equation (3.5) depends only on the resonance frequency suggesting that accelerometer size can be scaled without affecting the mechanical performance if the proof mass and spring constant are reduced proportionally. However, as was shown in Chapter 2, the mechanical vibrations due to thermal noise increase with decreasing mass size. As a result, large masses are desired for low noise sensors. The mechanical noise in accelerometers is further studied in Section 3.4.

Example 3.1: Accelerometer displacement
Problem: A capacitive accelerometer is to have a displacement of 0.5 μm at 2 G acceleration. Calculate the sensor resonant frequency.
Solution: From Equation (3.5) we obtain the resonant frequency as

$$f_0 = \frac{1}{2\pi}\sqrt{\left|\frac{\ddot{x}_f}{x}\right|} = \frac{1}{2\pi}\sqrt{\frac{2 \cdot 9.81 \text{ m/s}^2}{0.5 \ \mu\text{m}}} = 1.0 \text{ kHz}.$$

3.2.2 High-frequency response

Equation (3.5) shows that to obtain a sensitive accelerometer, the resonant frequency should be as low as possible. Taken to the extreme, the resonant frequency may be 1 Hz or lower and the sensor is operated above the natural frequency ($\omega \gg \omega_0$). These types of sensors are used as seismometers to measure the ground vibrations and earth quakes.

Substituting $\ddot{x}_f = s^2 x_f$ into Equation (3.4) and taking limit $\omega \gg \omega_0$ leads to

$$x \approx x_f \quad \text{for} \quad \omega \gg \omega_0. \tag{3.6}$$

The physical interpretation of Equation (3.6) is that at above resonance frequency, the proof mass essentially stays immobile and the difference of the mass and frame positions is simply $x = x_f - x_m \approx x_f$.

The mechanical resonance frequency of less than 1 Hz is obtainable with macroscopic devices that have large mass but is not easily achieved with small MEMS components. For this reason, the commercial MEMS accelerometers operate below the resonance frequency.

Example 3.2: A "MEMS seismometer" mass
Problem: A practical lower limit for a MEMS spring constant is around $k = 1$ N/m. Calculate the required proof mass dimensions to evaluate the feasibility of scaling down a macroscopic seismometer design with a resonant frequency of $f_0 = 0.2$ Hz.
Solution: With a spring constant $k = 1$ N/m, the required mass is

$$m = \frac{k}{\omega_0^2} \approx 0.63 \text{ kg}.$$

This corresponds to a cube of silicon with dimensions of

$$L = \left(\frac{m}{\rho}\right)^{1/3} \approx 6.5 \text{ cm},$$

which is not very micromechanical. If the resonance frequency was $f_0 = 100$ Hz, the mass would be $m = 2.5$ μkg and the size of mass would be 1 mm^3 – not quite micron-sized but typical for bulk micromachined accelerometers.

3.2.3   Time domain response

The time domain response is critical in many accelerometer applications. Ideally, the accelerometer output should follow the input (acceleration) instantaneously and without any error. However, as with any physical system, the accelerometer output will lag the change in acceleration. The mechanical response time is inversely related to the accelerometer resonance frequency ω_0. Moreover, the shape of the response depends on the damping: An under damped system shows significant overshoot and ringing and over damped systems are slow to respond. In commercial applications, a well-behaved response without ringing can be equally important as the fast response time. For example, in car stability control, under damping in the feedback control loop could result in oscillations with catastrophic consequences.

   The step response describes the accelerometer response after a change in acceleration. The step response is studied systematically in Appendix B and we will summarize the results here. As shown in Figure 3.3, over damped accelerometers are slow to respond. Under damped devices are fast but the step response exhibits significant overshoot and ringing. The optimal speed is obtained with critical damping ($Q = 0.5$) that offers the fastest step response without overshoot or ringing.

   In all cases, the displacement approaches the final displacement $x_{\text{final}} = F/k$. The transient error decays approximately as

$$err = |x - x_{\text{final}}| = e^{-t/\tau} \tag{3.7}$$

where τ is the time constant for the step response. From Appendix B, the time constant for under damped system ($Q > 0.5$) is $\tau_u \approx 2Q/\omega_0$ that describes how quickly the oscillations decay. For a critically damped ($Q = 0.5$) and over damped ($Q < 0.5$) systems, there is no closed form solution for the time constant. The settling times for the critical damped system are tabulated in Table B.1 on page 416. For over damped systems with $Q < 0.3$, the approximation $\tau_o \approx 1/\omega_0 Q$ can be used.

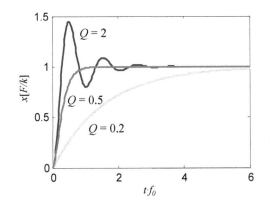

Figure 3.3: Accelerometer step responses for different quality factors (Figure B.2(c) from Appendix B reproduced here for convenience).

Example 3.3: Critically damped accelerometer settling time

Problem: A micromachined accelerometer has the resonant frequency $f_0 = 2.0$ kHz and the quality factor is $Q = 0.5$. How quickly does the accelerometer settle to within 1% of the final value?

Solution: For a critically damped system ($Q = 0.5$), the settling times are given in Table B.1 on page 416. The settling time to settle within 1% is

$$t_s = \frac{1.06}{f_0} \approx 0.5 \text{ ms.}$$

Example 3.4: Over damped accelerometer settling time

Problem: A micromachined accelerometer has the resonant frequency $f_0 = 2.0$ kHz and the quality factor is $Q = 0.1$. What is the time constant for the step response and how quickly does the accelerometer settle to within 1% of the final value?

Solution: For over damped critically damped system with $Q < 0.3$, the time constant is approximately

$$\tau \approx 1/\omega_0 Q \approx 80 \text{ } \mu\text{s.}$$

Solving Equation (3.7) for $err = 0.01$ gives the time to settle to within 1% of the final value as

$$t = -\log(0.01)\tau \approx 4 \text{ ms,}$$

which is almost ten times longer than for the critically damped accelerometer in Example 3.3.

3.3 Damping

The damping is controlled by the gas pressure inside the accelerometer package. The gas damping is covered in detail in Chapter 12 where the effect of device geometry and package pressure are analyzed. For now, it is sufficient to realize that the accelerometer damping can be adjusted with the device design. Often lowered package pressures are used to reduce the damping to an optimal level.

As shown in Figure 3.3, over damped accelerometers are slow to respond. Under damped devices are fast but the step response exhibits significant overshoot and ringing. The optimal speed is obtained with critical damping $(Q = 0.5)$ that offers the fastest step response without overshoot or ringing.

Practical limitations in device design may cause the actual damping to be above or below the critical damping level. For example, real devices have multiple degrees of freedom and over damping may be used to suppress unwanted vibration modes and resonances. Here the response speed is traded for greater stability. Another extreme is the surface micromachined accelerometers that are under damped. The gas damping is less significant for laterally moving structures. For example, surface micromachined accelerometers can have $Q > 10$ even at atmospheric pressure. The high quality factor results in ringing of the proof mass that is not desired. To filter this ringing from the output signal, the bandwidth of surface micromachined accelerometers is typically limited electrically to be significantly lower than the resonance frequency f_0.

Example 3.5: Damping coefficient
Problem: A micromachined accelerometer has the resonant frequency $f_0 = 2.0$ kHz and the mass $m = 0.5$ nkg. If the desired quality factor is $Q = 0.3$, what is the damping coefficient?
Solution: The damping coefficient is

$$\gamma = \frac{\omega_0 m}{Q} \approx 20.9 \ \mu\text{kg/s}.$$

3.4   Mechanical noise in accelerometers

The thermal noise induced mechanical vibrations set the lower limit for the measurable acceleration. In Chapter 2, we learned that noise force generator depends only on the dissipation and that small mechanical masses exhibit large noise induced vibrations.

For the noise analysis, it is helpful to model the mechanical system as a series connection of mass and spring as is shown in Figure 3.4. The mass m converts the acceleration into a force that is converted to a displacement by the spring k. The overall sensitivity $A = x/\ddot{x}_f$ is

$$A = A_1 A_2 = \frac{m}{k} = \frac{1}{\omega_0^2} \tag{3.8}$$

which is in agreement with Equation (3.5).

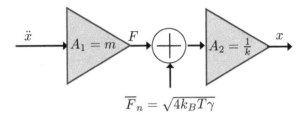

Figure 3.4: System level model for the accelerometer noise analysis.

The mechanical noise is modeled by the mechanical noise force generator $\overline{F}_n = \sqrt{4k_BT\gamma}$. As shown in Section 2.5, the input referred noise can be used to quantify the sensor noise performance as it gives a direct measure of the smallest measurable acceleration. The input referred noise equivalent acceleration spectral density is

$$\overline{\ddot{x}}_n = \frac{\overline{F}_n}{A_1} = \frac{\overline{F}_n}{m} = \frac{\sqrt{4k_BT\gamma}}{m} = \sqrt{\frac{4k_BT\omega_0}{mQ}}. \tag{3.9}$$

The noise equivalent acceleration given by Equation (3.9) is a measure of the smallest acceleration that can be measured: Acceleration smaller than $\overline{\ddot{x}}_n$ generate displacement that is below the thermal noise floor $\overline{x}_n$ for the mechanical vibrations.

Equation (3.9) suggests that the noise could be reduced by increasing the quality factor, increasing the mass, and reducing the resonant frequency ω_0. However, the benefit of increasing the quality factor is purely superficial. From the equipartition theorem, we know that the total noise energy integrated over all frequencies is constant. The rms-vibration amplitude given by Equation (2.24) is $x_{rms} = \sqrt{\frac{k_B T}{k}}$, which is independent of quality factor. Given the rms-displacement, the input referred rms-acceleration is

$$\ddot{x}_{rms} = \frac{x_{rms}}{A_1 A_2} = x_{rms}\omega_0^2 = \sqrt{\frac{\omega_0^2 k_B T}{m}}. \tag{3.10}$$

Equation (3.10) clearly shows that the quality factor does not affect the total input referred acceleration noise. In addition, the high-Q has the detrimental effect of increasing the step response time. Thus, increasing the proof mass and lowering the resonant frequency are the only effective methods to reduce the mechanical noise. This partially explains why there are no "nanomechanical" accelerometers on market.

Example 3.6: Accelerometer noise
Problem: A MEMS accelerometer has the mass $m = 0.5$ nkg and the mechanical resonant frequency $f_0 = 2$ kHz. Calculate the proof mass rms-noise displacement and the input referred rms-noise equivalent acceleration.
Solution: The element spring constant is $k = \omega_0^2 m = 12.67$ N/m and the rms-displacement due to thermal noise from Equation (2.24) is

$$x_{rms} = \sqrt{\frac{k_B T}{k}} \approx 2.3 \cdot 10^{-10} \text{ m.}$$

From Equation (3.10), the input referred noise equivalent acceleration is

$$\ddot{x}_{rms} = x_{rms}\omega_0^2 \approx 0.0362 \text{ m/s}^2 = 3.7 \text{ mG.}$$

Example 3.7: A Silicon accelerometer
Problem: Figure 3.5 shows a schematic of a bulk micromachined accelerometer. The mass dimensions are 1200 μm × 1200 μm × 550 μm and the silicon density is 2330 kg/m^3. The total spring constant for the four beams is $k = 40$ N/m and targeted quality factor is $Q = 0.2$. What is the
a) sensor resonance frequency?
b) damping coefficient?
c) proof mass displacement due to 1 G acceleration at low frequencies $(\omega \ll \omega_0)$?

d) noise induced displacement spectral density at low frequencies ($\omega \ll \omega_0$)?

e) proof mass rms-displacement due to noise?

f) noise equivalent acceleration (the spectral density and rms-noise)?

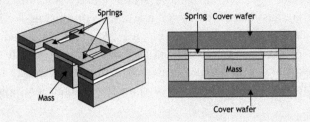

Figure 3.5: Silicon micromachined accelerometer. The proof mass displacement could be detected piezoresistively or capacitively.

Solution: a) The mass is $m = \rho V = 1.845\ \mu\text{kg}$ and the resonance frequency is $f_0 = \frac{1}{2\pi}\sqrt{\frac{k}{m}} = 737.7\ \text{Hz} \approx 740\ \text{Hz}$.

b) The damping coefficient is $\gamma = \frac{m\omega_0}{Q} = 0.0428\ \mu\text{kg/s} \approx 0.043\ \mu\text{kg/s}$.

c) The proof mass displacement due to the 1-G acceleration is $x = \ddot{x}_f / \omega_0^2 = 0.4565\ \mu\text{m} \approx 0.46\ \mu\text{m}$.

d) The noise induced displacement spectral density is

$$\overline{x}_n = \sqrt{\overline{x_n^2}} = \frac{\overline{F}_n}{k} = \frac{\sqrt{4 k_B T \gamma}}{k} \approx 0.67 \cdot 10^{-12}\ \text{m}/\sqrt{\text{Hz}}$$

e) The equipartition theorem states that the average potential energy due to thermal noise is $W = \frac{1}{2} k_B T$. Writing $W = \frac{1}{2} k x_{\text{rms}}^2 = \frac{1}{2} k_B T$ gives

$$x_{\text{rms}} = \sqrt{\frac{k_B T}{k}} = 10.22\ \text{pm} \approx 10\ \text{pm}.$$

f) The equivalent acceleration noise spectral density from Equation (3.9) is

$$\overline{\ddot{x}}_n = \sqrt{\frac{4 k_B T \omega_0}{m Q}} = 1.44 \cdot 10^{-5}\ \text{m/s}^2/\sqrt{\text{Hz}} \approx 1.47\ \mu\text{G}/\sqrt{\text{Hz}}.$$

and the rms-noise from Equation (3.10) is

$$\ddot{x}_{\text{rms}} = \sqrt{\frac{\omega_0^2 k_B T}{m}} \approx 2.20 \cdot 10^{-4}\ \text{m/s}^2 \approx 22\ \mu\text{G}.$$

3.5   Commercial devices

To exemplify what we have learned, we will analyze two commercial micromechanical accelerometers. The first device is a surface micromachined accelerometer that has a relatively small mass. The second device is a bulk micromachined
accelerometer that has 10^4 times larger mass. By comparing the intrinsic noise
for the two accelerometers, it is clear that a large mass is needed for low noise.
The low intrinsic noise however, does not guarantee low overall noise. For the
analyzed devices, the overall sensor noise performance including the noise from
circuitry is an order of magnitude worse than the mechanical noise limit.

3.5.1   Case study: A surface micromachined accelerometer

Figure 3.6 shows a schematic for a surface micromachined accelerometer. The
structure and design parameters are similar to ADXL50 accelerometer from
Analog Devices [30–32]. The proof mass is suspended by folded spring beams
and moves in plane above the die surface. The proof mass motion relative to the
substrate is measured with fixed sensing fingers that are anchored to the substrate. The capacitance measurement sensitivity is increased by measuring the
differential capacitance change $C_1 - C_2$ between two electrodes. The differential
capacitive measurement is analyzed in detail in Chapter 6.

   As the sensing element is relatively thin ($\sim$2 μm), several sensing fingers
are combined in parallel to increase the overall capacitance. Typical surface mi-
cromachine designs have 40-100 finger pairs but for illustrative purposes only
6 pairs are shown. Using the element values in Figure 3.6 and Equations (3.5)
and (3.9) we obtain $x/\ddot{x}_f = 0.4$ nm/G and $\bar{\bar{x}}_n = 0.2$ mG/$\sqrt{\text{Hz}}$ for the element sensitivity and noise, respectively. The intrinsic noise can be compared
to noise performance specifications of $\bar{\bar{x}}_n = 1$ mG/$\sqrt{\text{Hz}}$ in manufacturer's data
sheets [30]. The total noise including the measuring circuitry is a factor five
higher that the mechanical noise alone.

3.5.2   Case study: A bulk micromachined accelerometer

A typical bulk micromachined accelerometer is shown in Figure 3.7. The proof
mass moves in Z-direction in response to the Z-axis acceleration. The mass
displacement is detected by measuring the differential capacitance changes between the proof mass and the top and bottom electrodes.

   Representative device dimensions for a bulk micromachined accelerometer
are 1 mm$\times$1 mm$\times$0.38 mm. The mass and spring constant are $1 \cdot 10^{-6}$ kg
and 50 N/m, respectively. Using Equations (3.5) and (3.9), we obtain $x/\ddot{x}_f =$
0.2 μm/G and $\bar{\bar{x}}_n = 3$ μG/$\sqrt{\text{Hz}}$ for the element sensitivity and noise, respectively. The intrinsic noise can be compared to the noise performance specifi-
cations of $\bar{\bar{x}}_n = 20$ μG/$\sqrt{\text{Hz}}$ for a commercial accelerometer (SCA620 from

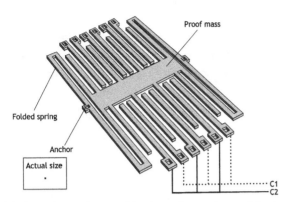

Parameter	Value	Units
Resonant frequency, f_0	25	kHz
Mass, m	0.2	nkg
Spring constant, k	5	N/m
Sense capacitance, C_0	0.1	pF
Quality factor, Q	5	

(a) A surface micromachined accelerometer and typical element parameters corresponding to ADXL50 accelerometer from Analog Devices.

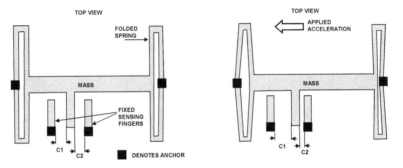

(b) The movement of the proof mass is detected by measuring the capacitance change between proof mass fingers and fixed sensing fingers. Typical designs have 40-100 finger pairs to increase total capacitance.

Figure 3.6: A schematic of typical surface micromachined accelerometer (After Analog Devices data sheet [30]).

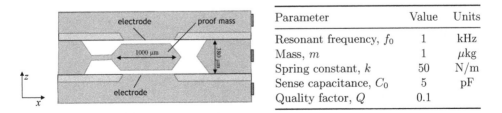

Parameter	Value	Units
Resonant frequency, f_0	1	kHz
Mass, m	1	μkg
Spring constant, k	50	N/m
Sense capacitance, C_0	5	pF
Quality factor, Q	0.1	

Figure 3.7: A bulk micromachined accelerometer and typical element parameters.

VTI Technologies [33]). We see that when noise from the measuring circuitry is included, the total noise is a factor of eight larger than the mechanical noise limit.

When comparing the surface and bulk micromachined accelerometers, it is evident that the large mass of the bulk micromachined device enables lower noise. In addition, the bulk micromachined accelerometer has a large sensitivity which further increases signal-to-noise ratio. This need for high sensitivity will be further investigated in Chapter 9 where the circuit noise is analyzed.

Key concepts

- Accelerometers consist of a proof mass and a spring. By measuring the proof mass displacement relative to the frame, the acceleration can be deduced.

- At low-frequencies ($\omega \ll \omega_0$), the inertial force $F = m\ddot{x}_f$ is balanced by the spring force $F = kx$. The accelerometer response $x/\ddot{x}_f = m/k = 1/\omega_0^2$ is inversely proportional to the resonance frequency squared.

- The mechanical noise sets the limit for the acceleration noise floor.

- The input referred noise equivalent acceleration spectral density is $\ddot{x}_n = \frac{F_n}{m} = \sqrt{\frac{4k_B T \omega_0}{mQ}}$.

- The input referred noise equivalent rms-acceleration is $\ddot{x}_{\mathrm{rms}} = \sqrt{\frac{\omega_0^2 k_B T}{m}}$.

- The system noise performance depends on both mechanical and electrical noise. In commercial devices, the electrical noise is typically larger than the fundamental mechanical noise limit.

Exercises

Exercise 3.1
Using your favorite search engine, find at least five MEMS accelerometer manufacturers.

Exercise 3.2
Calculate how big mass would be needed to obtain mechanical noise limit of $\ddot{x}_{\mathrm{rms}} = 1$ nG, 1 μG, and 1 mG if the mechanical resonance frequency is 100 Hz. If the mass is made of silicon, what would its physical dimensions be?

Exercise 3.3
Explain how the quality factor Q and the resonance frequency f_0 affect the accelerometer settling time.

Exercise 3.4
A capacitive accelerometer with 50 G full-scale acceleration is to have a maximum displacement of 1.5 μm. What is the lowest possible resonant frequency for the accelerometer?

Exercise 3.5
A micromachined accelerometer has the resonant frequency $f_0 = 100$ Hz and the quality factor is $Q = 0.2$. What is the time constant for the step response and how quickly the accelerometer settles to within 5% of the final value?

Exercise 3.6
A micromachined accelerometer has the resonant frequency $f_0 = 100$ Hz and the mass $m = 0.4$ μkg. If the desired quality factor is $Q = 0.2$, what is the damping coefficient?

Exercise 3.7
Consider the commercial accelerometer in Figure 3.7. What is the smallest proof mass size that could would lead to mechanical noise equal to the noise performance specifications of $\bar{\bar{x}}_n = 20$ μG/$\sqrt{\text{Hz}}$ in manufacturers data sheet? Assume that the resonant frequency is constant ($f_0 = 1$ kHz). Compare your results to the estimated actual mass $m = 1$ μkg.

Exercise 3.8
The low-G accelerometer ADXL05 from Analog Devices has a $\pm$5-G full scale range and the noise floor is 500 μG/$\sqrt{\text{Hz}}$ (12$\times$ less than ADXL50). Assume that the characteristics of ADXL05 are identical to those given in Figure 3.6 except for $k = 0.4$ N/m and $f_0 = 10$ kHz. Calculate the intrinsic acceleration noise spectral density and sensor element sensitivity (displacement for a 1-G acceleration). Compare numbers to ADXL50 in Figure 3.6.

Exercise 3.9
In this problem you are to explore noise in a micromachined accelerometer. Figure 3.8 shows an accelerometer fabricated of SOI (silicon on insulator) wafer. The plate dimensions are 200 μm $\times$ 200 μm $\times$ 10 μm and the silicon density is 2330 kg/m^3. The spring constant is $k = 0.08$ N/m and damping coefficient due viscous air damping is $\gamma = 4$ μkg/s. What is the
a) sensor resonance frequency?
b) noise induced displacement spectral density below the resonance frequency?
c) proof mass rms-displacement due to noise?
d) noise equivalent acceleration spectral density?
e) How should the sensor be modified to obtain noise floor of 1 μG/$\sqrt{\text{Hz}}$ if bandwidth is to remain constant?

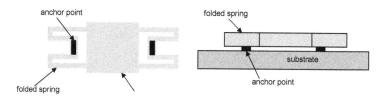

Figure 3.8: Figure for Exercise 3.9. A schematic view of SOI accelerometer (Left: top view, Right: side view).

Exercise 3.10

Figure 3.9 shows how the total output noise can be reduced electronically with a low pass filter. Although the total mechanical rms-noise is constant, we can limit the mechanical noise at the electronics output by increasing quality factor Q and low pass filtering the measured signal. After filtering, the total rms-noise is $\ddot{x}_{\text{rms}} = \overline{\overline{x}}_n \sqrt{BW}$, where BW is the filter bandwidth. Investigate whether this is a viable method make an accelerometer with performance "beyond noise floor" (total noise less than the rms-noise given by Equation (3.10)). Hint: take the sensitivity, the total thermal noise, and the settling time as the critical parameters. Compare the rms noise of the filtered accelerometer to that of critically damped accelerometer that has same mass but smaller resonance frequency corresponding to bandwidth of the filtered accelerometer. Are there other factors that should be considered for a fair judgment between the merits of electrical vs. mechanical bandwidth limiting?

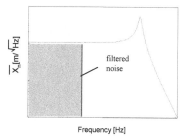

Figure 3.9: Figure for Excercise 3.10. The mechanical noise is electrically filtered so that the total measured rms-noise is less than $x_{\text{rms}} = \sqrt{\frac{k_B T}{k}}$.

4

Beams as micromechanical springs

Springs are the basic building blocks for many micromechanical devices. For example, the accelerometer proof mass is attached to the substrate by flexible springs and the optical micromirrors typically rotate on torsional springs. As MEMS springs are defined by optical lithography and etching, the microsprings do not resemble the macroscopic 3D coils. Instead, simple shapes such as beams and cantilevers are used.

Entire books are devoted to the beam analysis. The MEMS spring design, however, is remarkably simple and involves combinations of few "standard" structures. Here the design equations for the four most common springs (a rod extension, a cantilever, a guided beam, and a torsional beam) are given. Springs that are more complex can be analyzed as series and parallel combinations of the basic shapes. This simplistic approach is justified partly by the poor fabrication tolerances for the MEMS devices. Due to process variations, the spring constant may vary by more than $10 - 20\%$. In comparison, rough calculations presented here are typically accurate within a few percent. When higher accuracy is desired, numerical simulations such as finite element analysis are often used to refine the analytical results.

This chapter is organized as follows: First the Hooke's law for parallel and series springs is reviewed. This is followed by an introduction to the material properties. The elastic deformation theory is summarized and the stress-strain relationships are given. Next, the design equations for the common springs are provided in conjunction with several analysis examples. Finally, the analytical theories are compared to computer simulations.

4.1 Hooke's law for parallel and serial springs

Complex springs can often be broken into series and parallel connections of basic shapes. The ability to simplify complex springs into a connection of simple springs can greatly simplify the analysis effort. Here the Hooke's law for parallel and series springs is reviewed.

The well known relationship for the spring displacement x and force F is

$$F = kx, \tag{4.1}$$

where k is the spring constant. When two springs are connected in parallel as is illustrated in Figure 4.1(a), both springs stretch by the same amount x. The spring forces $F = k_1 x$ and $F = k_2 x$ sum

$$F = k_1 x + k_2 x = (k_1 + k_2)x. \tag{4.2}$$

From Equation (4.2), we recognize the total spring constant as

$$k_{\text{tot}} = k_1 + k_2 \quad \text{(parallel springs)} \tag{4.3}$$

or the sum of the two spring constants.

When springs are connected in series as illustrated in Figure 4.1(b), the same force acts on both springs and the displacements sum. The displacements are $x_1 = F/k_1$ and $x_2 = F/k_2$, and the total displacement is

$$x_{\text{tot}} = x_1 + x_2 = \frac{F}{k_1} + \frac{F}{k_2}. \tag{4.4}$$

From Equation (4.4), the effective spring constant is recognized as

$$\frac{1}{k_{\text{tot}}} = \frac{1}{k_1} + \frac{1}{k_2} \quad \text{(series springs)}. \tag{4.5}$$

Example 4.1: Serpentine spring
Problem: Figure 4.2 shows a mass supported by four serpentine springs. Each segment of the serpentine spring has the spring constant k. The connections between the segments can be assumed stiff in comparison. What is the total spring constant?

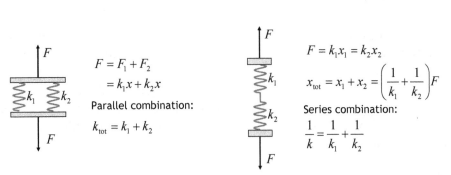

$$F = F_1 + F_2$$
$$= k_1 x + k_2 x$$

Parallel combination:

$$k_{\text{tot}} = k_1 + k_2$$

$$F = k_1 x_1 = k_2 x_2$$

$$x_{\text{tot}} = x_1 + x_2 = \left(\frac{1}{k_1} + \frac{1}{k_2}\right) F$$

Series combination:

$$\frac{1}{k} = \frac{1}{k_1} + \frac{1}{k_2}$$

(a) Two springs in parallel. Both springs stretch by the same amount and the forces sum.

(b) Two spring in series. The same force acts on both springs and the displacements sum.

Figure 4.1: Parallel and series connected springs.

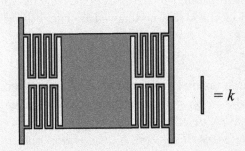

Figure 4.2: A mass connected with four serpentine springs each having six sections that have the spring constant k.

Solution: One serpentine spring is made of six beams connected in series. Each spring in series has the spring constant k. From Equation (4.5), the spring constant for one serpentine is

$$k_1 = \frac{k}{6}$$

and from Equation (4.3), the total spring constant for the four serpentines in parallel is

$$k_{\text{tot}} = 4k_1 = \frac{4}{6}k.$$

4.2   Material properties and theory of elasticity

Springs are structures that are designed to deform under external force. The magnitude of the deformation depends on material and geometry of the structure. Soft materials such as rubber deform more under the same force than hard material such as steel. Similarly, a coiled spring deforms more than a solid bar. Thus, to understand the spring behavior, we need to understand both the material properties and the effect of the spring geometry. This section will provide a summary of the stress-strain relationships in isotropic materials. This theory of elasticity is used in the following sections when the different spring geometries are analyzed.

4.2.1   Normal stress and strain

Figure 4.3(a) illustrates the normal stress. A force F is acting perpendicular on an area A. The mechanical stress is defined as force per unit area

$$T = \frac{F}{A}.$$ (4.6)

In addition to the magnitude, the stress depends on the applied direction of the force. The direction of the stress is denoted by the subscripts: For example,

$$T_{XX} = \frac{F_X}{A_X}$$ (4.7)

indicates that the area normal and force are in the X-direction. A positive force pulling the rod results in positive or *tensile* stress. A negative force pushes the rod resulting in negative or *compressive* stress.

When the normal stress T_{XX} acts on the end of the rod, its length changes by amount Δl due to elastic deformation. The longitudinal strain is defined as the relative change in the length

$$S_{XX} = \frac{\Delta l_X}{l_X},$$ (4.8)

where l_X is the initial length. Again, the subscripts indicate that the direction of elongation is in the X-direction.

The Young's modulus is a material dependent constant that does not depend on the shape or dimensions of the material. It is defined as the ratio of normal stress to longitudinal strain

$$E = \frac{T_{XX}}{S_{XX}}.$$ (4.9)

The Young's modulus characterizes the material stiffness for example in rod extension, beam bending, and membrane deflection. Consequently, it is one of the most used material constants in microsystem design.

As indicated in Figure 4.3(a), the rod also shrinks in lateral dimensions when it is elongated due to the normal stress. The transverse strain is the relative change in lateral dimensions

$$S_{YY} = S_{ZZ} = \frac{\Delta l_Y}{l_Y} = \frac{\Delta l_Z}{l_Z},$$ (4.10)

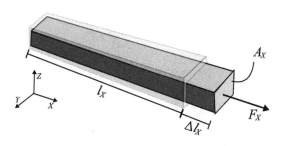
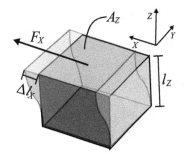

(a) Rod extension due to a normal force F_X acting at the end of the beam. Due to elongation, the rod also shrinks in lateral dimensions.

(b) Shear force acting in plane with the surface. The cube deforms but the volume does not change.

Figure 4.3: The schematic and spring constant for the three most common MEMS springs.

where l_Y and l_Z are the width and height of the rod, respectively.

The Poisson's ratio gives the ratio between transverse and longitudinal strains

$$\nu = -\frac{S_{YY}}{S_{XX}} = -\frac{S_{ZZ}}{S_{XX}}. \tag{4.11}$$

The negative sign is due to the rod shrinking in transverse directions when it is extended in the length direction. The Poisson's ration is between 0.1 and 0.4 for most materials.

Example 4.2: Stress in silicon ignot
Problem: A silicon ignot has the height $h = 0.8$ m, the diameter $D = 0.23$ m and it weighs approximately 80 kg. The ignot is hanging from its neck that has a cross sectional area $A = 3$ cm^2. What are the stresses and strains in the ignot's neck?
Solution: The force acting on the neck is $F = mG = 780$ N and the normal stress is

$$T_{XX} = \frac{F_X}{A_X} = 2.6 \text{ MPa}.$$

Since the next can freely shrink in the lateral dimensions, the transverse stresses are zero ($T_{YY} = T_{XX} = 0$). Using $E = 130$ GPa and $\nu = 0.3$ for silicon, the normal strain is

$$S_{XX} = \frac{T_{XX}}{E} = 2.0 \cdot 10^{-5}.$$

and the transverse strains are $S_{YY} = S_{ZZ} = -\nu S_{XX} = -6.0 \cdot 10^{-6}$.

4.2.2   Shear stress and strain

The shear stress is illustrated in Figure 4.3(b). The force F_X is acting in plane with the surface A_Z. Again, the subscript indicates the direction of the shear stress: T_{ZX} denotes stress in X-direction on area facing Z-direction.

When a shear stress $T_{ZX} = F_X/A_Z$ acts on a cube with the area A_Z and the height l_Z, the cube deforms by amount Δl_X. The shear strain is defined as the displacement per height

$$S_{ZX} = \frac{\Delta l_X}{l_Z}. \tag{4.12}$$

The shear modulus relates the shear stress T_{ZX} to shear strain S_{ZX}

$$G = \frac{T_{ZX}}{S_{ZX}}. \tag{4.13}$$

The shear modulus characterizes the material for example in torsional springs. For isotropic materials, the shear modulus is related to the Young's modulus and the Poisson's ratio by $G = \frac{E}{2(1+\nu)}$. Pure shear does not change the volume of the cube. Consequently, the shear modulus is significantly smaller than Young's modulus.

4.2.3   Material properties

Table 4.1 lists properties of selected materials. The single-crystal (SC) silicon has comparatively high Young's modulus, high yield strength, and a low thermal expansion coefficient. In addition, silicon has virtually no creep, meaning that the material properties and dimensions stay constant over time as opposed to materials such as plastic that slowly creep over time. In fact, single-crystal silicon wafers can have fewer than one defect per billion atoms making the man-made silicon nearly perfect material. The excellent mechanical properties, relatively low cost, and high machinability with wet and dry etching techniques have made silicon the dominant material for microsystems.

The crystalline silicon is anisotropic meaning that the elastic properties depend on the orientation of the structure with respect to crystal axis. The anisotropic elasticity is reviewed in Appendix E. Fortunately, the full anisotropic theory is seldom required as most of the MEMS designs are oriented along one of the three main crystalline directions shown in Table 4.1. When the anisotropic properties of silicon need to be incorporated into spring design, it may be desirable to resort to numerical simulations such as finite element modeling (FEM).

4.2.4   General definition of 3D strain

In our definition of the Young's modulus in Figure 4.3, we assumed that the stress and strain are constant in the stretched rod. The strain was defined in

Table 4.1: Young's modulus E, density ρ, thermal expansion coefficient α, and tensile yield strength T_{yield} of common materials. For single-crystal (SC) silicon, the elastic properties are given for the three most common crystalline directions.

Material	E[GPa]	ρ[kg/m^3]	α[10^{-6}/K]	T_{yield}[GPa]
Aluminum	69	2,690	23.0	0.4
Steel	200	7,900	17.3	2.1
SC Si [100]	130	2,330	2.33	7
SC Si [110]	170	2,330	2.33	7
SC Si [111]	189	2,330	2.33	7
Diamond	103.5	3,500	1.0	53
Polycrystalline Si	150-170	2330	2.33	103
Polypropylene (plastic)	1.5-2	850	100-200	0.01-0.04

terms of the total elongation Δl. For a non-uniform rod, the strain at each point varies as the rod's elastic modulus or cross-sectional area changes. To define the strain at each point, we need to define the infinitesimal strain. The general 3D strain theory is included here as a reference but is not required to understand the simple beam structures. Thus, the material in this section can be safely skipped.

Figure 4.4(a) illustrates the deformation for a small area element. The displacement $\vec{u}(X, Y, Z)$ indicates the displacement of the element at location (X, Y, Z). In general, the displacement is a vector with components in three directions. It is customary to denote these components with subscripts. For example, u_X denotes the displacement in the X-axis direction.

A simplified example is shown in Figure 4.4(b) where the displacement is only in X-direction. The original element length is ΔX, but due to deformation, the length changes by $u_X(X + \Delta X) - u_X(X)$. Notice that the element length changes only if the displacement in the locations X and $X + \Delta X$ differ. Constant displacement will result in a constant element displacement with no deformation. The strain is defined as the change in element length over the original element unit length

$$S_{XX} = \frac{u_X(X + \Delta X) - u_X(X)}{\Delta X} \approx \frac{\partial u_X}{\partial X}. \tag{4.14}$$

The most general definition for the strain is

$$S_{ij} = \frac{1}{2}\left(\frac{\partial u_i}{\partial X_j} + \frac{\partial u_j}{\partial X_i}\right), \tag{4.15}$$

where u is the displacement vector, X is coordinate, and the two indices i and j can range over the three coordinates axis X, Y, and Z. In the theory of

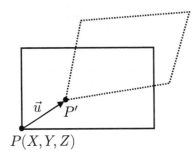
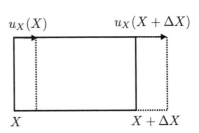

(a) Illustration of displacement and deformation. A point P at (X, Y, Z) is displaced by $\vec{u}$.

(b) Uniaxial strain in X-direction. The displacement at location X is $u_X(X)$.

Figure 4.4: 3D deformation due to location dependent displacement $\vec{u}$.

elasticity, numbers 1, 2, and 3 are sometimes used to denote the coordinate axis X, Y, and Z, respectively.

Using the general definition given by Equation (4.15), the strain S_{XX} is

$$S_{XX} = \frac{1}{2}\left(\frac{\partial u_X}{\partial X_X} + \frac{\partial u_X}{\partial X_X}\right) = \frac{\partial u_X}{\partial X}, \tag{4.16}$$

which agrees with Equation (4.14). As a final example of using Equation (4.15), the strain S_{XZ} is

$$S_{XZ} = \frac{1}{2}\left(\frac{\partial u_X}{\partial Z} + \frac{\partial u_Z}{\partial X}\right). \tag{4.17}$$

Example 4.3: Strain due to a sound wave
Problem: Calculate the strains due to an ultrasonic sound wave $u_X(X, t) = A\sin(kX - \omega t)$. Assume that $u_Y = u_Z = 0$.
Solution: The strain from Equation (4.15)

$$S_{XX} = \frac{1}{2}\left(\frac{\partial u_X}{\partial X_X} + \frac{\partial u_X}{\partial X_X}\right) = \frac{\partial u_X}{\partial X} = Ak\cos(kX - \omega t).$$

The other strains are zero.

4.3 Spring design equations

The three most common springs are shown in Figure 4.5: an extensional rod, a cantilever, and a torsional spring. Springs that are more complex can be

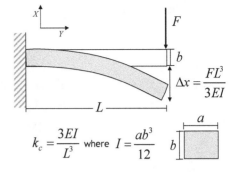

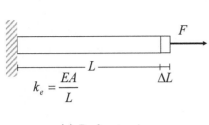

$$k_e = \frac{EA}{L}$$

(a) Rod extension.

$$k_c = \frac{3EI}{L^3} \quad \text{where} \quad I = \frac{ab^3}{12}$$

(b) Cantilever spring.

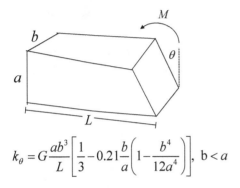

$$k_\theta = G\frac{ab^3}{L}\left[\frac{1}{3} - 0.21\frac{b}{a}\left(1 - \frac{b^4}{12a^4}\right)\right], \ b < a$$

(c) Torsional spring with a rectangular cross section.

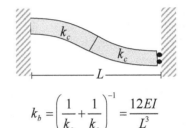

$$k_b = \left(\frac{1}{k_c} + \frac{1}{k_c}\right)^{-1} = \frac{12EI}{L^3}$$

(d) Guided beam spring is modeled as two cantilevers in series.

Figure 4.5: The schematic and spring constant for the three most common MEMS springs.

analyzed as a series and/or parallel combination of these basic structures. For example, a guided beam can be constructed from two cantilevers as shown in Figure 4.5(d). The spring constants shown in Figure 4.5 can be used to analyze the vast majority of practical MEMS devices. A more complete compilation of spring constants is given by Roark [34].

4.3.1   Rod extension

The simplest spring is a rod in extension shown in Figure 4.5(a). The spring constant is obtained from the relationship $T = ES$ between stress $T = F/A$ and strain $S = \Delta L/L$. Here E is the Young's modulus, F is the force acting on the rod, A is the cross sectional area, L is the beam length at rest, and ΔL is the change of length due to external force. Combining these expressions gives

$$F = \frac{EA}{L}\Delta L \equiv k_e\Delta L, \tag{4.18}$$

where we have recognized the spring constant for rod extension as

$$k_e = \frac{EA}{L}. \tag{4.19}$$

The rod in extension is a stiff structure and it is usually not suitable for sensors where resonance frequencies in the kilohertz range are desired. Possible applications for the extension spring are the high frequency resonators.

Example 4.4: Silicon rod spring
Problem: What is the spring constant for a silicon rod in extension with $E = 170$ GPa, $A = 2\ \mu\text{m} \times 2\ \mu\text{m}$, and $L = 50\ \mu\text{m}$? If a silicon mass with dimensions of $20\ \mu\text{m} \times 20\ \mu\text{m} \times 5\ \mu\text{m}$ is attached at the end of the beam, what is the resonance frequency?
Solution: Equation (4.19) gives

$$k_e = \frac{EA}{L} = 13.6\ \text{kN/m}.$$

The mass is $m = \rho V = 8.66$ pkg and the resonance frequency is

$$f_0 = \frac{1}{2\pi}\sqrt{\frac{k}{m}} = 8.6\ \text{MHz}.$$

4.3.2   Cantilever bending

The cantilever spring shown in Figure 4.5(b) is one of the most widely used microspring structures. The main advantage of the structure is that the beam can flex much more easily than it can extend. Thus, the cantilever can be used to make soft springs that are needed in sensors and actuators.

Figure 4.6 shows a cross section of a bending cantilever. The top surface of the beam is in tension and the lower surface is in compression so that the total stress is zero and the cantilever length is unchanged. The bending moment, however, is non-zero and the cantilever has curvature. The derivation of spring constant is more involved than for the rod and can be found in mechanical engineering textbooks [34, 35]. The results, however, are easy to use and are reviewed here.

The differential equation governing beam displacement is

$$EI\frac{\partial^2 x(Y)}{\partial Y^2} = M(Y), \tag{4.20}$$

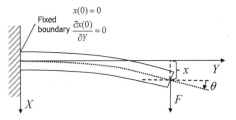
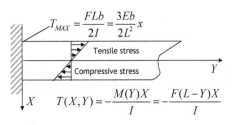

(a) Bending of the cantilever. Due to the bending moment, the cantilever is displaced by x and is in an angle θ relative to the rest position.

(b) Stress distribution in the cantilever beam. The maximum stress is located on the beam surface at the anchor.

Figure 4.6: Cantilever bending theory.

where I is the second moment of inertia that depends on the beam cross section, $x(Y)$ is the displacement at location Y along the beam length, and $M(Y)$ is the bending moment at location Y. MEMS beams are usually rectangular and the second moment of inertia is

$$I = \frac{ab^3}{12}, \tag{4.21}$$

where a is the transverse dimension and b is the thickness in the direction of cantilever bending. The second moment of inertias for other common geometries are tabulated in Appendix H.

Equation (4.20) is a second-order differential equation that is easily solved by integrating twice. The boundary conditions at the clamped end are $x(0) = 0$ and $\partial x(0)/\partial Y = 0$ which means beam cannot move or rotate (displacement x and angle $\theta = \partial x/\partial Y$ with respect to the Y-axis are both zero at $Y = 0$).

The spring constant depends on the location of the force acting on the cantilever. We will now derive effective spring constant for a cantilever with a length L and a point force F acting at beam end $Y = L$. Due to the point force, the moment acting on a beam at location Y is $M(Y) = F(L - Y)$. The beam displacement is obtained by integrating Equation (4.20) twice. Using the boundary conditions $x(0) = 0$ and $\frac{\partial x(0)}{\partial Y} = 0$ at the fixed end we obtain

$$x(Y) = \frac{F}{EI} \left(\frac{LY^2}{2} - \frac{Y^3}{6} \right). \tag{4.22}$$

From Equation (4.22), the beam tip displacement due to the force F acting on the tip is

$$x(L) = \frac{L^3}{3EI} F. \tag{4.23}$$

Comparing Equation (4.23) to Hooke's law $F = k_c x$, the effective spring constant for the cantilever is

$$k_c = \frac{3EI}{L^3}.$$ (4.24)

As indicated by Equations (4.24) and (4.21), the bending cantilever is sensitive to the length L and the thickness b but depends only linearly on beam width a.

The angle of the beam with respect to rest position θ shown in Figure 4.6(a) is

$$\theta(Y) = \frac{\partial x(Y)}{\partial Y} = \frac{F}{EI}\left[LY - \frac{Y^2}{2}\right].$$ (4.25)

The angle at the beam end is

$$\theta(L) = \frac{FL^2}{2EI}.$$ (4.26)

Example 4.5: Silicon cantilever spring
Problem: What are the bending spring constants for a silicon cantilever with $E = 170$ GPa, $h = 10$ μm, $w = 2$ μm, and $L = 50$ μm?
Solution: Equation (4.21) gives the second moment of inertias

$$I = \frac{ab^3}{12} = 1.66 \cdot 10^{-22} \text{ m}^4 \text{ and } 6.66 \cdot 10^{-24} \text{ m}^4$$

for bending in the h and w directions, respectively.
Equation (4.24) gives

$$k_c = \frac{3EI}{L^3} = 680 \text{ N/m and } 27.2 \text{ N/m}$$

for bending in the h and w directions, respectively.

Cantilever springs and its derivatives are useful for realizing piezoresistive sensors. For pure bending, the top surface of the cantilever is in tension while lower portion is in compressions, and the overall stress is zero as illustrated in Figure 4.6(b). The longitudinal stress at location (X, Y) is

$$T(X, Y) = -\frac{M(Y)X}{I}$$ (4.27)

where Y distance from the anchor along the beam length, X is the distance from the beam center where the stress is zero, and $M(Y)$ is the bending moment at location Y.

For a point force F at the beam end $Y = L$, the moment is $M = F(L - Y)$ and Equation (4.27) can be written as

$$T(X, Y) = -\frac{F(L - Y)X}{I}. \tag{4.28}$$

The maximum stress is located at the anchor point $(Y = 0)$ on the beam surfaces $(X = \pm b/2)$, and it is given by

$$|T_{MAX}| = \frac{FLb}{2I} = \frac{3Eb}{2L^2}x, \tag{4.29}$$

where x is the tip displacement from Equation (4.23).

Example 4.6: Stress due to a cantilever displacement
Problem: Calculate the longitudinal stress at the anchor location for a poly-crystalline silicon cantilever with the length $L = 100$ μm, the height $b = 2$ μm, and the tip displacement $x = 0.1$ μm. Assume $E = 150$ GPa.
Solution: The longitudinal stress at the beam surface near anchor from Equation (4.29) is

$$T_l = \frac{3Eb}{2L^2}x \approx 4.5 \text{ MPa}.$$

Example 4.7: Cantilever with a mass load
Problem: Figure 4.7 shows a simple accelerometer structure with a cantilever spring and a mass load. The center of the gravity for the mass is a distance $L_m/2 = 50$ μm from cantilever end. The cantilever dimensions are $L_c = 150$ μm, $b = 5$ μm, and $a = 20$ μm and the mass dimensions are $L_m = h_m = w_m = 100$ μm. Both the mass and the cantilever are made of single-crystal silicon with $E = 170$ GPa and $\rho = 2330$ kg/m^3. Calculate the cantilever tip displacement and the center of the mass displacement for a 10-G acceleration. You may ignore the mass of the cantilever.

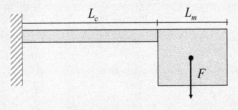

Figure 4.7: A mass loaded cantilever.

Solution: The mass is $m = \rho L_m^3 \approx 2.3$ nkg and the total force is $F = m\ddot{x} \approx 0.23$ μN. Since the force is not located at the cantilever tip, we need to rework our expression between the force and the cantilever tip displacement. With the force F at $L_c + k_m/2$, the moment at distance Y from the anchor is $M(Y) = F(L_c + L_m/2 - Y)$. By changing L to $L_c + L_m/2$, the cantilever displacement given by Equation (4.22) becomes

$$x(Y) = \frac{F}{EI}\left(\frac{(L_c + L_m/2)Y^2}{2} - \frac{Y^3}{6}\right).$$

Evaluating this at $Y = L_c$ gives the beam tip displacement $x \approx 11$ nm.

Due to the slope of the beam, the center of the mass will deflect more than the beam tip. The slope of the beam is

$$\theta(Y) = \frac{\partial x(Y)}{\partial Y} = \frac{F}{EI}\left[(L_c + L_m/2)Y - \frac{Y^2}{2}\right],$$

which at the beams end is $\theta(L) = 1.2 \cdot 10^{-4}$. Referring to Figure 4.8, the center of the mass is a distance $x_m = \theta L_m/2 = 6.1$ nm below the beam tip. The total center of the mass displacement is $x_{\text{tot}} = x + x_m = 17$ nm.

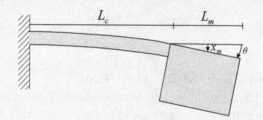

Figure 4.8: Deflection of the mass loaded cantilever.

4.3.3 Torsional springs

The Hooke's law for the torsion is $M = k_\theta \theta$ where M is the torsional moment acting on the beam and θ is the rotation angle in radians. For the rectangular beam in Figure 4.5(c), the spring constant is

$$k_\theta = G\frac{ab^3}{L}\left[\frac{1}{3} - 0.21\frac{b}{a}\left(1 - \frac{b^4}{12a^4}\right)\right] \quad \text{where } b < a. \tag{4.30}$$

As Equation (4.30) shows, the torsional spring is sensitive to the smallest lateral dimension b which can be oriented in the thickness or width direction. This makes the torsional springs well suited for making flexible springs on thick structures such as SOI where the thickness prevents using cantilevers or beams for out-of-plane displacements [36].

For silicon beams with length in [100]-crystal direction, the shear modulus is $G = 80$ GPa. As with the Young's modulus, the shear modulus for crystalline silicon depends on the device orientation with respect to crystal axis. As torsional twisting of beam results in shear along more than one axis, the analysis of torsional springs in other directions is complicated and the effective shear modulus depends on beam dimensions [37].

Example 4.8: Torsional accelerometer
Problem: Figure 4.9 shows a torsional accelerometer that measures motion in Z-direction [38]. The device is made of electroplated Ni ($E = 200$ GPa, $\nu = 0.31$, and $\rho = 8910$ kg/m^3). The torsional spring dimensions are $h = 5\ \mu$m, $w = 8\ \mu$m, and $L = 80\ \mu$m, and the plate thickness is 5 μm. The other dimensions are indicated in the figure. Due to asymmetry of the plates with respect to torsional springs, acceleration along Z-axis causes the plate to tilt. This plate movement is measured capacitively with electrodes placed under the accelerometer. What is the plate rotation and movement of the plate end due to a 1-G acceleration?

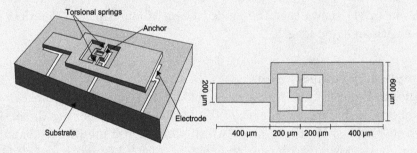

Figure 4.9: A torsional accelerometer [38].

Solution: From Equation (4.30), the spring constant for one beam is

$$k_\theta = G\frac{ab^3}{L}\left[\frac{1}{3} - 0.21\frac{b}{a}\left(1 - \frac{b^4}{12a^4}\right)\right] = 1.944 \cdot 10^{-7}\ \text{Nm}.$$

As there are two springs connected in parallel, the total spring constant is $k_{\text{tot}} = 2k_\theta = 3.888 \cdot 10^{-7}$ Nm. The moment due to acceleration $\ddot{x}$ acting on a mass dm at distance L from the axis of rotation is $dM = \ddot{x}Ldm$. The total moment acting on the accelerometer due to the 1-G acceleration is

$$M = \ddot{x}\rho w_R h \int_{L_1}^{L_2} L dL - \ddot{x}\rho w_L h \int_{L_1}^{L_2} L dL = \frac{1}{2}\ddot{x}\rho(w_R - w_L)h(L_2^2 - L_1^2).$$

Using $L_1 = 200$ μm, $L_2 = 600$ μm, $w_R = 600$ μm, and $w_L = 200$ μm, we get $M = 2.70 \cdot 10^{-11}$ Nm. The rotation due to the $\ddot{x} = 1$ G acceleration is

$$\theta = \frac{M}{k_{\text{tot}}} = 0.144 \text{ mrad}$$

and tip of the plate is displaced by

$$\Delta z = \theta L_2 = 86 \text{ nm}.$$

4.3.4 Guided beams

The guided beam shown in Figure 4.5(d) is the most used spring structure in surface micromachining. The free end of the beam is allowed to move but not to rotate. As a result, the beam bends into an S-shape. As illustrated in Figure 4.5(d), the spring constant can be derived by combining two cantilever springs with length $L/2$ in series to give

$$k_b = \frac{12EI}{L^3}. \tag{4.31}$$

Equation (4.31) shows that the guided beam is four times stiffer than a cantilever with an equal length.

Example 4.9: Folded beam
Problem: What is the spring constant for the surface micromachined accelerometer in Figure 3.6 on page 45. A close-up with spring dimensions is shown in Figure 4.10. If the beam width varies ± 100 nm due to manufacturing variances, how does the spring constant vary? Assume $E = 160$ GPa for polycrystalline silicon.

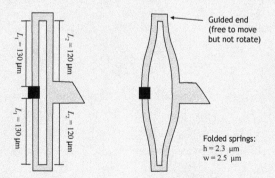

Figure 4.10: A close-up of the accelerometer in Figure 3.6 on page 45.

Solution: As shown in Figure 4.10, the folded beam can be analyzed as two guided beams in series. The small beam connecting the guided beams together can be neglected as its length is small and the spring constant is proportional to L^{-3}.

The guided beams have the spring constants

$$k_{b1} = \frac{12EI}{L_1^3} = 2.6 \text{ N/m}$$

and

$$k_{b2} = \frac{12EI}{L_2^3} = 3.3 \text{ N/m},$$

respectively where

$$I = \frac{ab^3}{12} = \frac{hw^3}{12}.$$

Note that the relatively small difference in beam length for springs k_{b1} and k_{b2} significantly changes the spring constant. The spring constant for the two springs in series is

$$k = \left(\frac{1}{k_{b1}} + \frac{1}{k_{b2}} \right)^{-1} = 1.5 \text{ N/m.}$$

Since there is a total of four folded beams attached to the mass, the total spring constant is

$$k_{\text{tot}} = 4k = 5.9 \text{ N/m.}$$

The estimated spring constant is close to actual value of 5 N/m on page 45. The range of spring constant for ± 100 nm variations is 5.2-6.6 N/m.

4.4 Computer simulations

In this chapter we have been analyzing the beams using simple expressions for the spring constants. Complex springs were analyzed using series and parallel combinations of the simple beams. Since the spring design is an integral part of designing many MEMS components, we will investigate the accuracy of the spring formulas using finite element modeling (FEM).

The FEM is the most accurate way to simulate three-dimensional physical systems. It is based on dividing the structure into small cells for which the approximate analytical solution is known. If the cell size is sufficiently small, the overall structure is accurately modeled even with the approximate solutions within a cell. The boundary conditions between the cells result in a large system of equations that is solved with a computer.

Figure 4.11 shows a finite element mesh for a simple guided beam. The beam is divided into small hexagonal bricks to model the entire structure. The FEM tool used allows the inclusions of the anisotropic silicon properties so the simulations can be carried out for a beam oriented in any direction. An example analysis is shown in Figure 4.11(b). One end of the beam is fixed and the other end is guided so it can move only in the beam width direction.

Table 4.2 shows a comparison of beam simulation and analytical solutions for a beam oriented in [100]-direction. The accuracy of the analytical results is excellent especially when we recall that the manufacturing tolerances will yield to variations greater than ten percent. However, both the analytical and FEM models are based on somewhat unrealistic fixed boundary conditions. In reality, the beam is going to be anchored to something that can also bend, stretch, and move.

(a) The mesh used to divide the beam into bricks.

(b) Simulated displacement for a guided displacement into the width direction.

Figure 4.11: Finite element model for a silicon beam with dimensions of 250 μm × 10 μm × 20 μm for the lenght, the width, and the height, respectively.

Table 4.2: Comparison of the finite element analysis and analytical beam equations for Figure 4.11.

Spring Symbol	Analytical	FEM	Units	Error
Rod extension, k_e	106.4	105.1	kN/m	1.3%
Guided beam in width direction, k_b	170.2	169.0	N/m	0.7%
Guided beam in thickness direction, k_b	681.0	667.0	N/m	2.1%
Torsional beam, k_θ	1.46	1.43	μNm	2.5%

Figure 4.12 shows a more realistic model where the beam is connected to blocks of silicon that form the anchors. Since the anchors are now elastic, we expect the beam to be less stiff than with the perfect boundaries. This is shown in Table 4.3 where the FEM calculated spring constants are consistently smaller

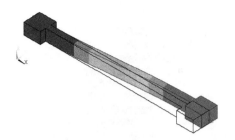

(a) Mesh for the model with anchors.

(b) Example simulations showing guided beam displacement in width direction.

Figure 4.12: A FEM model for the beam with anchors.

Table 4.3: Comparison of the finite element analysis and ideal analytical beam equations for a beam with realistic boundaries (Figure 4.12).

Spring Symbol	Analytical	FEM	Units	Error
Rod extension, k_e	106.4	94.6	kN/m	12.5%
Guided beam in width direction, k_b	170.2	153.9	N/m	11%
Guided beam in thickness direction, k_b	681.0	484.2	N/m	41%
Torsional beam, k_θ	1.46	1.42	μNm	3.3%

than the ideal analytical model. The accuracy for the beam displacement in the width direction and torsion is still good but the error for displacement in thickness direction is over 40%. The good agreement in the width direction and in torsion shows that the clambed boundary is a reasonable approximation as the anchor is much stiffer than the beam in this type of motion. For the displacement in the thickness direction, the beam and boundary are equally thick and the stiffness is comparable. Thus, the boundaries contribute more to the total compliance in this direction. As Example 4.10 shows, a good first approximation to account for the anchor elasticity is to increase the effective beam length by half of the beam width in the bending direction.

Figure 4.13 shows our final FEM example: A serpentine spring made of nine beams connected in series. The analytical solution is obtained by a simple series connection of the individual beam spring constants. Table 4.4 compares the analytical and FEM results. For the in-plane displacement and torsional beams, the analytical and FEM results are in good agreement but the analytical solution for out-of-plane displacement is off by a factor of two. This is due to the spring geometry that allows beams to rotate in addition to just bending in the thickness direction. The rotational movement lowers the overall beam stiffness in this direction.

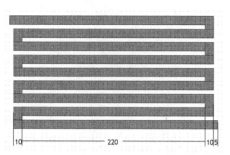

(a) Mask view and spring dimensions in micrometers.

(b) FEM model for the spring.

Figure 4.13: Finite element model for a serpentine spring.

Table 4.4: Comparison of the finite element analysis and analytical beam analysis for serpentine spring in Figure 4.13.

Spring Symbol	Analytical	FEM	Units	Error
Guided spring in width direction	22.3	21.8	N/m	2.3%
Guided spring in thickness direction	89.1	46.8	N/m	90%
Torsional beam	0.168	0.178	μNm	-5.5%

In summary, the simple analytical solutions can give excellent results even for complex serpentine springs. The main limitations of the simple spring constant approach covered in this chapter are that the possible anchor movements are not included in the spring models and the beam models are one dimensional, meaning that combined movement in two or more directions is not modeled. For example, combined spring torsion and bending is not modeled.

As stated before, the most accurate way to model any mechanical structure is to make a finite element model. In addition to giving an accurate spring constant, the FEM can be used to simulate, for example, the temperature affects and the effect of packaging stresses. Essentially all MEMS designs use FEM to verify the operation and to fine tune the design.

The drawback of FEM is that the model development is time consuming. The simple analytical models allow rough design with relative ease. Once the spring topology and rough dimensions have been chosen, it makes sense to go for more accurate computer analysis. Even after the construction of the finite element model, the analytical analysis is useful in providing guidance on what parameters to change and by how much to obtain the desired design. For example, if the FEM shows that a harder spring is needed, the analytical equations show that this is achieved by making the beams shorter or thicker.

Example 4.10: Effective beam length due to elastic anchors
Problem: Evaluate the bending spring constants for the anchored beam in Figure 4.12 but use the following "rule-of-thumb" to account for the anchor elasticity: An anchor is included in the spring constant by increasing the effective beam length by one half of the beam width in the bending direction.
Solution: For the bending in the width direction, $b = 10$ μm, and each anchor increases the effective length by $b/2 = 5$ μm. Total effective length is therefore $L_{eff} = L + b = 260$ μm. The spring constant is $k = 12EI/L_{eff}^3 \approx 151$ N/m. The difference between the calculated and the simulated value is just 1.6%.

For the bending in the thickness direction, $b = 20$ μm and the effective length is $L_{eff} = L + b = 270$ μm. The spring constant is $k = 12EI/L_{eff}^3 \approx 541$ N/m. The difference between the calculated and the simulated value is 11.6%.

Key concepts

- Most micromechanical springs are based on beam structures.

- Springs can be connected in series or parallel to form more complex structures.

- The two most common springs are the cantilever and the guided beam.

- Torsional springs are often used for out of plane rotating structures such as Z-axis accelerometers and micromirrors.

- Finite element modeling (FEM) is the most accurate way to analyze mechanical structures. Building and simulating FEM models is time consuming and should only be done after rough design has been done.

Exercises

Exercise 4.1
Calculate how much the single-crystal silicon spring constant changes if the spring is oriented to [110]- or [111]-direction instead of [100]-direction. What direction is preferred for making springs for accelerometers?

Exercise 4.2
Calculate the longitudinal stress at the anchor location for a silicon cantilever with the length $L = 150$ μm, the height $b = 4$ μm, and the tip displacement $x = 1$ μm. Assume $E = 170$ GPa and $\nu = 0.1$.

Exercise 4.3

A silicon rod is extended by pulling it from both ends. If the strain is 10^{-5}, by how much how much the rod volume changes? Assume $E = 130$ GPa and $\nu = 0.3$.

Exercise 4.4

Which one of the springs shown in Figure 4.14 is softer? Assume that the springs move in the direction of the force and the end points cannot rotate (guided springs).

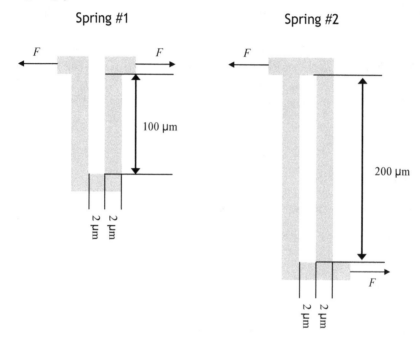

Figure 4.14: Two springs.

Exercise 4.5

Explain why modeling a cantilever beam with length L as a series connection of two cantilevers with spring lengths $L/2$ does not give correct result for the total spring constant.

Exercise 4.6

What are the bending spring constants for a silicon cantilever with $E = 170$ GPa, $h = 5$ μm, $w = 2$ μm, and $L = 100$ μm? How much the spring constants change if the dimensions change by ± 200 nm due to manufacturing variations.

Exercise 4.7

You are doing consulting work for an optical MEMS start-up company BS-MEMS, Inc. Their optical MEMS product line is not selling well, so the company wants to break into the lucrative aerospace and defense markets. You are reviewing their first design that is based on a micromirror product. The device is a fast response time accelerometer. The thickness of silicon structure is 3 μm, the mechanical quality factor is 300, distance to the substrate is 2.5 μm, the Young's modulus is 170 GPa, and the density is 2330 kg/m^3. To sense the acceleration, the capacitance between the structure and substrate is measured with an amplifier that is on a separate chip.

The specifications for the accelerometer are:

Maximum acceleration: ± 100 G

Noise limited resolution: 1 mG

Step response time: 100 μs to settle within 95%

Evaluate the design based on the performance criteria above. If required, recommend design changes.

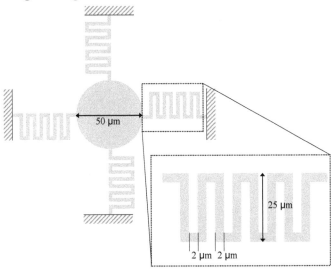

Figure 4.15: Figure for Exercise 4.7. Top view of (not so good) accelerometer.

Exercise 4.8

A silicon mass with dimensions of 100 μm $\times$ 100 μm $\times$ 50 μm is supported from one side by four cantilever beams with dimensions $h = 10$ μm, $w = 20$ μm, and $L = 200$ μm each. What is the effective spring constant in height direction assuming $Y = 170$ GPa? What is the center of the proof mass displacement and the maximum stress on the beams if it the device undergoes a 10-G acceleration?

Exercise 4.9

Again, consider the structure in Exercise 4.8 but now assume that the four beams are located so that there are two beams on the opposite sides. What is the center of the proof mass displacement and the maximum stress on the beams if it the device undergoes a 10-G acceleration?

5

Piezoresistive sensing

Electrical resistance changes due to mechanical stress. This effect occurs in all materials and it is called the piezoresistive effect. The piezoresistivity is utilized in many commercial devices including pressure sensors and accelerometers. In piezoresistive accelerometers, the acceleration is sensed by measuring the resistance change of piezoresistors embedded in the supporting springs. In pressure sensors, the deformation of pressure sensing membrane is measured by integrated piezoresistors.

The piezoresistive measurement offers two advantages that has made it popular for microsensors: the resistance measurement is easy to implement and piezoresistors are inherently shielded structures. In comparison, capacitive sensors require more complex measurement circuitry and hermetic packaging is needed to shield the device from environmental variations. Unfortunately, the resistance measurement is inherently noisy, has a large temperature dependency, and consumes a significant amount of dc power. Due to these drawbacks, the piezoresistive sensors have lost market share to capacitive sensors.

In this chapter, the piezoelectric effect and its application to microsensors are reviewed. The resistance change is analyzed and typical measurement configurations are covered. The noise, temperature dependency, sensitivity, and power consumption are discussed. The piezoresistive sensor performance is exemplified by analyzing the noise in a piezoresistive accelerometer.

5.1   Piezoresistive effect

When material is subjected to stress, its electrical resistance changes. The piezoresistive effect can be understood by considering resistance of a bar

$$R = \rho \frac{L}{A}, \tag{5.1}$$

where ρ is the resistivity, A is the cross sectional area, and L is the length. By deriving Equation (5.1), the fractional change in resistance is obtained as

$$\frac{\Delta R}{R} = \frac{\Delta \rho}{\rho} - \frac{\Delta A}{A} + \frac{\Delta L}{L}, \tag{5.2}$$

where the first term is due to the change in material resistivity $\Delta \rho$, the second term is due to the change in cross-sectional area ΔA, and the last term is due to the change in the length ΔL. As Equation (5.2) shows, the resistance changes are due to the geometrical and material changes. Hence, the piezoresistive effect occurs in all materials but is pronounced in materials that have a high resistivity change $\Delta \rho / \rho$.

The change in resistivity due to stress is described by the piezoresistivity coefficient π defined as the fractional change in resistivity ρ per unit stress

$$\pi = \frac{\Delta \rho / \rho}{T} = \frac{\Delta \rho / \rho}{ES}, \tag{5.3}$$

where T is stress, E is Young's modulus, and S is the strain. Another coefficient often used to characterize the piezoresistive effect is the gauge factor GF which is the fractional change in resistance per unit strain

$$GF = \frac{\Delta R / R}{S} = \frac{\Delta R / R}{\Delta L / L}. \tag{5.4}$$

Combining Equations (5.2), (5.3), and (5.4) and using $\frac{\Delta A / A}{\Delta L / L} = -2\nu$, we obtain

$$GF = 1 - \frac{\Delta A / A}{\Delta L / L} + \frac{\Delta \rho / \rho}{\Delta L / L} = 1 + 2\nu + E\pi, \tag{5.5}$$

where ν is the Poisson's ratio. The term $1 + 2\nu$ in Equation (5.5) is due to geometrical changes and $E\pi$ is due to material properties.

The gauge factor is a dimensionless figure-of-merit for strain gauges. As the Poisson's ratio for almost all materials is between 0 and 0.5, the gauge factor for purely geometrical changes ranges from 1 to 2. Table 5.1 shows that most metals have gauge factors within this range. Some metals such as nickel show larger gauge factors indicating significant resistivity change in addition to geometrical effects. Single-crystal (SC) silicon has gauge factors ranging from -102 to 135

Table 5.1: Gauge factors for selected materials [39].

Material	Gauge factor GF
Al	1.4
Cu	2.1
Ni	-12.62
Pt	2.60
Si(SC)	-102 to 135
Si(poly)	-30 to 40

and most of the change in resistance is due to material effects ($GF \approx E\pi$). The large gauge factors of crystalline and polycrystalline silicon makes them attractive for piezoresistive sensors.

Example 5.1: Piezoresistive coefficient in polycrystalline silicon
Problem: Estimate the piezoresistive coefficient in polycrystalline silicon with $GF = 30$ and $E = 150$ GPa.
Solution: As the gauge factor is large, the geometrical effects can be ignored. The piezoresistive coefficient is

$$\pi = \frac{GF}{E} \approx 20 \cdot 10^{-11} \text{ Pa}^{-1}. \tag{5.6}$$

5.2 Piezoresistive properties of silicon

The piezoresistive effect in silicon was discovered by Smith [40] in 1954. The effect has been utilized for example in atomic force microscope cantilevers [41], accelerometers [29], and pressure sensors [4]. As silicon is anisotropic material, its piezoresistive coefficients depend on crystalline orientation and direction of current with respect to stress. The formalism to calculate the piezoresistive coefficients for current and stress in any direction is covered in Appendix F. Most practical cases, however, have stress that is in the direction of the current (longitudinal stress T_l) and/or perpendicular to the current (transverse stress T_t). In these cases, fractional change in resistance can be written as

$$\frac{\Delta R}{R} = \pi_l T_l + \pi_t T_t, \tag{5.7}$$

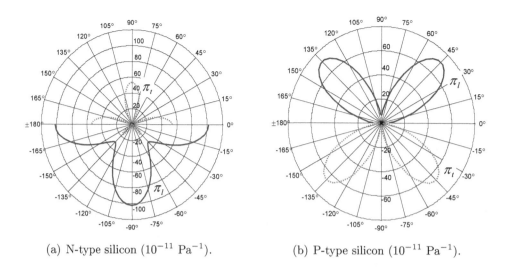

(a) N-type silicon (10^{-11} Pa^{-1}).   (b) P-type silicon (10^{-11} Pa^{-1}).

Figure 5.1: Longitudinal π_l (solid lines) and transverse π_t (dashed lines) piezoresistive coefficients for silicon for low doping concentrations at room temperature (After [42]).

Table 5.2: Longitudinal π_l and transverse π_t piezoresistive coefficients for silicon in selected orientations.

	n-type			p-type		
	[100]	[110]	[111]	[100]	[110]	[111]
π_l [10^{-11}Pa^{-1}]	-102.2	-31.2	-7.5	6.6	71.8	93.5
π_t [10^{-11}Pa^{-1}]	53.4	-17.6	6.1	-1.1	-66.3	44.6

where π_l is the longitudinal piezoresistive coefficient and π_t is the transverse piezoresistive coefficient. Figure 5.1 shows the longitudinal π_l and transverse π_t piezoresistive coefficients for silicon in different crystalline directions at low to modest doping concentrations (the dopant concentration $n < 10^{17}$ cm^{-3}). For higher concentrations, the piezoresistive effect decreases [42]. For convenience, the piezoresistance coefficient values for the most common orientation are also listed on Table 5.2.

The piezoresistive coefficients are temperature dependent. For doping concentrations less than 10^{17} cm^{-3}, the piezoresistive coefficients decrease as T^{-1}. Thus, the piezoresistive coefficients are obtained from

$$\pi(T) = \pi_{300K} \frac{300 \text{ K}}{T}, \tag{5.8}$$

where π_{300K} is the piezoresistive coefficient at 300 K and T is the temperature in Kelvins. At higher doping concentrations, the temperature dependency decreases [42].

In addition to the temperature dependency of the piezoresistive coefficients, the silicon resistivity is also temperature dependent. The silicon resistivity is a complex function of doping, carrier mobility, and temperature. A detailed model for n-type silicon is given in reference [43] and for p-type silicon in reference [44]. In general, the resistivity increases with temperature as the carrier mobility decreases. The dependency is not linear, but at room temperature, the temperature change is approximately 1-2%/K. This effect is much larger than the piezoresistive effect for reasonable stresses. Some form of temperature compensation is therefore needed to cancel out the temperature dependency.

Example 5.2: Piezoresistive effect in silicon
Problem: Consider a piezoresistor with the length $L = 50$ μm, the cross sectional area $A = 20$ μm^2, and the resistivity $\rho = 0.02$ Ωm at the room temperature. Calculate the resistance change due to a 10-MPa stress change in the direction of the current and due to a 5-K temperature change. Assume $\pi_l = 70 \cdot 10^{-11}$ Pa^{-1} and $\frac{\Delta\rho}{\rho} = 0.015\Delta T$ for a temperature change ΔT.
Solution: The resistance given by Equation (5.1) is

$$R = \rho\frac{L}{A} \approx 50 \text{ k}\Omega.$$

From Equation (5.7), the resistance change due to the stress is

$$\Delta R = \pi_l T_l R \approx 350 \ \Omega$$

or 0.7%. The resistance change due to temperature change ΔT is

$$\Delta R = 0.015\Delta T R \approx 3750 \ \Omega$$

or 7.5%. The change due to temperature is much greater than the piezoresistive effect.

Example 5.3: Piezoresistive accelerometer
Problem: Figure 5.2 shows a piezoresistive accelerometer which is similar to the first silicon accelerometer developed in the 1970s [29]. The structure is similar to Example 4.7 on page 61 except that there is a piezoresistor integrated on the beam. The cantilever dimensions are $L_c = 150$ μm, $b = 5$ μm, and $a = 20$ μm and the mass center of mass is $L_m/2 = 50$ μm from cantilever end. The mass

is $m = 2.3$ nkg and the force due to a 10-G acceleration is $F = m\ddot{x} \approx 0.23$ μN. Assume $E = 170$ GPa and $\pi_l = 20 \cdot 10^{-11}$. Calculate the relative piezoresistance change due to the 10-G acceleration. You can ignore the transverse stresses.

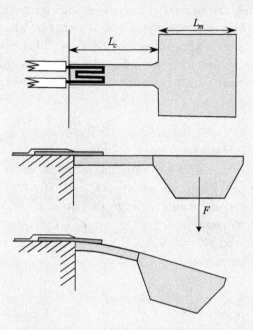

Figure 5.2: Piezoresistive accelerometer

Solution: From Equation (4.29), the stress in the piezoresistor is

$$T_{MAX} = \frac{F(L_c + L_m/2)b}{2I} \approx 550 \text{ kPa}.$$

The fractional change in resistance from Equation (5.7) is

$$\frac{\Delta R}{R} \approx \pi_l T_l \approx 1.1 \cdot 10^{-4}.$$

5.3   Piezoresistance measurement

Piezoresistive transducers translate the mechanical strain into a change in resistance. The task then is to carry out the resistance measurement to translate the change in resistance into electrical signal. The relative ease of measuring dc resistance is one of the main advantages of piezoresistive sensors. For example, we can apply known current through the resistor and measure the voltage drop to deduce the absolute value of the resistance.

While the absolute resistance measurement outlined above is possible, there are significant advantages to integrating reference resistors with the sensor element. The measurement change is carried out ratiometrically by comparing the ratio of sensing and reference resistances. When the *relative change* in resistance is measured, the temperature dependency of resistors is canceled in the first order and the measurement circuit is simplified.

5.3.1   Single-ended ratiometric measurement

Figure 5.3 shows a possible single ended measurement for detecting the fractional change in resistance $\Delta R/R$. During the fabrication process, two identical resistors are integrated with the sensor. The location of the resistors is chosen such that one of the sensors is subjected to stress changes while the other is at a constant stress. When a constant voltage V is applied to the series connected resistors, the change in resistance can be read by measuring resistor divided voltage

$$V_{\text{sig}} = \frac{R_1}{R_1 + R_2}V \approx \frac{1}{2}V + \frac{\Delta R}{4R_0}V, \qquad (5.9)$$

where $R_1 = R_0 + \Delta R$ and $R_2 = R_0$. The approximation in Equation (5.9) is valid for $\Delta R \ll R_0$. Using Equation (5.7), the signal voltage given by Equation (5.9) becomes

$$V_{\text{sig}} \approx \frac{1}{2}V + \frac{1}{4}(\pi_l T_l + \pi_t T_t)V. \qquad (5.10)$$

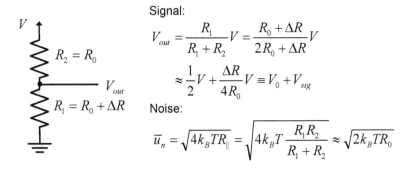

Signal:
$$V_{out} = \frac{R_1}{R_1 + R_2}V = \frac{R_0 + \Delta R}{2R_0 + \Delta R}V$$
$$\approx \frac{1}{2}V + \frac{\Delta R}{4R_0}V \equiv V_0 + V_{sig}$$

Noise:
$$\bar{u}_n = \sqrt{4k_B TR_\parallel} = \sqrt{4k_B T\frac{R_1 R_2}{R_1 + R_2}} \approx \sqrt{2k_B TR_0}$$

Figure 5.3: Single-ended ratiometric resistance measurement. Note that the resistor temperature dependency cancels to the first-order.

Equation (5.10) is a significant result as the signal voltage does not depend on the absolute value for the resistors R_1 and R_2. Also, as long as the resistors have the same temperature dependency and are at the same temperature, the temperature dependency of the resistors is canceled to the first-order. This is a major advantage for the ratiometric measurements as the changes in resistance, due to temperature changes, can be much larger than the changes due to stress. However, the fractional change in resistance $\frac{\Delta R}{4R_0}$ due to stress still shows the T^{-1} temperature dependency of piezoresistivity coefficients π_l and π_t given by Equation (5.8). This can be canceled only by measuring the temperature and compensating for the temperature induced change electronically.

An added complication is that in a realistic manufacturing environment, no two resistors are identical. This will cause sensor offset and the offset will also have non-zero temperature dependency. The compensation of the temperature induced errors is the biggest challenge in piezoresistive sensors. A brute force approach is to calibrate the sensor in multiple temperatures. This type of calibration is a slow and expensive process.

5.3.2   Differential ratiometric measurement

The single-ended ratiometric measurement shown in Figure 5.3 provides the advantage of being independent of the absolute resistance values. This independence reduces the sensor temperature dependency and the effect of processing variations. As shown by Equation (5.10), however, the single-ended signal has a large dc offset $V/2$ which makes the measurement of small resistance changes difficult. To appreciate this challenge, we will consider a typical signal with amplitude ranging from 0.05 mV to 5 mV (0.001% to 0.1% of 5 V bias). In order to digitize the signal, it should be amplified by roughly a factor of one hundred. This, however, would also amplify the offset which would become $100 V/2 \approx 250$ V which is clearly not possible. The dc offset has to be removed before the signal can be further processed.

The Wheatstone bridge shown in Figure 5.4 enables differential measurement of the piezoresistor output, thus removing the dc offset. The Wheatstone bridge can be understood as a combination of two voltage dividers. Each voltage divider consists of a piezoresistor and a reference resistor. By taking the difference in voltage divider outputs, the dc offset is canceled and the output is proportional to the relative resistance change. As the offset is removed at the sensor level, the measuring electronics is simplified.

As with the single-ended ratiometric measurement, the output is not dependent on the absolute resistance, and temperature dependency of resistors is canceled to the first-order. The added benefits of the differential measurement are increased sensitivity, increased linearity, and the removal of large dc offset.

As Figure 5.4 shows, the Wheatstone bridge may also be biased with a constant current I instead of the more common voltage bias. The benefit of the constant current bias is that the output is more linear in comparison to the voltage bias. Alternative ways to arrange the resistors and the biasing are tabled in reference [45].

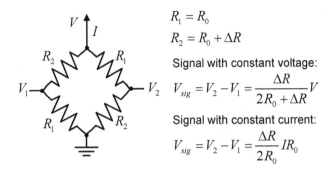

$$R_1 = R_0$$

$$R_2 = R_0 + \Delta R$$

Signal with constant voltage:

$$V_{sig} = V_2 - V_1 = \frac{\Delta R}{2R_0 + \Delta R} V$$

Signal with constant current:

$$V_{sig} = V_2 - V_1 = \frac{\Delta R}{2R_0} I R_0$$

Figure 5.4: Differantial ratiometric resistance measurement using the Wheatstone bridge.

Example 5.4: Piezoresistive silicon accelerometer

Problem: Figure 5.5 shows a piezoresistive bulk micromachined accelerometer. The proof mass is $m = 1.845$ μkg and the beam dimensions are $L = 280$ μm, $h = 4$ μm, and $w = 20$ μm. The total spring constant for the four beams combined is $k = 40$ N/m. Piezoresistors are integrated into the beams at the location of maximum stress and the current through the resistance is in the direction of stress (the transverse effects can be ignored). Assume that the piezoresistors are p-type silicon and $\pi_l = 70 \cdot 10^{-11}$ Pa^{-1}. In addition to the two piezoresistors, there are two reference resistors to form a Wheatstone bridge. If the bridge is biased with 2.5 V, what is the sensor sensitivity (mV/G) at the Wheatstone bridge output?

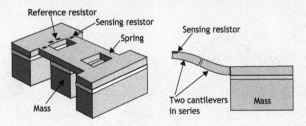

Figure 5.5: Silicon micromachined piezoresistive accelerometer. The proof mass displacement is detected with piezoresistors. The beam can be analyzed as two cantilevers in series.

Solution: For an acceleration $\ddot{x}$, the proof mass displacement at low frequencies is $|x| = m|\ddot{x}|/k = |\ddot{x}|/\omega_0^2$. For the analysis, the springs can be divided into two cantilevers and the cantilever tip placement is half of the proof mass displacement ($|x_c| = |x|/2$). The stress at the base of the cantilever is related to the tip displacement by Equation (4.29) giving

$$|T_{MAX}| = \frac{3Eb}{2L_c^2}|x_c| = \frac{3Eb}{2L_c^2}\frac{m|\ddot{x}|}{2k},$$

where $L_c = L/2$ is the length of the cantilever. From Equation (5.7), the change in resistance is

$$\Delta R = \pi_l T_l R_0 = \pi_l T_{MAX} R_0.$$

From Figure 5.4, the differential signal voltage from the Wheatstone bridge is

$$\Delta V = \frac{\Delta R}{2R_0 + \Delta R}V = \frac{\Delta R}{2R_0}V\left(1 - \frac{\Delta R}{2R_0} + \ldots\right) \approx \frac{\Delta R}{2R_0}V,$$

where the small terms $O(\Delta R^2/R_0^2)$ have been ignored. The sensitivity is

$$\left|\frac{\Delta V}{\ddot{x}}\right| \approx \frac{\pi_l}{2}\frac{3Eb}{2L_c^2}\frac{m}{2k}V = 10.4 \text{ mV/G}.$$

An alternative approach to find the maximum stress would have been to note that the force acting on the each beam is $F_{1/4} = m\ddot{x}/4$. This force must also be acting on each cantilever and from Equation (4.29), the stress is obtained as

$$T_{MAX} = \frac{F_{1/4}L_c b}{2I}$$

giving identical results.

Example 5.5: Piezoresistive pressure sensor

Problem: A typical use of piezoresistors in a pressure sensor is shown in Figure 5.6. The example piezoresistive pressure sensor makes creative use of the positive and negative piezoresistive coefficients in p-type silicon. When the diagram bends due to differential pressure across the membrane, there will be stress that will change the piezoresistor resistances. The stress will have two components shown: The stress normal to the edge is $T_N = 1.02pa^2/h^2$ where p is the pressure across the membrane, $2a$ is the membrane edge length, and h is the membrane thickness. The stress parallel to the membrane edge is $T_P = \nu T_N$ where ν is the Poison's ratio. Calculate the resistance change of the longitudinal (A) and transverse (B) resistors for pressure $p = 1$ kPa if the edge length is $2a = 2 \cdot 250$ μm and diaphragm thickness is $h = 10$ μm.

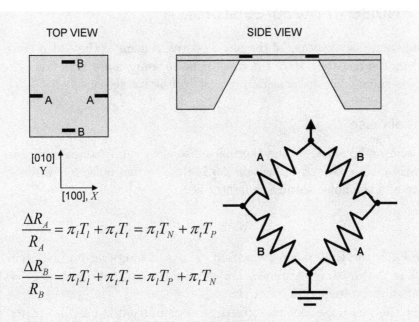

TOP VIEW   SIDE VIEW

[010] ↑
Y

[100], X

$$\frac{\Delta R_A}{R_A} = \pi_l T_l + \pi_t T_t = \pi_l T_N + \pi_t T_P$$

$$\frac{\Delta R_B}{R_B} = \pi_l T_l + \pi_t T_t = \pi_l T_P + \pi_t T_N$$

Figure 5.6: Example of piezoresistive pressure sensor. The membrane has p-type piezoresistors integrated in longitudinal and lateral configurations. As the sign of the pizoresistive coefficients differ, the sensitivity of the sensor is increased in the Wheatstone bridge configuration. (After [10]).

Solution: Both resistors are under longitudinal and transverse stress. For resistor 'A', the stresses are $T_l = T_N = 1.02pa^2/h^2 \approx 10.2$ MPa and $T_t = \nu T_N \approx 2.86$ MPa. The fraction change of resistance from Equation (5.7) is

$$\frac{\Delta R_A}{R_A} = \pi_l T_l + \pi_t T_t = \pi_l T_N + \pi_t \nu T_N \approx 0.0051,$$

where we have used $\pi_l = 70 \cdot 10^{-11}$ Pa^{-1} and $\pi_l = -70 \cdot 10^{-11}$ Pa^{-1}. For the resistors 'B', the current in X-axis direction. The longitudinal and transverse stresses are reversed and we have $T_l = \nu T_N$ and $T_t = T_N$. The fractional change of resistance is

$$\frac{\Delta R_B}{R_B} = \pi_l T_l + \pi_t T_t = \pi_l \nu T_N + \pi_t T_N \approx -0.0051.$$

Thus, resistance 'A' increases and resistance 'B' increases with pressure. This clever utilization of the positive and negative piezoresistive coefficients increases the sensitivity of the Wheatstone bridge.

5.4   Noise in piezoresistors

The inherent disadvantage of the piezoresistive sensing is the added noise introduced by the resistors. Here the resistor noise sources are reviewed. The noise
theory is exemplified by considering a piezoresistive accelerometer.

Thermal noise

The Johnson-Nyquist, or the thermal noise, sets the fundamental noise limit
for resistors. As covered in Section 2.3.1, the thermal noise has white spectral
density and the noise voltage generator is

$$\overline{v_n^2} = 4k_B T R, \tag{5.11}$$

where k_B is the Boltzmann's constant, T is the temperature in Kelvin units,
and R is the resistor resistance. For example, a 1-kΩ resistor at $T = 300$ K
exhibits noise voltage of 4 nV/$\sqrt{\text{Hz}}$.

The thermal noise in a piezoresistive measurement is best illustrated by an
example. Let us consider the noise voltage in Figure 5.3. The noise voltage at
the output due to piezoresistors is

$$\overline{v_n^2} = 4k_B T R_{||} = 2k_B T R_0, \tag{5.12}$$

where $R_{||} = R_1 || R_2$ is the parallel combination of the resistors $R_1 \approx R_2 \approx R_0$.
Equation (5.12) suggest that the noise voltage can be reduced by decreasing
the resistance values. This conclusion, however, is misleading as the power used
for the measurement

$$P_{\text{dc}} = \frac{V^2}{2R_0} \tag{5.13}$$

increases with decreasing R_0. By comparing the signal and noise powers,

$$P_{\text{sig}} = \frac{V_{\text{sig}}^2}{2R_0} = \left(\frac{\Delta R}{4R_0}\right)^2 \frac{V^2}{2R_0} = \left(\frac{\Delta R}{4R_0}\right)^2 P_{\text{dc}} \tag{5.14}$$

and

$$\overline{P}_n = \frac{\overline{v_n^2}}{R_{||}} = 4k_B T, \tag{5.15}$$

respectively, we find that the signal-to-noise ratio depends only on the fractional
sensitivity of the element $\frac{\Delta R}{4R_0}$ and measurement power P_{dc}. Thus, the only way
to increase the signal-to-noise ratio in piezoresistive measurements where the
noise is dominated by the thermal noise, is to increase the measurement power.
This is a major disadvantage for piezoresistive sensors especially in battery
powered systems.

Shot noise

Shot noise is due to random fluctuations of the electric current across a potential barrier. Shot noise is observed for example in vacuum tubes and semiconductor junctions. In vacuum tubes, the noise is due to random emissions of electrons from the cathode. In semiconductors, the shot noise is due to the random diffusion of electrons and holes across a junction.

The shot noise was first analyzed by Schottky in 1918 [46] and the equation for the noise current spectral density

$$\bar{i}_n = \sqrt{2q_e I} \tag{5.16}$$

is known as the Schottky formula. Here q_e is electron charge and I is current through the potential barrier. Unlike the Johnson-Nyquist noise, which is independent of the applied voltage and current, shot noise increases with current.

Ideal conductors (or resistors) do not exhibit the shot noise as there is no potential barrier. However, piezoresistors may exhibit shot noise due to poor contacts. For example, a metal-semicondutor contact can have a potential barrier on the order of ~0.2-0.4 V if proper manufacturing procedures have not been followed. In the case of bad contacts to the piezoresistor, the shot noise may be significant. For example, a 1-mA current through a 1-kΩ resistor with bad contacts exhibits 18 pA/$\sqrt{\text{Hz}}$ of shot noise. This is equivalent to the noise voltage $\bar{v}_n = i_n R = 18$ nV/$\sqrt{\text{Hz}}$ and four times larger than the thermal noise $\bar{v}_n = \sqrt{4k_B T R} = 4$ nV/$\sqrt{\text{Hz}}$.

Flicker noise (1/f-noise)

The flicker noise or 1/f-noise was introduced in Section 2.2.2. The flicker noise in resistors can be modeled with the experimental Hooges formula

$$\bar{v}_n = \sqrt{\frac{\alpha}{fN}} V, \tag{5.17}$$

where $\bar{v}_n$ is the noise voltage spectral density corresponding voltage fluctuation, V is the bias voltage across the resistor, N is the total number of carriers in the resistor, f is the frequency, and α is the Hooge's factor which is a material and process dependent parameter [47–49]. Typically, the Hooge's factor α varies between 10^{-7} and 10^{-3}. For a uniform bar shaped resistor, the total number of carriers can be written as $N = nlwh$, where n is the carrier density and l, w, and h is the piezoresistor length, width, and thickness, respectively.

Equation (5.17) has several important consequences:

1. The noise power increases as 1/f. At high frequencies, the 1/f-noise is below the Johnson-Nyquist noise but at low frequencies the 1/f-noise dominates. Due to the 1/f-noise, the piezoresistor noise in dc or low

frequency measurements can be orders-of-magnitude higher than expected from Equation (5.11) for the Johnson-Nyquist noise.

2. The $1/f$-noise is proportional to the bias voltage V. As a result, increasing the bias voltage has no effect on the signal-to-noise ratio when the resistor $1/f$-noise dominates.

3. The $1/f$-noise is inversely proportional to the resistor volume. Physically large resistors may be needed for low noise design.

4. Increasing the doping density increases the number of carriers and reduces the noise. High doping densities are used to improve noise performance but concentrations higher than 10^{17} cm^{-3} reduce the piezoresistor sensitivity. Optimal design maximizes the signal-to-noise ratio.

Example 5.6: Flicker noise in piezoresistor
Problem: A n-type piezoresistor with the carrier concentration $n = 10^{16}$ cm^{-3} and the mobility $\mu = 1250$ cm^2/s is used in a silicon sensor. Assuming Hooge's factor $\alpha = 3 \cdot 10^{-5}$, device dimensions $l = 40$ μm, $w = 5$ μm , and $h = 2$ μm, and the bias voltage $V = 1$ V, what is the $1/f$-corner frequency for the resistor? What is the rms-noise for 0.1 Hz to 1 kHz bandwidth? For semiconductors, the resistivity is $\rho = 1/q_e\mu n$.
Solution: The resistance is

$$R = \rho \frac{l}{wh} = \frac{1}{q_e\mu n}\frac{l}{wh} = 20 \text{ k}\Omega$$

and the number of carriers in the resistor is

$$N = nwlh = 4 \cdot 10^6.$$

Equating the thermal noise (Equation (5.11))

$$\overline{v_n^2} = 4k_BTR \approx 3.31 \cdot 10^{-16} \text{ nV}^2/\text{Hz}$$

and the $1/f$-noise (Equation (5.17))

$$\overline{v_n^2} = \frac{\alpha}{fN}V^2$$

gives the corner frequency

$$f_c = \frac{\alpha V^2}{N4k_BTR} = 22.6 \text{ kHz}.$$

From Equation (2.13), the rms-noise due to Flicker-noise is

$$v_{\mathrm{rms}} = \sqrt{\overline{v_n^2} f_c \ln \frac{f_H}{f_L}} \approx 25 \ \mu\mathrm{V}.$$

In comparison, the thermal noise over the same 1 kHz bandwidth is just 0.6 μV. Summing the noise powers, the total noise is $v_{\mathrm{rms}} = \sqrt{(25 \ \mu\mathrm{V})^2 + (0.6 \ \mu\mathrm{V})^2} = 25 \ \mu$V.

Example 5.7: Piezoresistive silicon accelerometer

Problem: The piezoresistive accelerometer in Figure 5.5 of Example 5.4 has piezoresistor dimensions of $l_r = 36 \ \mu$m, $h_r = 0.4 \ \mu$m, and $w_r = 6 \ \mu$m, the resistor carrier concentration is $n = 5 \cdot 10^{16}$ cm^{-3}, the resistivity is $\rho = 0.34 \ \Omega$cm, and the Hooge's factor is $\alpha = 10^{-5}$. What is the input referred rms-noise (rms-acceleration) for a $0.1 - 100$ Hz noise bandwidth if the resistor Flicker noise is ignored? What is the input referred rms-noise with Flicker noise included? The mass $(m = 1.845 \ \mu$kg) and damping $(Q = 0.2$, and $\gamma = \frac{m\omega_0}{Q} \approx 0.043 \ \mu$kg/s) of this accelerometer are identical to Example 3.7 on page 42.

Solution: Figure 5.7 shows the schematic representation of the signal and the noise in a piezoresistive accelerometer. The proof mass m converts the acceleration to a measurable force F. In addition, there is mechanical noise represented by the mechanical noise force generator $\overline{F}_n$ that adds to the signal force. From Equation (2.25) the mechanical rms-noise force is

$$F_{\mathrm{rms}} = \sqrt{4k_B T \gamma BW} = 2.7 \cdot 10^{-10} \ \mathrm{N}$$

where BW is the noise bandwidth. The input referred acceleration noise is $\ddot{x}_{\mathrm{rms}} = F_{\mathrm{rms}}/A_1 = F_{\mathrm{rms}}/m = 12 \ \mu$G.

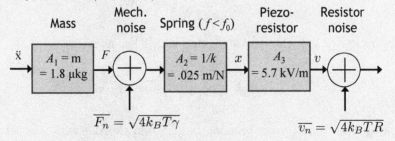

Figure 5.7: System level schematic of the noise in piezoresistive accelerometer.

The beams and piezoresistors convert the force to measurable voltage. From Example 5.4, the force to displacement transfer function is

$$A_2 = \frac{1}{k} \approx 0.025 \text{ m/N},$$

where we have assumed that $f \ll f_0$. The piezoresistors convert the displacement to measurable voltage with the conversion factor given by

$$A_3 = \frac{\pi_l}{2} \frac{3Eb}{2L_c^2} \frac{V}{2} \approx 5700 \text{ V/m},$$

where we have again used the results from Example 5.4.

Ignoring the flicker noise, the noise due to the piezoresistor is

$$v_{\text{rms}} = \sqrt{4k_B TRBW} \approx 0.29 \ \mu\text{V},$$

where the resistance

$$R = \rho \frac{l_r}{h_r w_r} = 51 \text{ k}\Omega$$

was used. The total noise at the output due to the mechanical and the electrical noise is

$$v_{\text{rms,tot}} = \sqrt{(A_2 A_3 F_{\text{rms}})^2 + v_{\text{rms}}^2} = 0.29 \ \mu\text{V}$$

and input referred acceleration noise is $\ddot{x}_{\text{rms}} = v_{\text{rms,tot}}/A_1/A_2/A_3 = 0.11$ mG. Compared to the noise equivalent acceleration of 12 μG for the mechanical noise alone, the noise due to resistors is ten times higher.

The power dissipation in the Wheatstone bridge is

$$P = \frac{V^2}{R} = 120 \ \mu\text{W}.$$

The number of carriers in the piezoresistor is $N = nl_r w_r h_r$ and from Equations (5.17) and (2.25), the Flicker noise corner frequency is obtained as

$$f_c = \frac{\alpha V^2}{N4k_B TR} \approx 17 \text{ kHz}.$$

From Equation (2.13), the rms-noise due to Flicker-noise is

$$v_{\text{rms},1/f} = \sqrt{v_n^2 f_c \ln \frac{f_H}{f_L}} \approx 26 \ \mu\text{V}.$$

We see that the Flicker noise is much larger than the thermal noise. The input referred noise with the Flicker-noise included is $\ddot{x}_{\text{rms},1/f} = v_{\text{rms},1/f}/A_1/A_2/A_3 = 10$ mG which is 1,000 larger than the intrinsic mechanical noise.

Key concepts

- Piezoresistive sensors are fairly easy to manufacture and the dc resistance measurement is simple to implement.

- Piezoresistors are inherently shielded structures.

- Resistive measurements are inherently noisy and consume power.

- Ratiometric measurements compare two or more resistors. The benefit of ratiometric measurement is that the resistor temperature dependency and the effect of process variations are canceled to the first order.

- Differential measurements have the advantage that the dc offset is canceled and the measurement circuitry is thus simplified.

Exercises

Exercise 5.1
A nickel strain gauge has the nominal resistance $R = 500$ Ω. Calculate the resistance when the strain gauge is elongated by 10^{-3}%.

Exercise 5.2
Explain why differential measurements are used in resistive sensing.

Exercise 5.3
Consider the Wheatstone bridge in Figure 5.6. Derive an expression for the differential output voltage if $R_A = R_0 + \Delta R$ and $R_B = R_0 - \Delta R$.

Exercise 5.4
Consider the piezoresistive accelerometer in Figure 5.8. The proof mass dimensions are $L = 1.2$ mm, $W = 1.2$ mm, and $H = 380$ μm. The mass is anchored with a beam with dimensions dimensions $L = 245$ μm, $w = 80$ μm, and $h = 6$ μm. Two 12.5 kΩ piezoresistors with $\pi_l = 70 \cdot 10^{-11}$ Pa^{-1} are integrated on the cantilever near the anchor and two identical resistor are integrated as a reference at a zero stress location. The four resistors are connected to form a Wheatstone bridge as shown in Figure 5.4.

a) Calculate the relative change in piezoresistances due to a 1-G acceleration in Z-direction. Assume $E = 170$ GPa and $\rho = 2330$ kg/m^3.

b) What is the signal voltage for the 1-G acceleration with 2.5 V bias over the Wheatstone bridge?

c) What is the dc power consumption due to the bias?

d) What is the signal-to-noise ratio for a 1-mG acceleration? Assume noise bandwidth from 0.1 Hz to 100 Hz and that the resistor $1/f$-corner frequency is $f_c = 2$ kHz.

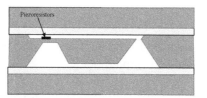

Figure 5.8: Piezoresistive accelerometer.

Exercise 5.5

Silicon resistivity can be expressed as $\rho = 1/q_e\mu n$, where q_e is the electron charge, μ is mobility of carriers, and n is the carrier concentrations? A n-type piezoresistor with carrier concentration $n = 10^{15}$ cm^{-3} and mobility $\mu = 1250$ cm^2/s is integrated on a silicon sensor. Assuming the Hooge factor $\alpha = 3 \cdot 10^{-5}$, the resistor dimensions $l = 20$ μm, $w = 5$ μm , and $h = 2$ μm, and the bias voltage $V = 1$ V, what is:

i. the resistance value?

ii. the $1/f$-corner frequency for the resistor?

iii. rms-noise over 0.1 Hz to 1 kHz bandwidth?

Exercise 5.6

Consider the piezoresistive accelerometer in Example 5.3. List as many ways as possible to improve the noise performance while maintaining the mechanical bandwidth.

Exercise 5.7

Investigate the use of piezoresistive microcantilever as an atomic force microscope. The cantilever must be able to detect pits that are 0.1 nm deep. Assume that the noise is limited by the resistor noise $v_{\text{rms}} = 20$ μV and use piezoresistive sensing with $n = 10^{15}$ cm^{-3}, $\mu = 1250$ cm^2/s, and $\pi = 70 \cdot 10^{-11}$ Pa^{-1}. Choose dimension for the cantilever to meet the displacement resolution specifications.

Exercise 5.8

Consider an atomic force microscope cantilever with the length L, the width W, and the height H that has piezoresistors integrated in the location of the maximum stress. Derive expressions for the piezoresistance change due to i. cantilever tip displacement x and due to ii. force F acting on the cantilever tip. Is it possible to optimize the cantilever sensitivity to both displacement and force?

6

Capacitive sensing

Capacitive sensors offer excellent noise performance and low power consumption. The basic principle is to measure the change in capacitance due to mechanical movement. The classic example is the capacitive accelerometer in which the acceleration is deduced by measuring the capacitance between the proof mass and the package (frame). Other applications include capacitive pressure sensors, microphones, and gyroscopes.

While the basic operation principle is simple, the practical designs must overcome several challenges, as the capacitance changes can be extremely small. For example, the capacitance change for bulk micromachined accelerometers would typically be around femtofarad for a 1-mG acceleration and the capacitance changes for surface micromachines are orders-of-magnitude smaller. Measuring these small capacitances accurately is not trivial.

In this chapter, the capacitive measurement principles are covered. The most common capacitive sensing configurations are reviewed, including the differential measurement and the rate-of-change measurement. The effect of parasitic capacitances is discussed and possible ways to minimize the parasitic effects are analyzed. The noise performance is illustrated with a bulk micromachined accelerometer example.

6.1 Capacitance measurement

Figure 6.1 shows the schematic of a micromachined capacitive accelerometer. The acceleration is deduced by measuring the capacitance between the proof mass and the package (frame). All capacitive measurements are based on de-

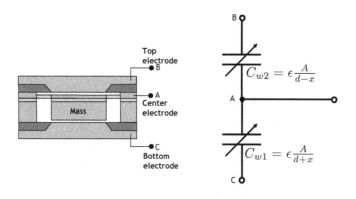

Figure 6.1: A bulk micromachined accelerometer. The proof mass movement x is measured by sensing the capacitance between the proof mass and top and bottom electrodes.

tecting the current through the capacitor given by

$$i = \frac{\partial(CV)}{\partial t} \tag{6.1}$$

$$= C\frac{\partial V}{\partial t} + V\frac{\partial C}{\partial t}, \tag{6.2}$$

where V is the voltage over the capacitor and C is the capacitance. The first term in Equation (6.2) represents the current due to the time varying voltage, which is familiar from the circuit analysis. The second term is the current due to the time variations of the capacitance value. In the normal circuit analysis, the second term is absent as capacitances are assumed to be constant.

In typical operation, one of the two terms in Equation (6.2) is significantly larger than the other. The capacitance measurement based on the first term is called the *displacement measurement*. Capacitance measurement based on the capacitance change is called the *rate-of-change* or *velocity measurement* as it depends on how quickly the capacitance changes. These two measurement principles are analyzed in the following sections.

6.1.1   Rate-of-change measurement

If the voltage $V = V_{\mathrm{dc}}$ over the capacitor is kept constant ($\frac{\partial V}{\partial t} = 0$), the current through the capacitor from Equation (6.2) is

$$i = V_{\mathrm{dc}}\frac{\partial C}{\partial t} \equiv i_{\mathrm{mot}}. \tag{6.3}$$

As the current through the capacitor due to capacitance variations will appear repeatedly, we define it as the *motional current* i_{mot}. Because the motional current is proportional to the capacitance rate-of-change $\frac{\partial C}{\partial t}$, the rate-of-change measurement cannot be used to determine a static displacement and the signal is small at low frequencies. Due to this limitation, the rate-of-change measurement is rarely used for accelerometers; however, the rate-of-change measurement finds applications in microphones and resonators that operate at higher frequencies.

Parallel plate capacitor is the most common MEMS capacitor geometry. As the name implies, a parallel plate is formed by two plates that are parallel to each other and separated by a small gap. The capacitance is given by

$$C = \epsilon \frac{A}{d - x}, \tag{6.4}$$

where ϵ is the permittivity, A is the electrode area, d is the initial electrode gap, and x is the movement of the capacitor plate from the initial position. The motional current is obtained as

$$i_{\text{mot}} = V_{\text{dc}} \frac{\partial C}{\partial t} = V_{\text{dc}} \frac{\partial C}{\partial x} \frac{\partial x}{\partial t} = V_{\text{dc}} \epsilon \frac{A}{(d-x)^2} \frac{\partial x}{\partial t} \approx V_{\text{dc}} \epsilon \frac{A}{d^2} \frac{\partial x}{\partial t} = V_{\text{dc}} \epsilon \frac{A}{d^2} \dot{x}, \tag{6.5}$$

where the approximation is valid when $x \ll d$. As Equation (6.5) shows, the motional current is proportional to the *velocity* of the capacitor plate.

Example 6.1: A capacitive microphone
Problem: As an example of the rate-of-change sensing, Figure 6.2 shows a schematic and electrical equivalent for a capacitive microphone that is based on measuring the capacitance change between a moving upper plate and fixed lower plate [50]. The upper plate moves due to the sound pressure and the fixed lower plate has holes to reduce the air damping between the plates. The measured capacitance change is $\Delta C = 7$ fF/Pa. Calculate the sensor sensitivity (V/Pa) for $V_{\text{dc}} = 5$ V dc bias assuming $R = \infty$. What is the low frequency limit for the measurement if $R = 100$ MΩ?

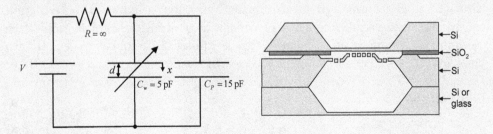

Figure 6.2: An equivalent circuit diagram and schematic view of bulk micromachined capacitive microphone [50].

Solution: The capacitance between the plates is $C_w = C_0 + \Delta C$ where the C_0 is the capacitance with no sound signal and ΔC is the capacitance change due to sound pressure. The motional current given by Equation (6.3) is

$$i_{\text{mot}} = V_{\text{dc}}\frac{\partial(C_0 + \Delta C)}{\partial t} = V_{\text{dc}}\frac{\partial \Delta C}{\partial t} = V_{\text{dc}}j\omega\Delta C, \qquad (6.6)$$

where we have used $\frac{\partial \Delta C}{\partial t} = j\omega\Delta C$. As shown in Figure 6.2, the large resistor prevents the current from flowing into the bias supply, and the motional current is divided between the work and parasitic capacitances, C_w and C_P, respectfully. The output voltage with the 5-V bias is

$$v = \frac{i_{\text{mot}}}{j\omega(C_w + C_P)} = \frac{\Delta C}{C_w + C_P}V_{\text{dc}} = 1.8 \text{ mV/Pa.} \qquad (6.7)$$

This theoretical sensitivity is in good agreement with the reported experimental sensitivity of 1.6 mV/Pa [50]. The measured microphone noise was dominated by the preamplifier noise.

In this example we assumed that the dc blocking resistor resistance was much higher than the capacitor impedances. As the frequency is decreased, the capacitive impedance increases as $1/\omega$. Thus, there is a practical low frequency limit for the capacitive measurements: Below the -3 dB cutoff frequency most motional current will flow through the resistor and not the capacitors. The output voltage becomes $v \approx i_{\text{mot}}R$, which decreases with frequency as i_{mot} decreases with frequency. For example, assuming $R = 100$ MΩ, the -3 dB cutoff frequency is $f_{-3\text{ dB}} = 1/(2\pi R(C_P + C_w)) \approx 80$ Hz.

6.1.2   Displacement measurement

The displacement measurement is based on measuring the regular ac current through the capacitor. The measurement frequency is assumed to be significantly higher than the frequency of capacitance variations. For example, the mechanical movement in accelerometers is limited to frequencies less than a few kilohertz and the frequency of the measurement voltage can be 100 kHz or more. In this case, we can approximate the capacitance C as a constant and Equation (6.2) becomes

$$i \approx C\frac{\partial V}{\partial t} = C\dot{V}, \qquad (6.8)$$

where $\dot{V}$ is the time derivative of the voltage. In frequency domain we have $i = sCV = j\omega CV$.

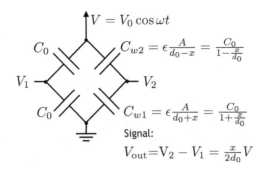

$$V = V_0 \cos \omega t$$

$$C_{w2} = \epsilon \frac{A}{d_0 - x} = \frac{C_0}{1 - \frac{x}{d_0}}$$

$$C_{w1} = \epsilon \frac{A}{d_0 + x} = \frac{C_0}{1 + \frac{x}{d_0}}$$

Signal:

$$V_{\text{out}} = V_2 - V_1 = \frac{x}{2d_0}V$$

Figure 6.3: Capacitive Wheatstone bridge.

Equation (6.8) is the normal ac current through a capacitor that is familiar from the circuit analysis. It can be measured directly to deduce the capacitance; however, as with the resistance measurement covered in Chapter 5, there are significant benefits to measuring the capacitance change ratiometrically by comparing two or more capacitors. By looking at the difference in the capacitance changes, the effect of environmental variations can be minimized and the measuring circuitry is simplified.

Figure 6.3 shows a capacitive Wheatstone-bridge similar to the resistive Wheatstone bridge of Figure 5.4. For the two parallel plate capacitors moving in opposite directions, the changing capacitances are $C_{w1} = \epsilon A/(d + x)$ and $C_{w2} = \epsilon A/(d - x)$ where ϵ is the permittivity, A is the capacitor area, d is the initial gap, and x is the capacitor plate displacement. The Wheatstone bridge output is the difference between the left arm voltage $V_1 = V/2$ and right arm voltage $V_2 = C_{w2}/(C_{w1} + C_{w2})V$. We have

$$V_{\text{out}} = V_2 - V_1 = \frac{x}{2d}V, \qquad (6.9)$$

which is proportional to the displacement x. As with the resistive measurement, the removal of common mode signal is a significant benefit of the differential measurement. Also, with the measurement configuration shown in the Figure 6.3, the measurement signal is a linear function capacitor displacement even though the capacitance itself is a nonlinear function of the displacement.

The challenge in measuring small capacitances is that any two conductors separated by a dielectric (air or solid) will form a capacitance. Examples 6.2 and 6.3 illustrate the typical parasitic capacitances and how they can affect the measured signal voltage. If the parasitic capacitances are fixed, the effect is to reduce the sensor sensitivity. However, if the parasitic capacitances vary for example due to temperature or humidity variations, it may be impossible to differentiate between changes in the sensor capacitor and parasitic capacitances. Methods to reduce the parasitic effects are covered in the next section.

Example 6.2: Parasitic capacitances in two chip accelerometer
Problem: Figure 6.4 shows a two chip solution for a capacitive microsensor. The MEMS sensor chip is wire bonded to an integrated circuit. The work capacitances are $C_{w1} = C_{w2} = 0.35$ pF and the capacitance change due to acceleration is $\Delta C/|\ddot{x}| = 2$ fF/mG. (1) Estimate the bond pad to ground capacitances $(A = (80\ \mu\text{m})^2$, $d = 2\ \mu\text{m}$, $\epsilon_R = 7)$. (2) If a finger is brought within 1 mm of the MEMS device, what is the capacitance from MEMS to finger and how are the measurement affected?

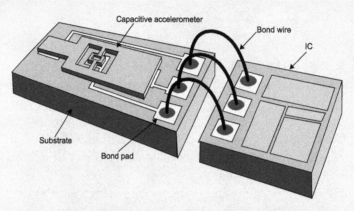

Figure 6.4: Two chip solution for integration of MEMS and IC.

Solution: (1) The bond pad capacitance is

$$C = \epsilon_R \epsilon_0 \frac{A}{d} = 0.2\ \text{pF}. \tag{6.10}$$

The bond pad capacitance is of similar size as the work capacitance and it can decrease the sensitivity of the sensor. Moreover, the bond pad capacitance can have a sizable temperature dependency which may introduce variations in the sensor output.

(2) An order-of-magnitude estimate for the capacitance from MEMS device to finger is obtained by using the parallel plate approximation. Assuming the MEMS device area is $A = (400\ \mu\text{m})^2$, the distance $d = 1$ mm gives $C = 14$ fF from device to finger. Similar femtofarad capacitances are expected from other portions of the chip to finger. For example, finger in proximity of the bond wires will change the capacitances between the bond wires. If the measurement circuitry cannot separate between the capacitance change due to the finger proximity and the work capacitance changes, the accelerometer will give erroneous readings. For example, with the sensitivity $\Delta C/|\ddot{x}| = 2$ fF/mG, the $C = 14$ fF capacitance change due to finger has potential to change the accelerometer output by 7 mG.

Example 6.3: Sensitivity reduction due to parasitics
Problem: Figure 6.5 shows a capacitive Wheatstone bridge with parasitic capacitances. Calculate the bridge output voltage $V_{\text{sig}} = V_2 - V_1$ with and without the parasitic capacitances C_P. Assume $C_0 = 100$ fF, $\Delta C = 1$ fF, $C_P = 1$ pF, and $V = 1$ V.

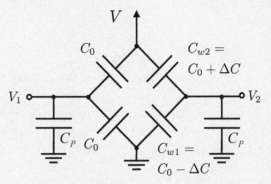

Figure 6.5: A capacitive Wheatstone-bridge with parasitic capacitances.

Solution: The bridge output voltage is

$$V_{\text{sig}} = V_2 - V_1 = \frac{\Delta C}{2C_0 + C_P}V.$$

With the parasitics, we have $V_{\text{sig}} \approx 0.83$ mV and without parasitics we have $V_{\text{sig}} \approx 5$ mV. The parasitics reduce the bridge sensitivity by $6\times$.

6.2   Minimizing the effect of parasitic capacitances

As Example 6.2 illustrates, the typical capacitance due to connections can be of similar magnitude as the sensor capacitance itself. It is impossible to avoid the parasitic effects completely but a good design can minimize the problems caused by the parasitics. The two approaches used in parallel are the reduction of the magnitude of parasitic by sensor design and minimizing the effect by circuit design techniques, such as bootstrapping. In this section, the common design techniques are reviewed.

6.2.1   Single-chip integration

An effective way to reduce the parasitic capacitances between the sensor and the amplifier is to integrate the MEMS elements and electronics on a single die. This eliminates both the bond pads and the bond wires. In addition, the "single-chip" integration can be cost effective as the assembly cost is reduced. Unfortunately, the MEMS and IC processes usually have conflicting process requirements which may result in compromises in the sensor and/or IC design. For example, the thickness of surface micromachined accelerometers may be limited to a few micrometers as larger thickness can interfere with the lithography process for the integrated circuit. The economics of the single-chip integration are investigated in Chapter 24.

6.2.2   Physical separation

As the capacitance is inversely proportional to the distance, an effective way to decrease the parasitic capacitance is to increase the physical separation between capacitor electrodes. Figure 6.6(b) illustrates how this can be achieved by removing a portion of the substrate under the MEMS device. Alternatively, a portion of the conductive substrate can be replaced with a dielectric material as shown in Figure 6.6(c).

The physical separation is often used for high frequency components such as capacitors and inductors because at high frequency even a small capacitance has a low impedance. For example, the impedance of a 1-pF capacitor at 1 GHz is only 160 Ω. Thus, even small parasitics can lead to significant losses at radio frequencies.

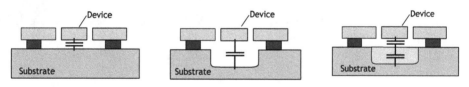

(a) Parasitic capacitance between the MEMS device and the conductive substrate.

(b) Portion of the substrate can be removed to reduce the parasitic capacitance.

(c) Dielectric well under the MEMS device reduces the parasitic capacitance.

Figure 6.6: Increasing physical separation between the device and the substrate reduces the parasitic capacitance.

6.2.3   Shielding and grounding

The unshielded MEMS devices are subjected to external disturbances. To eliminate these variations, the device can be surrounded with a grounded conductive shield. Some form of shielding of the sense capacitors is a requirement for stable operation of capacitive sensors. In this regard, the capacitive sensors are much less forgiving than the piezoresistive sensors.

Surrounding the sensor with an ungrounded conductive body is not sufficient to shield the device. Anything conductive with a capacitive coupling path to the sensor must be electrically grounded. Figure 6.7 illustrates the importance of grounding all conductive bodies with capacitive coupling path to the sensor. When the conductive body such as the sensor substrate is not grounded, the floating conductor forms a current path between the sensor electrodes. Electrically, this parasitic capacitance is parallel to the measurement capacitance and it is impossible to distinguish these in the electrical measurement. At best, the capacitance to the ungrounded bodies, such as substrate and package cover, is stable and the parasitic capacitance will reduce the sensor sensitivity as is illustrated in Example 6.4. At worst, the capacitances are unstable and the drift in parasitics, for example due to change in humidity or temperature, will directly change the sensor output.

By grounding the conductive bodies around the device, the direct path between the measurement electrodes is eliminated. Electrically, the parasitic capacitors appear as added capacitors to the ground which *can* be electrically discerned from the work capacitance. Circuit techniques that minimize the effect of the grounded parasitic capacitors are discussed in the following sections.

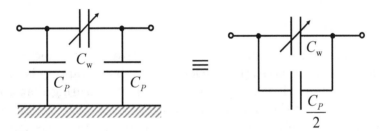

(a) Floating conductor near the working capacitor C_w acts as a short circuit between input and output. Any changes in the parasitic capacitance C_P directly affects the measurement.

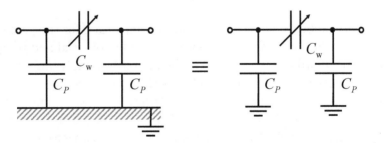

(b) By grounding the conductor, the parasitic capacitances are not in parallel with the work capacitance.

Figure 6.7: Conductors near the sensed capacitor should not be floating.

Example 6.4: The effect of grounding on sensor sensitivity
Problem: Calculate the relative capacitance change between the input and output port for the two cases in Figure 6.7 (1) when the C_w changes by 1% and (2) when the C_P changes by 1%. Assume $C_P = 0.3$ pF and $C_w = 100$ fF.
Solution: (1) The capacitance change due to the work capacitance variation is $\Delta C_f = 0.01 C_f = 1$ fF. For Figure 6.7(a), total capacitance between the input and output ports is $C_w + C_P/2$. The relative change in capacitance is $\Delta C_f/(C_w + C_P/2) = 0.004$. For Figure 6.7(b) with the parasitics grounded, total capacitance between the input and output ports is just C_w. The relative change in capacitance is $\Delta C_f/C_w = 0.01$.
(2) The capacitance change due to the parasitic capacitance change is $\Delta C_P = 0.01 C_P = 3$ fF. For Figure 6.7(a), the relative change in capacitance between the input and output is $\Delta C/(C_w + C_P/2) = 0.012$. This is larger than the change due to work capacitance change! For Figure 6.7(b) with the parasitics grounded, the change in C_P does not affect the capacitance between the input and output.

6.2.4   Bootstrapping

Bootstrapping illustrated in Figure 6.8 is an effective way to reduce the parasitic capacitances and resistances [51]. The idea behind bootstrapping is to eliminate the current flow through the parasitic capacitors by ensuring that there is no voltage difference over them. This is achieved by measuring the voltage at the sense electrode and applying exactly the same potential to the second capacitor electrode. As there is no potential difference over the capacitance, there is no current flow and the effect of parasitics is eliminated. The second electrode where the sense electrode potential is applied is often referred to as the "guard" electrode.

The bootstrapping is an active method to compensate for the parasitics as it requires an amplifier to measure the sense electrode potential and to apply this potential to the guard electrode. Hence, the bootstrapping is limited by the accuracy of the unity gain amplifier which sets the guard electrode potential. The bootstrapping and the effect of amplifier accuracy is illustrated in Example 6.5.

A common example of laboratory use of bootstrapping are the triaxial cables that have a third "guard" electrode to eliminate cable capacitance and leakage currents. The main challenge in using bootstrapping for microdevices is that it is difficult or impossible to completely surround the measurement electrode with a guard electrode. For example, to eliminate the parasitic capacitance to the substrate, a guard electrode is needed between the sense electrode and

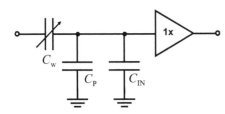
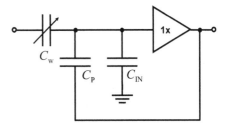

(a) The parasitic capacitance C_P in parallel with the amplifier input capacitance C_{IN} directly affects the voltage measured by the amplifier.

(b) The feedback ensures that there is no voltage difference over the parasitic capatance C_P.

Figure 6.8: Bootstrapping is an active way to compensate the parasitic capacitances.

substrate. This added electrode may not be possible without significant changes in the manufacturing process.

Example 6.5: The effect of amplifier imperfections in bootstrapping
Problem: The amplifier in Figure 6.8 has gain $A = 0.999\angle - 3°$ and the parasitic capacitance is $C_P = 1$ pF. What is the effective parasitic impedance at 1 MHz with C_P connected to ground and with C_P bootstrapped? How does this compare to an almost ideal amplifier that does not have phase lag and has the gain of 0.999?
Solution: With the C_P connected to the ground, the impedance is

$$Z = \frac{1}{j\omega C_P} = -j \cdot 160 \text{ k}\Omega = 160\angle - 90° \text{ k}\Omega.$$

The voltage at the amplifier input is v_A and the voltage at the amplifier output is Av_A. With the feedback connected, voltage over the capacitor C_P is the difference beween the amplifier input and output voltages

$$v = v_A - Av_A,$$

The current through the capacitor is

$$i = \frac{v}{Z} = \frac{v_A(1 - A)}{Z},$$

where $Z = 1/j\omega C$. The effective input impedance is the input voltage v_A divided by the current

$$Z_{eff} = \frac{v_A}{i} = \frac{Z}{1 - A} = 3\angle - 177° \text{ M}\Omega.$$

The impedance is seen to increase by a factor of twenty and hence the effect of C_P is reduced by 20×.

For an almost idea amplifier with $A = 0.999$ and no phase shift, we obtain

$$Z_{eff} = \frac{v}{i} = \frac{Z}{1 - A} = 160\angle - 90° \text{ M}\Omega,$$

which is a factor of thousand higher than for the capacitor without bootstrapping.

6.2.5 Current measurement

The bootstrapping covered in the previous section eliminated the parasitics by ensuring that there is no voltage difference over them. An alternative way to achieve this is to keep the both sides of the parasitic at the ground potential. Since the amplifier input is held at the ground potential, we cannot measure the amplifier input voltage (which is always zero). We can, however, measure the current flow through the capacitor as shown in Figure 6.9. The current meter has very low impedance and it keeps the node 'A' at the ground potential. The effect is the same as with the bootstrapping: As there is no voltage difference over the parasitic capacitor, no current flows through the parasitic capacitor and it will not affect the measurement.

Figure 6.9(b) shows a typical implementation for the current amplifier that is based on an operational amplifier that converts the current to voltage. This configuration is often referred to as a transimpedance amplifier as it converts current to voltage. The operation of the transimpedance amplifier is analyzed in detail in Chapter 8.

Example 6.6: Bulk micromachined capacitive accelerometer
Problem: Example 3.7 on page 42 covered the mechanical noise in a typical bulk micromachined accelerometer. The proof mass for the accelerometer is $m = 1.845$ μkg and the total spring constant for the four beams combined is $k = 40$ N/m. The proof mass displacement due to a 1-G acceleration is $x = \ddot{x}m/k = 0.4565$ μm ≈ 0.46 μm and the noise induced rms-displacement density is $x_{\text{rms}} = 10$ pm.

The proof mass movement is sensed using the differential measurement shown in Figure 6.10. The current meter grounds the node 'A' and measures the difference of currents i_1 and i_2 going through the capacitors C_{w1} and C_{w2}, respectively. The capacitances have the nominal gap $d = 2.5$ μm and the area

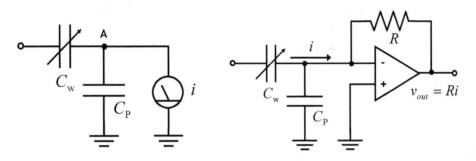

(a) Current meter grounds the node 'A'. As voltage over C_P is zero, there is no current through C_P. The current through C_w is sensed by current meter.

(b) A typical implementation for the current meter is a current-to-voltage converter or transimpedance amplifier that is based on an operational amplifier.

Figure 6.9: Current measurement can be used to eliminate the effect of parasitic capacitance C_P.

is $A = 1200\ \mu$m $\times$ 1200 μm. The amplitude of the ac measuring voltage is $V = 0.1$ V and the measuring signal frequency is $f = 100$ kHz.

a) Calculate the relative change in capacitances due to a 1-G acceleration in Z-direction.

b) What is signal current for the 1-G acceleration?

c) What is the power consumption due to ac measurement signal dissipated in charging and discharging the capacitances?

d) What is the signal current change for the acceleration corresponding to a 100-μG acceleration?

e) What is the rms-noise current due to mechanical noise?

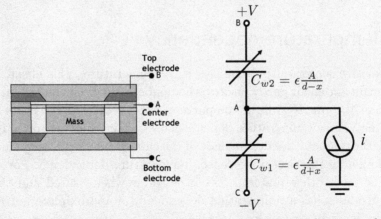

Figure 6.10: Capacitive accelerometer with electrode on top and bottom of the proof mass. The proof mass movement is sensed with a current amplifier.

Solution: a) From the calculated displacement $|x| = m\ddot{x}/k = 0.46$ μm, the relative changes in the capacitance are $d/(d - |x|) = 1.223$ and $d/(d + |x|) = 0.846$.

b) The current through C_{w1} toward node 'A' is $i_1 = -j\omega C_1 V$ and the current through C_{w2} toward node 'A' is $i_1 = j\omega C_1 V$. The signal current is

$$i_{\text{sig}} = i_2 + i_1 = j\omega C_2 V - j\omega C_1 V = j2\omega V C_0 d\frac{x}{d^2 - x^2} = 0.12 \ \mu\text{A}.$$

Here $C_0 = \epsilon_0 A/d = 5.1$ pF and $\omega = 2\pi f$ where $f = 100$ kHz is the frequency of the measurement voltage V.

c) The energy used for charging one capacitor is $E = \frac{1}{2}C_0 V^2$. The power wasted in charging and discharging the two work capacitances is $P = 2\frac{1}{2}C_0 V^2 f = 5$ nW.

d) Scaling the result from part b) gives the signal current $i_{\text{sig},100 \ \mu\text{G}} = 12$ pA corresponding to the 100-μG acceleration.

e) The rms-noise current due to mechanical noise is $i_{\text{rms}} = 2\omega V C_0 x_{\text{rms}}/d = 2.6$ pA which is more than $4\times$ smaller than the current due to the 100-μG acceleration. In comparison, the piezoresistive accelerometer had input referred noise $\ddot{x}_{\text{rms}} \approx 10$ mG.

The full noise analysis will have to include the noise from the electronics. But we notice that the capacitive sensing can reach much better noise performance at a fraction of the power consumed by piezoresistive sensors. In addition, capacitive accelerometers can have a larger bandwidth for a given sensitivity and the temperature performance is significantly better than for the piezoresistive accelerometers [52]. Due to the advantages, the capacitive accelerometers have largely replaced the piezoresistive accelerometers in commercial applications.

6.3 Temperature dependency

The capacitive sensor output changes with temperature. This effect, however, can be small: For an air gap capacitor, the temperature coefficient of capacitance is less than 10 ppm/K [53]. In comparison, the piezoresistive sensors analyzed in Chapter 5 show temperature dependencies greater than 300 ppm/K.

Since the temperature dependency of the capacitors itself is very small, other effects dominate the capacitive sensor temperature dependency. For example, silicon like most other materials becomes softer when heated and the silicon Young's modulus has a temperature dependency of approximately -60 ppm/K. This effect is illustrated in Example 6.7.

Another often more significant effect is the thermal expansion of dissimilar sensor materials. For example, silicon has a relatively low thermal expansion coefficient in comparison to packaging materials which results in significant temperature dependent stresses. This effect can cause changes in the capacitance gap, which can results in temperature offsets greater than 100 ppm/K. This effect is further discussed in Chapter 11 where the sensor specifications are covered.

Example 6.7: Temperature dependency of silicon spring constant
Problem: Silicon Young's modulus has the temperature dependency of $\frac{\Delta E/E}{\Delta T} = -60$ ppm/K. How much does a silicon beam spring constant change if the temperature increases by $\Delta T = 50$ K?
Solution: The Young's modulus changes by

$$\frac{\Delta E}{E} = \frac{\Delta E/E}{\Delta T}\Delta T = -3{,}000 \text{ ppm}$$

or 0.3%. Since the spring constant is proportional to the Young's modulus, it changes by the same amount.

6.4   Demodulation

As capacitors do not transmit dc current, the capacitive displacement measurement is carried out with ac excitation at frequency ω. The typical measurement frequencies range from 10 kHz to 1 MHz. The sensor output is therefore a small current at a high frequency. Measurement of for example 1 fA current at 100 kHz is not trivial and somehow this ac current has to be converted to a dc signal. This process is called demodulation.

Figure 6.11 shows a schematic representation for the synchronous demodulation where the sensor output signal is multiplied with the measurement signal. First, the sensor signal is amplified with an amplifier and the amplifier output is

$$v_{\text{amp}} = A\Delta C(t)\cos(\omega t + \theta), \tag{6.11}$$

where A is a constant term, $\Delta C(t)$ is the differential capacitance change, ω is the capacitance measurement frequency, and θ is a phase lag due to amplifier non-idealities. The amplifier output is then multiplied with the original measurement signal using a mixer. Using the trigonometric identity $\cos(\omega t + \theta) \cdot \cos(\omega t) = \frac{1}{2}\left[\cos(\theta) + \cos(2\omega t + \theta)\right]$, mixer output is

$$v_{\text{mixer}} = \frac{AV_0\Delta C(t)}{2}\left[\cos(\theta) + \cos(2\omega t + \theta)\right]. \tag{6.12}$$

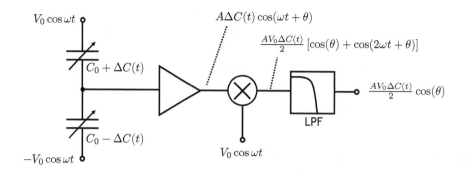

Figure 6.11: Synchronous signal demodulation.

Finally, the high frequency component $\cos(2\omega t + \theta)$ is removed with a low pass filter (LPF) and the filter output is

$$v_{\text{out}} = \frac{AV_0 \Delta C(t)}{2} \cos(\theta), \qquad (6.13)$$

which is proportional to the capacitance change $\Delta C(t)$. The $\cos(\omega t)$ modulation has been removed and the signal is at dc.

The phase shift θ in Equation (6.13) is undesirable for two reasons: First, the phase shift decreases the sensitivity. Second, if the phase shift is not constant but varies for example with temperature, the demodulator gain will drift. The phase shift should be minimized which requires that the amplifier in Figure 6.11 has a high bandwidth.

The synchronous demodulation covered in this section has good performance and it is used in accelerometers and gyroscopes to measure small signals. The mixer, however, consumes significant amount of power. In Chapter 10 we present a simple alternative to the capacitance measurement, the switched capacitor circuit, that can be used to measure capacitances with minimal power consumption.

Key concepts

- Capacitive sensing offers excellent noise performance and low power consumption.

- Capacitive sensing requires electrical and physical shielding.

- Differential measurements are used to obtain a signal that is proportional to the change in capacitance.

- As the measured capacitance changes are extremely small, care must be taken to avoid parasitic capacitances. Shielding, bootstrapping and current sensing can be used to mitigate the parasitic capacitances.

- Demodulation is needed to convert the ac measurement signal to dc output.

Exercises

Exercise 6.1
A micromechanical resonator is vibrating at the frequency $f_0 = 20$ kHz with vibration amplitude $x_0 = 10$ nm. The vibrations are measured using the rate-of-change measurement with a parallel plate capacitor that has the area $A = 1,000\ \mu\text{m}^2$ and the gap $d = 1\ \mu\text{m}$. Calculate the motional current if the bias voltage over the capacitor is $V_{\text{dc}} = 5$ V.

Exercise 6.2
Explain why: 1. Reference capacitors and 2. differential measurement are used in capacitive sensing.

Exercise 6.3
List methods to reduce the effect of parasitic capacitances in capacitive measurements. Comment on advantages and disadvantages of each method.

Exercise 6.4
Derive the output voltage $V_{\text{out}} = V_2 - V_1$ vs. displacement x relationships for the capacitive Wheatstone bridges shown in Figure 6.12.

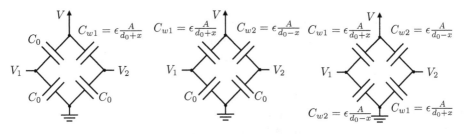

Figure 6.12: Capacitive Wheatstone bridges

Exercise 6.5
Calculate the relative sensitivity $\frac{\Delta C/C}{\bar{x}}$ (Units: 1/G) and the absolute capacitance change $\frac{\Delta C}{\bar{x}}$ (Units: F/G) due to acceleration for the surface and bulk micromachined accelerometers in Figures 3.6 and 3.7, respectively. Assume electrode gap $d = 2.2\ \mu\text{m}$ for both accelerometers. What are the signal currents due to a 1-G acceleration if the measuring circuit in Figure 6.10 is used? Assume that the measurement voltage and frequency are $V = 0.1$ V and $f = 200$ kHz, respectively.

Exercise 6.6
Consider the capacitive sensor with a sensing capacitance C_w, reference capacitance C_0, and parasitic capacitances shown in Figure 6.13. The capacitance is measured by applying voltage V_{AC} to one side of the sensing capacitance and measuring the current on the other side of the capacitance. Which parasitic capacitances affect the output voltage?

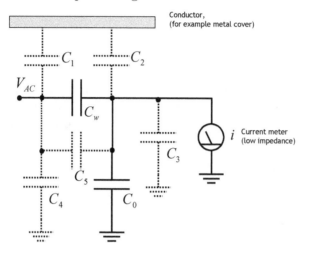

Figure 6.13: Capacitive sensor.

Exercise 6.7
Consider the Exercise 6.6 but assume that the conductor above the work capacitance is grounded. Which parasitic capacitances affect the output voltage?

Exercise 6.8
Using a surface micromachining process with a 2-μm sacrificial layer thickness and a 3-μm structural layer thickness, design a differential X-axis accelerometer (acceleration in plane with the substrate) that has the electrode capacitances $C_{w1} = C_{w1} = 0.2$ pF, the sensitivity to acceleration is $\Delta C/|\ddot{x}| = 1$ fF/G, and the maximum range is ± 5 G. Optimize your design for the smallest rectangular footprint. Assume that the capacitor gap is $d = 2$ μm, the smallest mechanical structure width is $w = 2$ μm, and the Young's modulus is $E = 150$ GPa.

Exercise 6.9
Consider the capacitive bridge in Figure 6.5. Assume $C_{w1} = \epsilon A/(d-x)$, $C_{w2} = \epsilon A/(d+x)$, and $C_0 = \epsilon A/d$, where $A = (400 \ \mu\text{m})^2$ and $d = 2$ μm. Plot the bridge output as a function of displacement with $C_P = 0$ F and $C_P = 500$ fF. How does the parasitic capacitance C_P affect the sensitivity and linearity?

7

Piezoelectric sensing

Certain materials generate an electrical field when subjected to mechanical deformation. Alternatively, the same materials deform when an electrical field is applied. This effect was first discovered in 1880 by Pierre and Jacques Currie and is named piezoelectricity; the prefix is derived from the Greek work *piezein* which means "to press".

Piezoelectric sensors include record player pickups, microphones, and accelerometers. Piezoelectric actuators have been used for example in adaptive optics where small but accurately controlled displacements are needed. Several applications utilize the dual nature of piezoelectric materials as sensors and actuators. The piezoelectric underwater transducers or sonars were developed during World War II to detect submarines. Today, similar transducers are used for medical imaging. Finally, piezoelectric resonators are needed in practically all communication systems.

The piezoelectric sensing offers the advantage of being self-generating: Unlike capacitive or piezoresistive sensing, no bias voltage or current is needed. The drawback of piezoelectric sensing is that the signals are proportionate to the *change* in strain. Thus, the piezoelectric sensing is not suitable for static measurements. For example, piezoelectric accelerometers can only measure change in acceleration but not dc acceleration. The piezoelectric materials can also be difficult to process and may not be compatible with the other clean room processes.

This chapter focuses on the use of piezoelectric materials for microsensor applications. The piezoelectric actuation is covered in Chapter 16. First, we cover the relationship between the stress, the strain, and the electric field in piezoelectric materials. Next, the two most common transducer configurations, longitudinal and transverse transducers, are analyzed. The operation of the transducers is exemplified with macroscopic and micromechanical analysis examples.

7.1 Piezoelectric effect

The piezoelectric materials are characterized by the constitutive equations for the stress

$$T = ES - e\mathcal{E}. \tag{7.1}$$

and the electric displacement

$$D = \epsilon\mathcal{E} + eS. \tag{7.2}$$

Here T is the stress, E is the Young's modulus, S is the strain, ϵ is the permittivity, e is the piezoelectric coefficient, $\mathcal{E}$ is the electric field, and D is the electric displacement. Equations (7.1) and (7.2) can be compared to the corresponding equations $T = ES$ and $D = \epsilon\mathcal{E}$ for non-piezoelectric materials. We see that the piezoelectric effect couples the normally independent stress and electric field equations.

Table 7.1 shows the properties for selected piezoelectric materials. Aluminum nitride (AlN) and zinc oxide (ZnO) are typical piezoelectric thin-films and both have been used for microdevices. Although the coupling coefficient for these materials is not the highest of all materials, they have high quality factors which makes them especially suited for high frequency resonators [54–56]. Lead zirconium titanate oxides (PZTs) are piezoelectric ceramics that have exceptionally large piezoelectric coupling coefficients. Several different formulations of PZT have been developed and the material is used extensively for macroscopic sensors and actuators [57]. It is also possible to deposit PZT in a thin-film form but the material is not compatible with IC processes and therefore is not allowed in most clean rooms. Crystalline quartz is a unique material as the resonant frequencies of quartz resonators are insensitive to the temperature. Although quartz crystals have relatively weak coupling coefficients, the insensitivity to temperature variations has made quartz resonators invaluable as frequency and time references [58]. Finally, PVDF is an example of a piezoelectric plastic. The coupling coefficient is relatively small but the softness of the plastic makes it attractive for many applications.

Piezoelectric materials are anisotropic and accurate analysis requires representing Equations (7.1) and (7.2) in a matrix form [59] resulting in a total of twelve coupled equations making piezoelectric materials intimidating when encountered for the first time. Fortunately, the majority of piezoelectric sensors fall under the two cases shown in Figure 7.1. In both cases, the piezoelectric material is sandwiched between two electrodes and there is an electric field $\mathcal{E}_3$ between the electrodes. As is standard for the piezoelectric materials, the axis

Table 7.1: Selected piezoelectric materials.

Material	ϵ_R	E[GPa]	ρ[kg/m^3]	e_{33}[C/m^2]	e_{31}[C/m^2]
AlN	8.5	400	3,200	1.55	-0.48
ZnO	8.5	210	5,600	1.14	-0.61
PZT-4	1,300	110	7,500	15.1	-5.2
PZT-5A	1,700	110	7,750	15.8	-5.4
Quartz (SiO$_2$)	4.52	107	2,650	0.17	–
PVDF (plastic)	13	3	1,880	-0.10	0.06

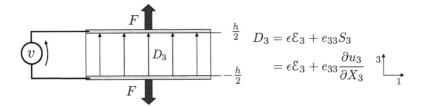

(a) Longitudinal configuration. The force is in the direction of electric field.

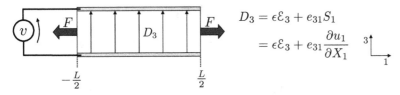

(b) Transverse configuration. The force is perpendicular to the electric field.

Figure 7.1: The two common configurations for a piezoelectric sensor. The piezoelectric material is sandwiched between two electrodes.

directions is represented with numbers 1 to 3. In the longitudinal configuration of Figure 7.1(a), the strain is parallel to the electric field and other strains are zero. Figure 7.1(b) shows the transverse sensor configuration where the strain is perpendicular to the electric field.

7.1.1 Longitudinal transducer

For the longitudinal configuration of Figure 7.1(a), the electric field is in the direction of stress. To calculate the signal current due to a deformation, we will start by rewriting Equation (7.2) to explicitly show the material directions:

$$D_3 = \epsilon \mathcal{E}_3 + e_{33} S_3. \tag{7.3}$$

Here e_{33} is the piezoelectric coefficient that couples the strain S_3 and the electric displacement D_3 that are aligned along the 3-axis. Other strains and electric fields are zero ($S_1 = S_2 = 0$ and $D_1 = D_2 = 0$). As shown in Appendix G, Equation (7.3) is a good approximation even if the other strains are not zero.

Substituting $S_3 = \frac{\partial u_3}{\partial X_3}$ where $u_3(X_3)$ is the material displacement in the 3-direction at the location X_3 and integrating Equation (7.3) from the bottom electrode ($X_3 = -h/2$) to the top electrode ($X_3 = h/2$) gives

$$\int_{-\frac{h}{2}}^{\frac{h}{2}} D_3 dX_3 = \int_{-\frac{h}{2}}^{\frac{h}{2}} \epsilon \mathcal{E}_3 dX_3 + \int_{-\frac{h}{2}}^{\frac{h}{2}} e_{33} \frac{\partial u_3}{\partial X_3} dX_3$$

$$\Leftrightarrow D_3 h = \epsilon \mathcal{E}_3 h + e_{33} \left[u_3 \left(\frac{h}{2} \right) - u_3 \left(-\frac{h}{2} \right) \right]. \tag{7.4}$$

Next, we rewrite Equation (7.4) in terms of the voltage over the electrodes $v = \mathcal{E}_3 h$:

$$D_3 = \epsilon \frac{v}{h} + \frac{e_{33}}{h} \left[u_3 \left(\frac{h}{2} \right) - u_3 \left(-\frac{h}{2} \right) \right]. \tag{7.5}$$

The charge on the electrode is obtained by multiplying the dielectric displacement D_3 with the electrode area A_3:

$$q = A_3 D_3. \tag{7.6}$$

Finally, we obtain the current across the electrodes by taking the time derivative of the charge:

$$i = \dot{q} = \epsilon \frac{A_3}{h} \dot{v} + \frac{e_{33} A_3}{h} \left[\dot{u}_3 \left(\frac{h}{2} \right) - \dot{u}_3 \left(-\frac{h}{2} \right) \right]. \tag{7.7}$$

Using $\dot{v} = sv$, $\dot{u}_3 = su_3$, and $C = \epsilon A/h$, we can write Equation (7.7) as

$$i = sCv + s\frac{e_{33} A_3}{h} \left[u_3 \left(\frac{h}{2} \right) - u_3 \left(-\frac{h}{2} \right) \right] \equiv i_{\text{ac}} + i_{\text{mot}}. \tag{7.8}$$

The first term in Equation (7.8) is the normal ac current through the capacitance and the second term is the motional current due to displacement velocities $\dot{u}_3$ at $h/2$ and $-h/2$. Equation (7.8) shows that only the electrode displacements $u_3(h/2)$ and $u_3(-h/2)$ are needed to calculate the motional current. This is a large simplification over the Constitutive Equation (7.2) that requires the knowledge of strain everywhere within the sample.

If the strain in the transducer is uniform, we can use $S_3 = \left(u_3 \left(\frac{h}{2} \right) - u_3 \left(-\frac{h}{2} \right) \right) /h$ to write Equation (7.8) as

$$i = sCv + se_{33} A S_3 \qquad \text{(uniform strain)}. \tag{7.9}$$

Example 7.1: Macroscopic piezoelectric accelerometer

Problem: Figure 7.2 shows an example of an accelerometer that is manufactured using traditional machining [60]. A steel proof mass is mounted onto a PZT4 plate. Acceleration in Z-direction will result in the proof mass exerting force to the PZT plate. Calculate the output current for a 1-G acceleration at 1 Hz, 10 Hz, and 100 Hz. Assume $m = 0.025$ kg for the proof mass and $A = 2$ cm^2 for the PZT plate area.

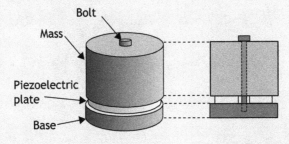

Figure 7.2: Macroscopic piezoelectric accelerometer.

Solution: The inertial force acting on the piezoelectric "spring" is $F = m\ddot{x}$. The stress on the piezoelectric layer is $T_3 = F/A$. We assume that the voltage over the electrodes is small and the effect of the electric field to the strain can be ignored. Equation (7.1) can then be approximated by $T_3 = ES_3$ and the strain is

$$S_3 = \frac{T_3}{E} = \frac{m\ddot{x}}{EA} \approx 1.1 \cdot 10^{-8}.$$

The piezoelectrically generated current from Equation (7.9) is

$$i_{\text{mot}} = e_{33}AsS_3.$$

Using $s = j\omega$ and $e_{33} = 15.1$ C/m^2 from Table 7.1, we obtain $|i_{\text{mot}}| = 0.21$ nA, 2.1 nA, and 21 nA for $f = 1$ Hz, 10 Hz, and 100 Hz, respectively.

The error caused by neglecting the effect electric field on stress in Equation (7.1) is investigated in Example 7.3.

7.1.2   Transverse transducer

The second sensor configuration shown in Figure 7.1(b) has the strain perpendicular to the electric field. We rewrite Equation (7.2) as

$$D_3 = \epsilon\mathcal{E}_3 + e_{31}S_1 \qquad (7.10)$$

to explicitly show the dependence of electric displacement D_3 on the direction of the electric field and strain. The piezoelectric coefficient e_{31} couples the orthogonal electrical displacement D_3 and strain S_1 that depends only on location along the 1-axis. Other strains and electric fields are zero ($S_2 = S_3 = 0$ and $D_1 = D_2 = 0$). Again, as shown in Appendix G, the error from ignoring the stresses S_2 and S_3 is small.

The analysis for the transverse actuator is similar to the longitudinal transducer. We use $S_1 = \frac{\partial u_1}{\partial X_1}$ and integrate Equation (7.10) from the left boundary to the right boundary:

$$\int_{-\frac{L}{2}}^{\frac{L}{2}} D_3 dX_1 = \int_{-\frac{L}{2}}^{\frac{L}{2}} \epsilon \mathcal{E}_3 dX_1 + \int_{-\frac{L}{2}}^{\frac{L}{2}} e_{31} S_1 dX_1 \tag{7.11}$$

$$\Leftrightarrow DL = \epsilon \mathcal{E}_3 L + e_{31} \left[u_1 \left(\frac{L}{2} \right) - u_1 \left(-\frac{L}{2} \right) \right].$$

Noticing the similarity between (7.4) and (7.11), we proceed as before to obtain

$$i = \dot{q} = \epsilon \frac{A_3}{h} \dot{v} + \frac{e_{31} A_3}{L} \left[\dot{u}_1 \left(\frac{L}{2} \right) - \dot{u}_1 \left(-\frac{L}{2} \right) \right] \tag{7.12}$$

$$= sCv + s \frac{e_{31} A}{L} \left[u_1 \left(\frac{L}{2} \right) - u_1 \left(-\frac{L}{2} \right) \right] \equiv i_{\text{ac}} + i_{\text{mot}}.$$

The first term in Equation (7.12) is the normal ac current through the capacitance and the second term is the motional current due to the displacement velocities $\dot{u}_1$ at $L/2$ and $-L/2$. Again, only the displacements at the electrode edges are needed to calculate the current.

If the strain in the transducer is uniform, we can use $S_1 = (u_1(L/2) - u_1(-L/2))/L$ to write Equation (7.12) as

$$i = sCv + se_{31} AS_1 \qquad \text{(uniform strain)}. \tag{7.13}$$

7.2 Sensing circuits

The piezoelectric sensors generate current that is proportional to the rate of change for the strain. The two ways to measure this signal are the current measurement and the voltage measurement shown in Figure 7.3. In both circuits, the resistance R sets the low frequency response of the sensor system.

7.2.1 Current measurement

In the current measurement shown in Figure 7.3(a), the transimpedance amplifier maintains the piezoelectric sensor at a virtual ground potential. The voltage v over the electrodes in Equations (7.8) and (7.12) is therefore zero and only the motional current is measured. The transimpedance amplifier is analyzed in detail in Section 8.1.3. Looking ahead, the parallel combination of R and

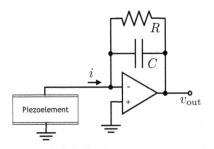 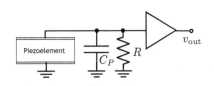

(a) The current is measured with a tran-simpedance amplifier.

(b) The signal current is measured with a voltage amplifier.

Figure 7.3: The common ways to measure piezoelectric signals.

C converts the motional current into voltage and the output voltage from the amplifier is

$$v_{\text{out}} = -i_{\text{mot}} R || \frac{1}{sC} = -i_{\text{mot}} \left(\frac{1}{R} + sC \right)^{-1} \tag{7.14}$$

$$\approx -\frac{i_{\text{mot}}}{sC} \text{ for } \omega \gg \frac{1}{RC}. \tag{7.15}$$

As the motional current proportional to the frequency, the voltage given by Equation (7.15) does not depend on frequency. Below the -3-dB cut-off frequency, the output voltage is $v_{\text{out}} = -i_{\text{mot}} R$ which decreases with frequency as the motional current decreases with frequency.

7.2.2 Voltage measurement

In the voltage measurement shown in Figure 7.3(b), the piezoelectric current is converted to the voltage by the load capacitance C_P, the resistance R, and the sensor capacitance C. Referring to Figure 7.3(b), the current through the sensor is $i_{\text{ac}} + i_{\text{mot}}$ and the currents through the load resistor and capacitor are i_R and i_{C_P}, respectively. The sum of all the currents is zero and we have

$$i_{\text{ac}} + i_{\text{mot}} + i_R + i_{C_P} = 0. \tag{7.16}$$

Substituting $i_{\text{ac}} = sCv$, $i_R = v/R$, and $i_{C_P} = sC_P v$ to Equation (7.16) and solving for v we obtain

$$v = -\frac{i_{\text{mot}}}{sC + \frac{1}{R} + sC_P} \tag{7.17}$$

$$\approx -\frac{i_{\text{mot}}}{s(C + C_P)} \text{ for } \omega \gg \frac{1}{R(C + C_P)}. \tag{7.18}$$

The physical interpretation of Equations (7.17) and (7.18) is that the motional current is converted to voltage by the piezoelement capacitance in parallel with the load impedances.

Example 7.2: Bulk piezoelectric accelerometer
Problem: The bulk piezoelectric accelerometer of Example 7.1 is measured with a voltage pick-up configuration in Figure 7.3(b). Assume $R = 100$ MΩ and $C_P = 0.1$ nF for the circuit and use $h = 1$ mm for the PZT plate thickness. What are the -3-dB cut-off frequency and the signal voltage above the -3-dB cut-off frequency for a 1-G acceleration?
Solution: The element capacitance is

$$C = \epsilon \frac{A}{h} \approx 2.3 \text{ nF}.$$

In this case, the element capacitance dominates the total capacitance. The -3 dB frequency is

$$f_{-3 \text{ dB}} = \frac{1}{R(C + C_P)} \approx 4.2 \text{ Hz}.$$

Following Example 7.1, the piezoelectrically generated current is

$$i_{\text{mot}} = e_{33} s A S_3 = e_{33} s \frac{m\ddot{x}}{E}.$$

When $f \gg f_{-3 \text{ dB}}$, the signal voltage from Equation (7.18) is

$$v = -\frac{i_{\text{mot}}}{s(C + C_P)} = -\frac{e_{33} m\ddot{x}}{E(C + C_P)} \approx -14.0 \text{ mV}.$$

Example 7.3: Including the electric field in the stress analysis
Problem: Recalculate Example 7.1 without ignoring the effect of the electrical field. Assume that the sensing circuit of Example 7.2 is used.
Solution: Combining Equations (7.1) and (7.9) gives

$$i_{\text{mot}} = sCv + se_{33}A\left(\frac{T_3}{E} + \frac{v}{Eh}\right) = s\left(C + \frac{e_{33}A}{Eh}\right)v + se_{33}A\frac{T_3}{E}.$$

Mathematically, the effect of electric can be modeled by replacing the physical capacitance C with $C' = C + \frac{e_{33}A}{Eh}$. From Example 7.2, we have

$$v = -\frac{i_{\text{mot}}}{s(C' + C_P)} = -\frac{e_{33}m\ddot{x}}{E(C' + C_P)} \approx -13.9 \text{ mV}.$$

We see that that in this example, the effect of ignoring the electric field in stress calculations is about 1%. This shows that the piezoelectric coupling effects are fairly small and the cross coupling of stress and the electric field can be ignored in the first-order estimations.

7.3   Case study: A micromechanical accelerometer

Since the macroscopic piezoelectric accelerometers are a commercially available product [60], it is natural for us to use the accelerometer as a test case to investigate the relative merits of piezoelectric sensing in micro-scale. Figure 7.4 shows a possible implementation for a MEMS accelerometer. The device resembles the piezoresistive accelerometer that was already analyzed in Example 5.4 on page 81. The only difference is that instead of the piezoresistor implanted into the beams, the acceleration is sensed using piezoelectric ZnO thin-film deposited onto the beams. To simplify the analysis, we assume that the piezoelectric film is sufficiently thin so as not to affect the beam spring constant.

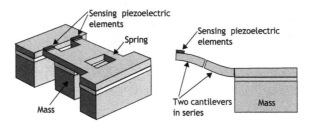

Figure 7.4: Micromachined piezoelectric accelerometer.

As the beam bends, the piezoelectric film is strained. This strain is picked up by the electrodes on top of and under the film. The stress in the thin-film from Example 5.4 and Equation (4.29) is

$$T = \frac{F_{1/4}L_c b}{2I},$$

where $F_{1/4} = m\ddot{x}/4$ is the force acting on one beam and $L_c = L/2$ is half of the total beam length. For a 1-G acceleration, our example accelerometer has

$T = 12$ MPa. Using $S = T/E$ and Equation (7.13), we obtain the motional current

$$i_{\mathrm{mot}} = e_{31} As\frac{T}{E}.$$

The piezoelectric coefficient for ZnO from Table 7.1 is $e_{31} = -0.61$. Assuming $A = 30 \ \mu\mathrm{m} \times 20 \ \mu\mathrm{m}$ for the electrode area and $\omega = 2\pi 10$ Hz for the frequency, we obtain $|i_{\mathrm{mot}}| = 1.6$ pA for the 1-G acceleration.

As the output is current, it is natural to compare the piezoelectric sensing with the capacitive sensing that also outputs current. Neither the capacitive sensing nor the piezoelectric transduction adds noise to the signal which is an advantage in comparison to the piezoresistive sensing. In addition, the piezoelectric accelerometer is self-generating but signal current is very small. In comparison, the capacitive accelerometer with a similar geometry (Example 6.6 on page 102) gives signal current of $i = 120$ nA for the 1-G acceleration. Even with piezoelectric materials with higher piezoelectric coefficient such as PZT, it is clear that the signal from piezoelectric transduction is inferior to the capacitive accelerometer. Similar results have been obtained experimentally. For example, a surface micromachined piezoelectric thin-film accelerometer has a reported sensitively of only $q/\ddot{x} = 0.2$ fC/G [61].

The relatively small output current explains why there are no commercially available micromachined piezoelectric accelerometers. Part of the reason for the small current is that the accelerometers operate at low frequencies. As the motional current for a given strain increases with frequency, we expect the piezoelectric transduction to work better at higher frequencies. The market concurs with this conclusion. Piezoelectric transduction is used commercially in resonating structures such as gyroscopes, reference oscillators, and RF filters where the operation frequency is higher and the ability to both sense and actuate is especially advantageous. Piezoelectric sensing has also been researched for use in micromachined microphones [62–64]. The piezoelectric actuation is covered in Chapter 16 and its applications in resonators and gyroscopes are covered in Chapter 21 and 22, respectively.

Key concepts

- Piezoelectric sensing is self-generating. The piezoelectric element generates output without bias voltage or current.

- The piezoelectric sensor output is proportional to the rate-of-change. Piezoelectric sensors cannot be used for measuring dc signals.

- Piezoelectric materials require special processing reducing their attractiveness for microsensors.

- Piezoelectric materials can be used for both sensing and actuation. This dual nature is used extensively in resonators.

Exercises

Exercise 7.1
Consider the piezoelectric accelerometer in Figure 7.2 but assume $m = 0.010$ kg for the proof mass and $A = 1.0$ cm^2 for the PZT5 plate area. Calculate the output current for a 1-G acceleration at 1 Hz, 10 Hz, and 100 Hz.

Exercise 7.2
A piezoelectric force sensor consists of a PZT4 plate with the area $A = 1.0$ cm^2 and the thickness for $d = 1$ mm. The PZT output is measured with the sensing circuit shown in Figure 7.3(b). Calculate the output voltage for a 100-N force if $C_P = 100$ nF. Assume that the resistance R in Figure 7.3(b) is large.

Exercise 7.3
Plot the magnitude of the frequency response on a logarithmic scale for the sensor in Exercise 7.2 from 0.01 Hz to 100 Hz if $R = 100$ MΩ. The Y-axis should be in units units of V/N.

Exercise 7.4
Piezoelectric vibration sensors detect vibrations in mechanical structures. The vibration sensors are used for example to detect vibrations caused by leaking valves. Calculate the sensor output current if vibration at 1 kHz frequency causes a longitudinal strain $S_3 = 10^{-5}$ in a PZT4 plate that is 0.3 mm thick and has the area $A = 25$ mm^2.

Exercise 7.5
Plot the magnitude of the frequency response on a logarithmic scale for the accelerometer in Example 7.2 from 0.01 Hz to 100 Hz. The Y-axis should be in units units of V/N.

8

Signal amplification

The small signals from micromechanical sensors are easily buried under noise. To rescue the signal, it needs to be strengthened with an amplifier. After the signal has been boosted by the amplifier, the noise from subsequent stages is comparatively small. The amplifier is also needed to isolate the sensor from the environment. For example, the femto-farad capacitance variations in capacitive sensors are easily lost to parasitic capacitances. The first amplifier is therefore an integral part of sensor design and should be considered early on when specifying the sensor performance.

In this chapter, the signal amplification is covered with the focus on modern operational amplifier circuits. The goal is to give the sensor designer a sufficient understanding of the amplifier properties and the inherent limitations. This will enable intelligent design trade-offs between the mechanical sensor element and the amplifier to obtain optimal performance.

We start the chapter by reviewing the operational amplifiers. The operational amplifiers are versatile electronic circuits that are the basis of the modern amplifiers. Common operational amplifier configurations are covered including the differential amplifier and transconductance amplifier. Next, we review the transistors that are the basic building blocks of all circuits. We will exemplify the transistor usage by presenting a simplified operational amplifier circuit. This chapter is complemented by Chapter 9 that focuses on amplifier noise.

8.1 Operation amplifiers

Operational amplifiers or "op amps" are an essential part of analog electronics and form the basis for precise sensor systems. The op amp can be loosely defined as a high gain differential amplifier. The actual value for the gain depends on the application but is typically more than 10^4. This is usually more than the needed signal amplification and op amp circuits use negative feedback to lower the gain to $1 - 100$ range.

Figure 8.1 shows an op amp circuit that has unity feedback from output to the negative input. The op amp amplifies the difference between the positive v_+ and negative v_- inputs by a large factor A. The amplifier output, however, is fed back to the negative input, and the resulting negative feedback reduces the overall gain. Solving for the output voltage gives

$$v_{\text{out}} = \frac{A}{1 + A} v_+ \approx v_+, \tag{8.1}$$

where the approximation assumes that the op amp gain A is large.

Equation (8.1) illustrates one of the most useful aspects of the operational amplifiers, namely that the total gain is largely independent of the op amp gain. Operational amplifiers are therefore versatile components and the designer can easily adjust the amplifier gain by simply changing the feedback configuration.

An additional benefit of the negative feedback is that small variations in op amp performance do not significantly affect the system performance. As the table in Figure 8.1 illustrates, the op amp gain can vary by orders of magnitude with little change in the overall gain. This robustness to gain variations is very important as the gain of transistors can vary by tens of percents. It is therefore difficult to make a transistor amplifier with precisely controlled gain, but it is relatively easy to make a large gain amplifier by combining several transistors. This gain can then be traded for accuracy with negative feedback.

As the actual gain of the op amp has little effect on the circuit gain, the op amp circuit can be analyzed more easily by the two "Golden Rules":

1. No current will flow into the op amp inputs.

2. The input voltages will be equal.

The use of the Golden Rules is illustrated in the following sections where typical op amp amplifiers are analyzed.

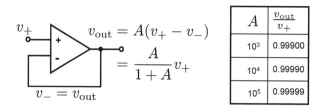

A	$\frac{v_{\text{out}}}{v_+}$
10^3	0.99900
10^4	0.99990
10^5	0.99999

Figure 8.1: Unity gain operational amplifier.

8.1.1 Inverting amplifier

Figure 8.2 shows an inverting op amp circuit. The positive input is connected to the ground. From the Golden Rule 2, the negative input (node 'A') is also at ground potential. The current flowing through Z_1 is

$$i_1 = \frac{v_{\text{in}} - v_A}{Z_1} = \frac{v_{\text{in}}}{Z_1}. \tag{8.2}$$

As this current does not flow into the op amp (Golden Rule 1), it must flow through Z_2 and $i_2 = i_1$. The output voltage is

$$v_{\text{out}} = -i_2 Z_2 = -i_1 Z_2 = -\frac{Z_2}{Z_1} v_{\text{in}}. \tag{8.3}$$

Equation (8.3) shows that the inverting amplifier voltage gain is $A_v = -Z_2/Z_1$. The gain is thus determined by the impedance ratio and not by the op amp.

The input impedance is also obtained using the Golden Rules: Since the node 'A' is at ground potential the input impedance for the circuit is simply

$$Z_{\text{in}} = \frac{v_{\text{in}}}{i_{\text{in}}} = \frac{v_{\text{in}}}{i_1} = Z_1. \tag{8.4}$$

As the impedance is set by Z_1, the inverting amplifiers have relatively low input impedance. This is usually not desirable when measuring microdevices and the inverting amplifiers are seldom connected directly to the sensor element.

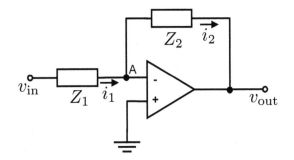

Figure 8.2: Inverting amplifier.

Example 8.1: Inverting amplifier
Problem: Calculate the gain for the inverting amplifier in Figure 8.3 for low frequencies ($\omega \ll 1/R_1 C_1$ and $\omega \ll 1/R_2 C_2$) and high frequencies ($\omega \gg 1/R_1 C_1$ and $\omega \gg 1/R_2 C_2$).

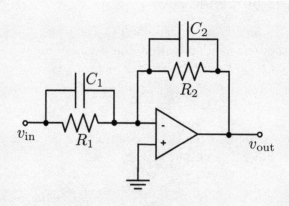

Figure 8.3: Inverting amplifier with capacitive and resistive feedback.

Solution: Substituting

$$Z_1 = R_1 \| \frac{1}{j\omega C_1} = \frac{1}{\frac{1}{R_1} + j\omega C_1}$$

and

$$Z_2 = R_2 \| \frac{1}{j\omega C_2} = \frac{1}{\frac{1}{R_2} + j\omega C_2}$$

to Equation (8.3) gives the voltage gain

$$A_v = \frac{v_{\text{out}}}{v_{\text{in}}} = -\frac{R_2}{R_1} \frac{1 + j\omega C_1 R_1}{1 + j\omega C_2 R_2}.$$

At low frequencies this simplifies to

$$A_v = -\frac{R_2}{R_1}$$

and at high frequencies this simplifies to

$$A_v = -\frac{C_1}{C_2}.$$

8.1.2 Non-inverting amplifier

Figure 8.4 shows the non-inverting op amp circuit. For given output voltage v_{out}, the voltage at node 'A' is

$$v_A = \frac{Z_1}{Z_1 + Z_2} v_{\text{out}}. \tag{8.5}$$

But from the Golden Rule 2, the input node potentials must be equal

$$v_A = v_{\text{in}}. \tag{8.6}$$

Combining Equations (8.5) and (8.6), we obtain the voltage gain

$$A_v = \frac{v_{\text{out}}}{v_{\text{in}}} = 1 + \frac{Z_2}{Z_1}. \tag{8.7}$$

Since no current flows into the op amp input (Golden Rule 1), the amplifier input impedance is ideally infinite. The non-inverting amplifier is therefore well suited for measuring signals without loading the signal source. For example, the non-inverting amplifier could be used to measure a capacitive or resistive voltage divider, as the amplifier would not affect the voltage division.

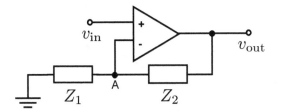

Figure 8.4: Non-inverting amplifier.

Example 8.2: Non-inverting amplifier
Problem: Calculate the gain for the amplifier in Figure 8.4 if $Z_1 = 100 \ \Omega$ and $Z_2 = 1 \ \text{k}\Omega$.
Solution: From Equation (8.7), we have

$$A_v = 1 + \frac{Z_2}{Z_1} = 11.$$

8.1.3 Transimpedance amplifier

Figure 8.5 shows a transimpedance amplifier circuit that transforms the input current i_{in} into a voltage v_{out}. Using the Golden Rules, the current i_{in} flows through the Z (Golden Rule 1) and input voltage v_{in} is zero (Golden Rule 2). We obtain

$$v_{out} = -Zi_{in}. \tag{8.8}$$

Equation (8.8) shows that the impedance Z transforms the current i_{in} into voltage v_{out} – hence the name transimpedance amplifier.

Since $v_{in} = 0$ (Golden Rule 2), the input impedance is

$$Z_{in} = \frac{v_{in}}{i_{in}} = 0. \tag{8.9}$$

Equation (8.9) shows that the transimpedance amplifier is a perfect voltage to current converter as it does not restrict the current flow. The transimpedance amplifiers are used extensively for measuring small currents for example from capacitive sensors and photo diodes.

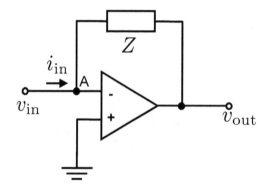

Figure 8.5: Transimpedance amplifier that converts current to voltage.

Example 8.3: Transimpedance amplifier
Problem: The current from a capacitive sensors ranges from 1 pA to 1 nA. Size the feedback impedance Z in the transimpedance amplifier so that the maximum output voltage is 1 V.
Solution: From Equation (8.8), we have

$$Z = \left| \frac{v_{out}}{i_{in}} \right| = \frac{1 \text{ V}}{1 \text{ nA}} = 1 \text{ G}\Omega.$$

This impedance is unrealistically large. A more realistic circuit would first use a smaller transimpedance (~ 10 MΩ) and then further amplify the signal for example with an inverting or non-inverting amplifier.

8.1.4   Differential amplifier

Differential amplification is desirable for measuring small relative changes in piezoresistors or capacitors. Figure 8.6 shows a differential amplifier that can be used to amplify the signal for example from a resistive or a capacitive Wheatstone bridge. Again, we recall that no current flows into the op amp inputs. Using voltage division, the voltage at node 'A' is

$$v_A = \frac{Z_2}{Z_1 + Z_2} v_{\text{in},+}. \tag{8.10}$$

The current from the negative input to node 'B' is

$$i_{\text{in},-} = \frac{v_{\text{in},-} - v_B}{Z_1} = \frac{1}{Z_1} v_{\text{in},-} - \frac{Z_2}{Z_1(Z_1 + Z_2)} v_{\text{in},+}, \tag{8.11}$$

where we have used $v_A = v_B$ (Golden Rule 2). The same current flows from node 'B' to the output and the output voltage is

$$v_{\text{out}} = -i_{\text{in},-} Z_2 + v_B = -i_{\text{in},-} Z_2 + v_A = (v_{\text{in},+} - v_{\text{in},-}) \frac{Z_2}{Z_1}, \tag{8.12}$$

where we have again used the Golden Rule $v_A = v_B$ and Equation (8.10).

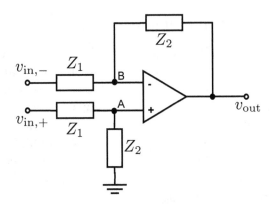

Figure 8.6: Differential amplifier amplifies the difference two inputs $v_{\text{in},+}$ and $v_{\text{in},-}$.

It is useful to rewrite the Equation (8.12) in terms of common mode voltage $v_{cm} = (v_{in,+} + v_{in,-})/2$ and differential voltage $v_{diff} = v_{in,+} - v_{in,-}$

$$v_{out} = v_{diff} \frac{Z_2}{Z_1}. \tag{8.13}$$

As Equation (8.13) shows, the ideal differential amplifier output is directly proportional to the *difference* in input voltages and the common mode signal v_{cm} is not amplified. These results only hold in case of perfect matching of the impedances Z_1:s and Z_2:s. Any real amplifier will have a finite common mode rejection ratio (CMRR)

$$CMRR = \frac{v_{diff}}{v_{cm}}. \tag{8.14}$$

For example, the impedance match of 0.01% results in CMRR of 80 dB which is sufficient for most applications.

Using the Golden Rules, the input impedance for a common mode signal is

$$Z_{cm} = \frac{Z_1 + Z_2}{2}. \tag{8.15}$$

For a differential signal, the input impedance is

$$Z_{diff} = 2Z_1. \tag{8.16}$$

When connecting the difference amplifier to a signal source such as the Wheatstone bridge, the finite input impedance of the amplifier will load the Wheatstone bridge output. The differential amplifier in Figure 8.6 is therefore suited for signal sources that have a low output impedance. The finite input impedance is a major shortcoming of the single op amp differential amplifier. The instrumentation amplifier covered in the next section uses three op amps to give better performance than is obtained with just a single op amp differential amplifier.

8.1.5   Instrumentation amplifier

The instrumentation amplifier is a differential amplifier with a high input impedance. A typical instrumentation amplifier is shown in Figure 8.7 where two op amps have been added to the input of a differential amplifier to increase the input impedance. The input op amps in Figure 8.7 are connected as unity followers but additional amplification can be obtained by converting the buffering op amps to non-inverting amplifiers [65, 66]. The differential amplifier connected after the buffers provides the common mode rejection ratio and the differential gain. As with the single op amp differential amplifier, the components require tight matching.

The instrumentation amplifier is a versatile amplifier and is used for wide variety of sensors including chemical and optical devices. Although the circuit

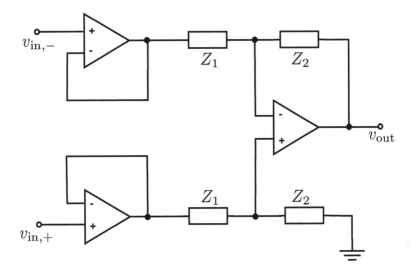

Figure 8.7: Instrumentation amplifier is a differential amplifier with high input impedance.

could be implemented with operational amplifiers and bulk resistors, the component matching requirements make this approach uneconomical. A dedicated IC that has laser-trimmed resistors provides very good matching at lower cost than can be obtained with a circuit based on discrete op amps and resistors. Further information on the instrumentation amplifier can be found in electronics textbooks [65–68].

8.2   Transistor amplifiers

As almost all analog circuit functions can be realized with op amps, a MEMS component or system designer may be tempted to leave transistors to the "circuit people". This ignorance is partially justified as the designer of mechanical sensors is seldom asked to design the measurement circuit or vice versa. However, a good MEMS component designer must have an understanding of the trade-offs within the system in order the find the optimal design. For example, the same design goal might be reached by increasing the sensor element sensitivity, reducing the amplifier noise, or both. The person making the right decisions at the system level is a well-paid individual!

A working knowledge of amplifier circuits is relatively easy to obtain. The op amp circuit may contain 30+ transistors, but as we learned in Chapter 2, the first amplifier stage is the most important in setting the noise performance. We will therefore focus on the input stage of the op amp to gain an understanding of the performance trade-offs between size, power consumption, and noise.

Today, practically all sensor amplifiers are based on the complementary metal-oxide-semiconductor (CMOS) technology. The CMOS transistors have a high input impedance and are economical to manufacture. Different process variants are available; for example, non-volatile memory is often added for sensor calibration.

In this section, we will review the basic operation principles of transistor amplifiers. Further information about transistor circuit design can be found in references [66, 68, 69].

8.2.1 Common source amplifier

The common source CMOS amplifier in Figure 8.8 is an ubiquitous circuit block in analog electronics. The transistor has three terminals; gate (G), drain (D), and source (S). In the common source configuration, the source is the common potential for the input and output. When voltage is applied to the gate, electrons can flow from source to the drain. As the electrons are negative charge carriers, the current flows from drain to source.

Transistors require power or bias to operate. In Figure 8.8(a), the bias is provided by a current source that supplies constant current from the power supply. We will omit the details of the current supply design and instead focus on the transistor as it dominates the noise-power trade-offs. The current source design is covered for example in references [68, 69].

The common source amplifier operates in what is referred to as a saturation region, where the gain for given bias current is the largest. The small signal model for the amplifier in the saturation region is shown in Figure 8.8(b). The

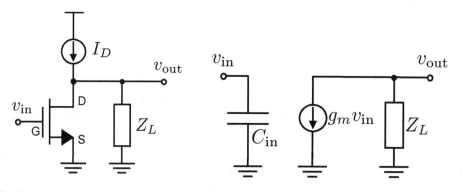

(a) Common source CMOS amplifier circuit. The bias current I_D flows through the transistor from drain (D) to source (S).

(b) Small signal model for the common source amplifier.

Figure 8.8: CMOS common source amplifier and small signal model.

transistor acts as a voltage controlled current source. The small signal current due to an input voltage v_{in} is

$$i = g_m v_{\text{in}},\tag{8.17}$$

where g_m is the small signal transconductance given by

$$g_m = \sqrt{2\mu_n C_{ox}\frac{W}{L}I_D}.\tag{8.18}$$

Here μ_n is the mobility of electrons, C_{ox} is the capacitance per unit area, W is the transistor width, L is the transistor length, and I_D is the dc bias current trough the transistor.

The transistor dimensions W and L have a minimum size that depends on the used technology. This minimum size is typically indicated in the process name. For example, CMOS035 indicates that the CMOS process has the minimum transistor length $L = 0.35$ μm. The constant $\mu_n C_{ox}$ also depends on the used technology. Typical parameters for transistors made in 0.35 micron process are shown in Table 8.1.

Table 8.1: Typical parameters for transistors made in 0.35 μm process.

Parameter	Symbol	Value
Threshold voltage	V_T	0.4 V
Oxide capacitance	C_{ox}	4 fF/μm^2
Electron mobility	μ_n	350 cm^2/Vs
Mobility×capacitance	$\mu_n C_{ox}$	140 μA/V^2

To obtain voltage amplification, the drain is connected to a load impedance Z_L that converts the current i to voltage

$$v_{\text{out}} = -iZ_L = -g_m v_{\text{in}} Z_L \equiv A_v v_{\text{in}},\tag{8.19}$$

where A_v is the small signal voltage gain $A_v = -g_m Z_L$.

As shown in Figure 8.8(b), the input impedance for the amplifier is capacitive and depends on the size of the transistor. The dominant capacitance is from gate to source and this is given by

$$C_{\text{in}} = C_{ox}WL.\tag{8.20}$$

Example 8.4: Common source amplifier
Problem: Calculate the frequency response of the common source amplifier in Figure 8.9. Simplify the expression for low frequencies, high frequencies, and determine the frequency at which the gain has dropped by 3 dB. Evaluate the gain at dc and at 10 MHz.

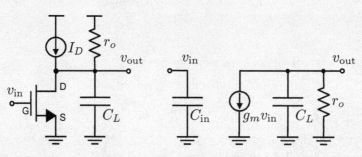

Figure 8.9: A common source amplifier and the small signal model ($g_m = 1$ mS, $C_L = 2$ pF and $r_o = 100$ kΩ).

Solution: The load seen by the transistor is

$$Z_L = r_o || \frac{1}{j\omega C_L} = \frac{r_o}{1 + j\omega C_L r_o}.$$

The gain $A_v = -g_m Z_L$ therefore decreases with frequency. The -3 dB frequency where the gain has dropped to half is

$$f_{\text{-3 dB}} = \frac{1}{2\pi C_L r_o} = 800 \text{ kHz}.$$

At dc, the Z_L simplifies to

$$Z_L = r_o$$

and at high frequencies this becomes

$$Z_L \approx \frac{1}{j\omega C_L}.$$

We obtain the voltage gain at dc as $A_v \approx -g_m r_o = -100$ and 10 MHz we have $A_v \approx -g_m / j\omega C_L = j8$.

Equation (8.18) indicates that we can adjust the transconductance by changing the transistor dimensions or the bias current. However, the inherent assumption in Equation (8.18) is that the transistor is biased to operate in so called "saturation region". The bias current in saturation region is

$$I_D = \frac{1}{2}\mu_n C_{ox} \frac{W}{L}(V_{GS} - V_T)^2, \tag{8.21}$$

where V_{GS} is the bias voltage from gate to source and V_T is the threshold voltage, typically about 0.4 V for a 0.35 micron process. In the interest of simplicity, we'll cut corners a little and require that $V_{GS} = 2V_T$. While the V_{GS} could in fact vary more, the range is quite limited especially in modern processes that have low supply voltages.

With the assumption of $V_{GS} = 2V_T$, the bias current from Equation (8.21) is

$$I_D = \frac{1}{2}\mu_n C_{ox}\frac{W}{L}V_T^2. \tag{8.22}$$

Equations (8.18) and (8.22) can be combined as

$$g_m = \mu_n C_{ox}\frac{W}{L}V_T = 2I_D/V_T. \qquad \text{(if } V_{GS} = 2V_T\text{)} \tag{8.23}$$

Equation (8.23) shows that in order to increase the transconductance, the transistor current and width must both be increased. This gives a powerful motivation for integrated circuits: It is very difficult to obtain very large gains from a single transistor, but by cascading the transistor amplifiers as illustrated in Example 8.5, very large gains can be easily obtained.

Example 8.5: Cascading amplifiers
Problem: Show that the amplifier gain can be increased with no increase in power consumption by connecting two transistor amplifiers in series as shown in Figure 8.10.

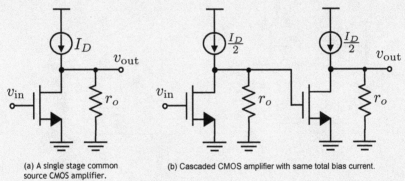

(a) A single stage common source CMOS amplifier.

(b) Cascaded CMOS amplifier with same total bias current.

Figure 8.10: Comparison of a single common source amplifier and two amplifiers cascaded.

Solution: Figure 8.10 shows two amplifiers with the same total power consumption. Assuming $I_D = 200~\mu$A, $V_T = 0.4$, and $r_o = 100$ kΩ, the transconductance for the first amplifier from Equation (8.23) is $g_m = 2I_D/V_T = 1.0$ mS and

$$A_{v1} = -g_{m1}r_o = -100.$$

For the second amplifier, the total gain is obtained by multiplying the gain from individual stages. Since the bias current for the transistors is halved, we have $g_{m2} = g_{m1}/2$ and

$$A_{v2} = g_{m2}r_o g_{m2}r_o = (g_{m2}r_o)^2 = \frac{(g_{m1}r_o)^2}{4} = 2{,}500.$$

This is the basis of op amp circuits: It is difficult to make an accurate transistor amplifier. But by cascading amplifiers, it is easy to make amplifiers with a large gain.

8.2.2   Differential pair

The differential pair shown in Figure 8.11(a) is a transistor circuit that amplifies the difference between two inputs. As illustrated in Figure 8.11(b), the circuit is the basis for the op amp input stage. After the differential pair, the op amp usually has additional gain stages but the latter stages are less critical for noise and common mode rejection performance. Once we understand the differential pair operation, we are well on our way to understanding op amp circuits.

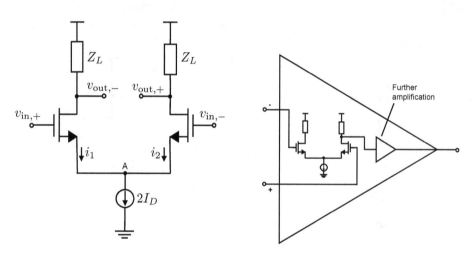

(a) Differential pair is formed by two source followers tied together.

(b) Differential pair is used as the first op amp amplifier stage.

Figure 8.11: Differential pair.

Figure 8.11(a) shows a simplified schematic for the differential pair. The easiest way to understand the differential pair is to note that it is based on two common source amplifiers with the sources connected together. The current through the left transistor is $i_1 = g_m v_{in,+}$ and the current through the right transistor is $i_2 = g_m v_{in,-}$. For a differential input $v_{in,-} = -v_{in,+}$, the currents i_1 and i_2 cancel at node 'A' and the potential at node 'A' is unchanged. Due to the current change $i_1 = g_m v_{in,+}$, the voltage at the negative output is $v_{out,-} = -g_m Z_L v_{in,+}$. Similarly, te positive output voltage is $v_{out,+} = -g_m Z_L v_{in,-}$ and the differential output voltage is

$$v_{\text{out}} = v_{\text{out},+} - v_{\text{out},-} = g_m Z_L (v_{\text{in},+} - v_{\text{in},-}) = g_m Z_L v_{\text{diff}}. \qquad (8.24)$$

Equation (8.24) shows that the gain for the differential pair equals the common source amplifier gain. As there are now two transistors each consuming bias current I_D, the total power consumption is twice as much for the common source amplifier with equal gain.

The most useful property of the differential amplifier is its ability to reject common mode signals. To understand this behavior, consider what happens when voltages at the both inputs are increased by an equal amount. If the sources were not connected to the same bias current source, the current through the transistors would increase and the voltages at the output nodes would decrease. However, as both transistors are connected to the same current source, the net current through the transistors is always the same. The current source effectively prevents common mode signals from affecting the transistor currents and the output voltage remains unchanged. This ideal picture assumes a perfect current source and perfect component matching. The quantitative analysis of common mode rejection of the differential pair is involved and depends on how the current source and the load resistors are implemented. Typically the CMRR ranges from 70 to 100 dB.

Key concepts

- Modern signal amplification is based on operational amplifiers that provide high gain. Combined with negative feedback, the operational amplifiers can be used for example as amplifiers, differential amplifiers, and transimpedance amplifiers.

- Single transistor amplifiers have limited gain and vary in characteristics. The transistor weaknesses can be can be overcome by combining a large number of transistors together. Integrated circuit technology enables integration of large number of transistors on a single-chip at a low cost.

- Today, the CMOS technology is the most cost effective integrated circuit technology. The trade-offs in CMOS circuits can be understood by analyzing the common-source amplifier.

- The common-source amplifier transconductance g_m is proportional to the bias current I_D. The voltage gain is $A_v = -g_m Z_L$. To obtain large gain from a single common-source amplifier, large bias current is needed. A power efficient way to increase the gain is to combine multiple gain stages together.

- Operational amplifier input is based on a differential pair. To increase the gain, the differential pair may be combined with additional transistor stages.

Exercises

Exercise 8.1
Plot the magnitude of gain for the inverting amplifier in Figure 8.3 from DC to 10 MHz if $R_1 = 10$ kΩ, $R_2 = 40$ kΩ, $C_1 = 100$ pF, and $C_2 = 200$ pF.

Exercise 8.2
Calculate the gain for the amplifier in Figure 8.4 if $Z_1 = 400$ Ω and $Z_2 = 2$ kΩ.

Exercise 8.3
The transimpendance amplifier in Figure 8.5 has $Z = 500$ kΩ. What is the input current if the output voltage is 0.5 V?

Exercise 8.4
Explain why an instrumentation amplifier is better suited for measuring a Wheatstone bride than a differential amplifier.

Exercise 8.5
An instrumentation amplifier with gain $A = 100$ and 80 dB CMRR is used to measure a Wheatstone bridge. The differential input voltage $v_2 - v_1$ is zero but the common mode voltage is $v_2 = v_1 = 2.5$ V. What is the amplifier output voltage?

Exercise 8.6
The frequency response of an op amp with two amplification stages is given by

$$A(s) = \frac{A_0}{(s + \omega_{P1})(s + \omega_{P2})},$$

where A_0 is the low frequency gain and ω_{P1} and ω_{P2} are the first and second pole, respectively. As illustrated in Figure 8.12, the gain of the op amp circuits is controlled by applying negative feedback. Defining β as the ratio of op amp output applied to the negative input, the closed loop gain is

$$A_{CL}(s) = \frac{A(s)}{1 + A(s)\beta} = \frac{A_0 \omega_{P1} \omega_{P2}}{s^2 + s(\omega_{P1} + \omega_{P2}) + \omega_{P1}\omega_{P2} + A_0\beta\omega_{P1}\omega_{P2}}.$$

This expression is identical to Equation (B.6) on page 410 if we write $m = 1/A_0/\omega_{P1}/\omega_{P2}$, $\omega_0^2 = \omega_{P1}\omega_{P2} + A_0\beta\omega_{P1}\omega_{P2}$, and $Q = \omega_0/(\omega_{P1} + \omega_{P2})$. i. What is the low frequency gain A_{CL} if A_0 is large? ii. How does the closed loop step response change with feedback β? When is the response over damped and when is it under damped? Assume $A_0 = 10^5$, $\omega_{P1} = 10^3$ Hz, and $\omega_{P2} = 10^6$ Hz.

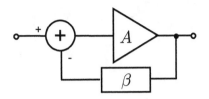

Figure 8.12: Amplifier with negative feedback.

Exercise 8.7

Calculate the low frequency gain and -3 dB frequency for the common source amplifier in Figure 8.9 if $I_D = 300\ \mu\text{A}$, $C_L = 0.5$ pF, and $r_o = 100$ kΩ. For the transistor, use parameters in Table 8.1 and assume $V_{GS} = 2V_T$.

9

Amplifier noise

The purpose of a sensor amplifier is to amplify the sensor signal to rescue it before it is corrupted by noise. The amplifier will add some noise of its own, but once the signal has been boosted, the noise from subsequent signal processing is usually small in comparison. The noise performance of the first amplifier stage is therefore critical for microsystem design.

In this chapter, the noise in amplifiers is reviewed. We will start by analyzing the noise in MOS-transistors, as the CMOS is a basis of most commercial microsensor electronics. Next, the noise in common source amplifiers and differential pairs are analyzed. The results are used to gain insights in operational amplifier noise trade-offs. Finally, the chapter is concluded by a full noise analysis of piezoresistive and capacitive accelerometers.

9.1 Noise in transistors

CMOS is the dominant technology for microsensor electronics and understanding the noise in CMOS circuits is necessary for optimizing the microsystems. The CMOS transistor operation was reviewed in Section 8.2. In this section, we review the noise analysis of CMOS amplifiers.

Figure 9.1 shows the noise sources in the common source amplifier analyzed in Section 8.2.1. The two main noise sources in the transistor are the channel noise and the $1/f$-gate noise. The channel noise current is

$$\overline{i_n^2} = 4k_B T \gamma_c g_m, \tag{9.1}$$

where the parameter γ_c is a process and transistor size dependent with a typical value of $2/3$ and g_m is the transistor transconductance given by Equation (8.18).

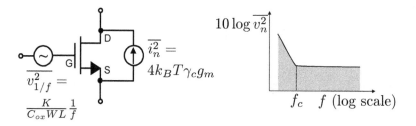

Figure 9.1: Noise sources in transistor. Due to $1/f$-noise, the transistor noise at low frequencies is significantly higher than the thermal noise limit due to the channel noise.

Equation (9.1) closely resembles the transistor thermal noise $\overline{i_n^2} = 4k_BTR$. This can be understood by noting that the transistor can be viewed as a voltage-controlled resistor where the gate voltage controls the channel current.

The $1/f$-gate noise or "flicker" noise is

$$\overline{v_{1/f}^2} = \frac{K}{C_{ox}WL}\frac{1}{f}, \tag{9.2}$$

where K is a process-dependent constant on the order of 10^{-25} V^{2}F [68]. The gate noise is caused by random trapping and release of electrons. The trapping and release occur more frequently at low frequencies and the gate noise has $1/f$-spectral characteristics. As Equation (9.2) shows, the $1/f$-noise can be decreased by increasing the transistor area WL. In low noise applications, the first amplifying transistor may have to be very large.

Using Equations (8.17) and (9.2), the drain current induced by the $1/f$-gate noise is $\overline{i_{1/f}^2} = g_m^2\overline{v_{1/f}^2}$. Equating this with the thermal channel noise current $\overline{i_n^2}$ given by Equation (9.1) gives the $1/f$-corner frequency for the transistor

$$f_c = \frac{K}{C_{ox}WL}\frac{g_m}{4k_BT\gamma_c}. \tag{9.3}$$

By combining Equations (8.23) and (9.3), the corner frequency can be written as

$$f_c = \frac{K\mu}{L^2}\frac{V_T}{4k_BT\gamma_c}, \tag{9.4}$$

where it is assumed that $V_{GS} = 2V_T$. Equation (9.4) shows that long transistors are needed to obtain a low $1/f$-corner frequency. This is further exemplified in Example 9.1.

Example 9.1: Transistor $1/f$-corner frequency
Problem: Based on the transistor parameters in Table 8.1 on page 131, calculate the $1/f$-corner frequency for an amplifier with $L = 0.3$ μm, 3 μm, and 30 μm. Assume $K = 10^{-25}$ V^{2}F.
Solution: From Equation (9.4) and $K = 10^{-25}$ V^{2}F, we get $f_c = 1.4$ MHz, 14 kHz, and 140 Hz for $L = 0.3$ μm, 3 μm, and 30 μm, respectively. We note that the sub-micron transistor has excessive noise at low frequencies and it is not suitable for low noise applications below 1 MHz. By increasing the transistor size by a factor of 100, the $1/f$-noise corner can be dropped to less than 1 kHz.

This example shows that to obtain good noise performance at low frequencies, large transistor size is needed. It is common to have transistors larger than several thousand square microns at the input of a low noise amplifier.

9.1.1 Noise in common source amplifier

The common source amplifier is the most used transistor amplifier structure. The operation of the common source amplifiers was reviewed in Section 8.2.1 on page 130. Here we will analyze the noise in the common source amplifier.

Figure 9.2(a) shows the noise model for a common source amplifier. The total noise current through a load impedance Z_L is

$$\overline{i_{n,\text{out}}^2} = g_m^2 \overline{v_{1/f}^2} + \overline{i_n^2} = g_m^2 \frac{K}{C_{ox}WL}\frac{1}{f} + 4k_BT\gamma_cg_m. \qquad (9.5)$$

From Equation (9.5), the noise voltage at the amplifier output is

$$\overline{v_{n,\text{out}}^2} = \overline{i_{n,\text{out}}^2}Z_L^2 = \left(g_m^2 \frac{K}{C_{ox}WL}\frac{1}{f} + 4k_BT\gamma_cg_m\right)Z_L^2. \qquad (9.6)$$

The first term in Equation (9.6) is due to the gate $1/f$-noise and the second term is due to the thermal channel noise.

Equation (9.6) shows that the noise at the amplifier output increases with increasing transconductance g_m. However, no benefit will be gained by decreasing g_m as this will also lower the amplifier gain. As was shown in Section 2.5, the input referred noise is a useful tool for quantifying the amplifier noise performance as the input referred noise factors in the amplifier gain.

Figure 9.2(b) shows the input referred noise for the common source amplifier. To account for any source impedance Z_S, the input referred noise representation requires a voltage source and a current source at the amplifier input. The input referred noise voltage due to the channel noise is

$$\overline{v_{n,\text{in}}^2} = \frac{4k_BT\gamma_c}{g_m} \qquad (9.7)$$

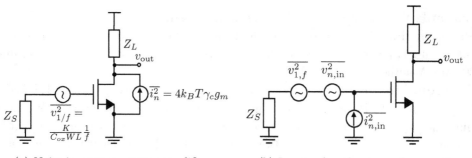

(a) Noise in common source amplifier.   (b) Input referred noise and current.

Figure 9.2: Noise and input referred noise in common source amplifier.

and the input referred current is

$$\overline{i_{n,\text{in}}^2} = (\omega C_{\text{in}})^2 \frac{4k_B T \gamma_c}{g_m}, \tag{9.8}$$

where $C_{\text{in}} = WLC_{ox}$ is the transistor input capacitance. We see that increasing g_m directly decreases the input referred noise.

The input referred noise generators given by Equations (9.7) and (9.8) both represent the transistor channel noise and are correlated. We therefore sum the noise contributions of $\overline{v_{n,\text{in}}}$ and $\overline{i_{n,\text{in}}}$.

In addition to the channel noise, the complete input referred noise model shown in Figure 9.2(b) has the $1/f$-noise applied to the gate. The gate flicker noise is given by Equation (9.2). As the channel noise and gate flicker noise are not correlated, the noise powers are summed to obtain the total input referred noise.

Example 9.2: Proof of input referred noise model

Problem: Show that the input referred noise generator Equations (9.7) and (9.8) give the correct results for the total noise at the amplifier output (Equation (9.6)) regardless of the source impedance Z_S. For simplicity, you can ignore the $1/f$-noise and focus on the channel noise.

Solution: The noise current and voltage in Figure 9.2(b) are divided between the source impedance Z_S and the transistor input impedance $Z_{IN} = 1/sC_{IN}$. The noise at the gate due to the noise voltage is obtained by voltage division $\overline{v_{n,\text{in}}} \frac{Z_{IN}}{Z_{IN}+Z_S}$. The noise at the gate due to the noise current is $\overline{i_{n,\text{in}}} \frac{Z_{IN}Z_S}{Z_{IN}+Z_S}$ where $\frac{Z_{IN}Z_S}{Z_{IN}+Z_S}$ is the total impedance seen by the noise current generator.

Remembering that the $\overline{v_{n,\text{in}}}$ and $\overline{i_{n,\text{in}}}$ are correlated, we sum gate voltages (not powers) due to the noise voltage and current generators. The total noise at the transistor gate is

$$\overline{v_{n,G}} = \overline{v_{n,\text{in}}}\frac{Z_{IN}}{Z_{IN} + Z_S} + \overline{i_{n,\text{in}}}\frac{Z_{IN}Z_S}{Z_{IN} + Z_S} = \frac{\overline{v_{n,\text{in}}} + \overline{i_{n,\text{in}}}Z_S}{1 + Z_S C_{IN}s}. \tag{9.9}$$

From Equations (9.7) and (9.8), the input noise current is $\overline{i_{n,\text{in}}} = (\omega C_{IN})^2 \overline{V_{n,\text{in}}}$ and the gate noise from Equation (9.9) is

$$\overline{v_{n,G}} = \overline{V_{n,\text{in}}}. \tag{9.10}$$

The noise at the output is $\overline{v_{n,\text{out}}} = -g_m Z_L \overline{v_{n,G}}$ which leads to

$$\overline{v_{n,\text{out}}^2} = g_m^2 Z_L^2 \overline{v_{n,G}^2} = 4k_B T \gamma_c g_m Z_L^2. \tag{9.11}$$

This is identical to Equation (9.6) without the flicker noise component, proving that the input noise current and voltage generator give the correct output noise, regardless of input impedance Z_S.

Example 9.3: Common source amplifier noise

Problem: Find the input referred noise for the amplifier in Figure 9.3 and compare this to the noise at the gate due to the source resistance $R_S = 10 \text{ k}\Omega$. For the transistor, use the parameters in Table 8.1 and assume $W = 100 \ \mu\text{m}$ and $L = 1 \ \mu\text{m}$. What is the bias current through the transistor? Assume an operating frequency of $f = 1$ MHz and the $1/f$-noise can be ignored.

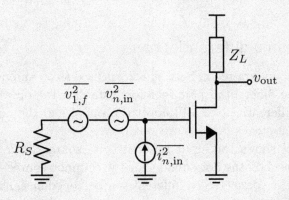

Figure 9.3: Noise in common source amplifier connected to resistive source.

Solution: The transistor bias current and transconductance from Equations (8.22) and (8.23) are

$$I_D = \frac{1}{2}\mu_n C_{ox}\frac{W}{L}V_T^2 = 1.1 \text{ mA}$$

and

$$g_m = \mu_n C_{ox}\frac{W}{L}V_T = 2I_D/V_T = 5.6 \text{ mS}.$$

Here we have assumed that $V_{GS} = 2V_T$.

From Equations (9.10) and (9.7) we get the

$$\overline{v_{n,G}} = \sqrt{\frac{4k_B T\gamma_c}{g_m}} = 1.4 \text{ nV}/\sqrt{\text{Hz}}$$

for the total input referred noise at the transistor gate. The noise at the gate due to source resistor is

$$\overline{v_{n,R_S}} = \left|\frac{1/sC_{IN}}{R_S + 1/sC_{IN}}\right|\sqrt{4k_B T R_s} = 13 \text{ nV}/\sqrt{\text{Hz}}.$$

We see that the noise due to source resistance dominates. A larger amplifier noise can therefore be tolerated without affecting the overall noise and the transistor power consumption can be significantly reduced. For example, $W = 25\ \mu$m and $L = 5\ \mu$m would give $I_D = 56\ \mu$A and $\overline{v_{n,G}} = 6.3 \text{ nV}/\sqrt{\text{Hz}}$ which would bring the total input referred noise to $\sqrt{(6.3 \text{ nV}/\sqrt{\text{Hz}})^2 + (13 \text{ nV}/\sqrt{\text{Hz}})^2} = 14 \text{ nV}/\sqrt{\text{Hz}}$. This is just a 1 nV/$\sqrt{\text{Hz}}$ more than the resistor noise alone. Compared to the original $I_D = 1.1$ mA, the power consumption is reduced by close to 20×.

9.1.2   Noise in a differential pair

The differential pair covered in Section 8.2.2 is used as op amp inputs. The noise properties of the differential pair therefore establish the op amp noise performance as the differential pair will boost the signal and the noise contribution of the subsequent stages is therefore small.

Figure 9.4(a) shows the input referred noise sources in a differential pair. The model is based on the noise model for the common source amplifier. As the two inputs may be connected to different source impedances, both the positive and the negative input need to have independent noise sources. Ignoring the noise due to the loads Z_L and the current source, the noise voltage and current generators are given by Equations (9.7) and (9.8).

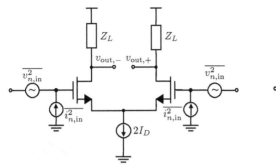
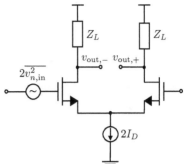

(a) Differential pair showing input referred noise sources for both inputs.

(b) Simplified noise model for $|Z_{IN}| > |Z_S|$.

Figure 9.4: Input referred noise in differential pair.

This idealistic model will underestimate total noise as noise sources other than the input transistors are ignored. But in a well designed amplifier, the input transistors are the main noise sources, and the simple model is a fairly accurate first-order estimate. Thus, the simple noise model allows us to gain design insights into noise-power performance without dwelling too deep into the details of the circuit design. More complete noise analysis that accounts for the current source noise is found in references [68, 69].

A significant simplification is possible when the source impedances are much lower than the transistor input impedance $|Z_{IN}| = 1/\omega C_{IN}$. In this case, the current noise sources can be ignored and a simpler noise model shown in Figure 9.4(b) can be used. The noise in this case can be represented by a single voltage noise source which has twice the power of the uncorrelated noise voltage sources.

9.2   Operational amplifier noise

We are now ready to give a simplified noise model for an operational amplifier. We start by noting that the input of the op amp is a differential pair. After the differential pair, there can be additional amplifier stages. The noise contribution of the latter stages, however, is reduced by the differential pair gain. This justifies the op amp noise model shown in Figure 9.5 where a differential pair is followed by an ideal noiseless op amp [70]. The model in Figure 9.5 is easy to use and is the basis for a noise model for commercial operational amplifiers. For off-the-shelf devices, the input referred noise voltages and currents are usually given in the data sheet provided by the chip manufacturers.

As stated before, the idealistic model accounts only for the noise due to input transistors. However, in a well designed op amp, the input transistors are

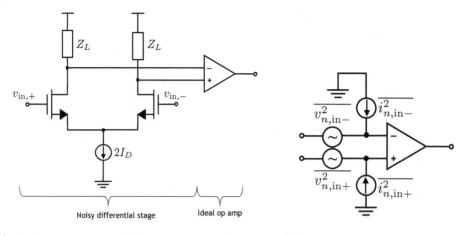

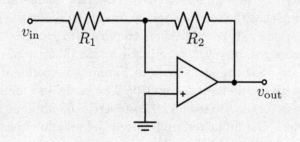

(a) Ideal op amp is preceded with a noisy differential pair.

(b) Input referred noise for the op amp.

Figure 9.5: Operational amplifier noise model.

the dominant noise sources and Figure 9.5 can be used to quantify the trade-off between the noise, transistor size, and power consumption. Low noise op amps require large differential pairs which are readily observed by studying the data sheets for low noise amplifiers. The commercial low noise op amps can have input capacitances greater than 10 pF indicating a large differential pair at the input.

Example 9.4: Noise in an inverting amplifier
Problem: Figure 9.6 shows an inverting amplifier. Assuming that the op amp noise is dominated by the input stage differential amplifier ($W = 100$ μm, $L = 5$ μm, $2I_D = 450$ μA), calculate the total noise at the output for a 1-kHz bandwidth around 1 MHz. Ignore the $1/f$-noise and assume the impedance of the voltage source connected to R_1 is small.

Figure 9.6: Inverting amplifier ($R_1 = 1$ kΩ and $R_2 = 4$ kΩ).

Solution: The differential pair transistor transconductance from Equation (8.23) is

$$g_m = 2I_D/V_T = 1.1 \text{ mS}.$$

The input referred op amp noise voltage from Equation (9.7) is

$$\overline{v_{n,\text{in}}^2} = \frac{4k_B T \gamma_c}{g_m} = 9.9 \cdot 10^{-18} \text{ V}^2/\text{Hz}$$

or $\overline{v_{n,\text{in}}} = 3.1 \text{ nV}/\sqrt{\text{Hz}}$. The input referred noise current from Equation (9.8) is

$$\overline{i_{n,\text{in}}^2} = (\omega C_{\text{in}})^2 \frac{4k_B T \gamma_c}{g_m} = 1.6 \cdot 10^{-27} \text{ A}^2/\text{Hz}$$

or $\overline{i_{n,\text{in}}} = 39 \text{ fA}/\sqrt{\text{Hz}}$. Here C_{in} is the transistor input capacitance given by $C_{\text{in}} = C_{ox}WL = 2 \text{ pF}$ and $\omega = 2\pi f \approx 6.28 \text{ MHz}$.

To calculate the total noise at the output, we will calculate the noise contributions from the individual noise sources and sum the total noise. At first, the analysis may appear involved but in the end all the steps taken are simple.

Figure 9.7 shows the contribution of the opamp noise voltage $\overline{v_{n,\text{in}-}^2}$. Since the amplifier input is connected to a low impedance source, R_1 is grounded for the noise analysis. As both op amp inputs are at ground potential, the noise voltage at node 'A' is $\overline{v_A} = \overline{v_{n,\text{in}-}}$ and the current through R_1 is

$$\overline{i_{n,R_1}} = \frac{\overline{v_A}}{R_1}.$$

As this current must flow through R_2, the noise at the output is

$$\overline{v_{\text{out,in}-}} = \overline{v_A} + \overline{i_{n,R_1}}R_2 = \overline{v_{n,\text{in}-}}\frac{R_1 + R_2}{R_1} = 16 \text{ nV}/\sqrt{\text{Hz}}.$$

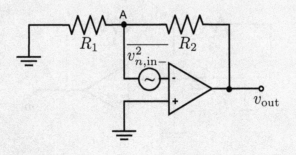

Figure 9.7: Inverting amplifier noise due to $\overline{v_{n,\text{in}-}}$.

The noise due to the op amp noise voltage $\overline{v_{n,\text{in}+}^2}$ is obtained similarly as is shown in Figure 9.8. As there is no voltage difference between op amp inputs, the noise voltage at node 'A' is $\overline{v_A} = \overline{v_{n,\text{in}+}}$. The current through R_1 is $\overline{v_A}/R_1$ and this current must also flow through R_2. The noise at the output is

$$\overline{v_{\text{out},v-}} = \overline{v_A} + \overline{i_{n,R_1}}R_2 = \overline{v_{n,\text{in}+}}\frac{R_1 + R_2}{R_1} = 16 \text{ nV}/\sqrt{\text{Hz}}.$$

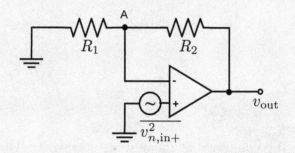

Figure 9.8: Inverting amplifier noise due to $\overline{v_{n,\text{in}+}}$.

Figure 9.9 shows the noise contribution from the input referred noise current. As the voltage over R_1 is zero, no current is flowing through it. Hence, the noise current must flow through R_2. The noise at the output is

$$\overline{v_{\text{out},i-}} = \overline{i_{n,\text{in}-}}R_2 = 0.2 \text{ nV}/\sqrt{\text{Hz}}.$$

As the impedance at the positive input is zero, the noise current $\overline{i_{n,\text{in}+}}$ flows directly to ground and does not contribute to the output noise. We see that due to relatively low frequency and low source impedances, the contribution of the op amp noise currents is small.

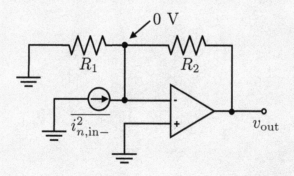

Figure 9.9: Inverting amplifier noise due to $\overline{i_{n,\text{in}-}}$.

The noise current due to R_1 is $\overline{i_{n,R_1}} = \sqrt{4k_BT/R_1} = 4.1$ fA/$\sqrt{\text{Hz}}$. As shown in Figure 9.10, voltage over R_1 is zero and no current flows through it. Hence, the resistor noise current flows through R_2. The noise at the output is

$$\overline{v_{\text{out},R1}} = \overline{i_{n,R_1}}R_2 = 16 \text{ nV}/\sqrt{\text{Hz}}.$$

Similarly the noise due to R_2 shown in Figure 9.11 is

$$\overline{v_{\text{out},R_2}} = \overline{i_{n,R_2}}R_2 = 8.1 \text{ nV}/\sqrt{\text{Hz}},$$

where $\overline{i_{n,R_2}} = \sqrt{4k_BT/R_2} = 2.0$ fA/$\sqrt{\text{Hz}}$.

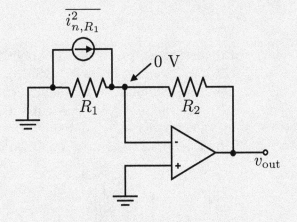

Figure 9.10: Inverting amplifier noise due to R_1.

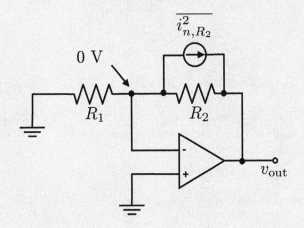

Figure 9.11: Inverting amplifier noise due to R_2.

The total noise at the output is obtained by summing up the individual noise contributions. Recalling that the noise powers from uncorrelated sources add and noise voltages from correlated sources add, we have

$$\overline{v_\text{out}} = \sqrt{\overline{v^2_{\text{out},v+}} + (\overline{v_{\text{out},v-}} + \overline{v_{\text{out},i-}})^2 + \overline{v^2_{\text{out},R1}} + \overline{v^2_{\text{out},R2}}} = 29 \text{ nV}/\sqrt{\text{Hz}}.$$

The rms-noise over 1 kHz bandwidth is

$$v_\text{rms} = \overline{v_\text{out}}\sqrt{BW} = 0.9 \ \mu\text{V}.$$

Example 9.5: Noise in inverting amplifier with capacitive feedback
Problem: Figure 9.12 shows an inverting amplifier with capacitive feedback. Assuming that the op amp noise is dominated by the input stage differential amplifier ($W = 20 \ \mu$m, $L = 5 \ \mu$m, and total bias current $2I_D = 90 \ \mu$A), calculate the total noise spectral density at the output at 100 kHz. Ignore the $1/f$-noise and assume the impedance of the source connected to C_1 is small.

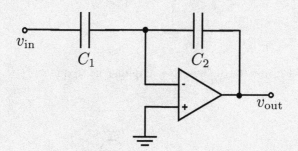

Figure 9.12: Inverting amplifier with capacitive feedback ($C_1 = 12$ pF and $C_2 = 1$ pF).

Solution: The analysis is almost identical to Example 9.4. The differential pair transistor transconductance from Equation (8.23) is

$$g_m = 2I_D/V_T = 0.22 \text{ mS}.$$

The input referred op amp noise voltage from Equation (9.7) is

$$\overline{v^2_{n,\text{in}}} = \frac{4k_BT\gamma_c}{g_m} = 4.9 \cdot 10^{-17} \text{ V}^2/\text{Hz}$$

and $\overline{v_{n,\text{in}}} = 7.0 \text{ nV}/\sqrt{\text{Hz}}$. The input referred noise current from Equation (9.8) is

$$\overline{i^2_{n,\text{in}}} = (\omega C_{\text{in}})^2 \overline{v^2_{n,\text{in}}} = 3.1 \cdot 10^{-30} \text{ A}^2/\text{Hz},$$

or $\overline{i_{n,\text{in}}} = 1.8 \cdot 10^{-15} \text{ A}/\sqrt{\text{Hz}}$, where $C_{\text{in}} = C_{ox}WL = 0.4 \text{ pF}$ is the transistor input capacitance.

The noise at the output due to noise voltage at the positive or negative input is

$$\overline{v_{\text{out},v}} = \overline{v_{n,\text{in}}}\frac{C_1 + C_2}{C_1} = 91 \text{ nV}/\sqrt{\text{Hz}}.$$

The noise at the opamp output due to the noise current at the negative input is

$$\overline{v_{\text{out},i}} = \overline{i_{n,\text{in}}}\frac{1}{\omega C_2} = \overline{v_{n,\text{in}}}\frac{C_{\text{in}}}{C_2} \approx 2.8 \text{ nV}/\sqrt{\text{Hz}}.$$

The output noise due to the noise current at the positive input is zero.

Thus, for the values in this example, the contribution of the input referred noise currents is small and we can use the simplified model in Figure 9.5(b). The total output noise voltage for the two inputs is

$$\overline{v_{\text{out}}} = \sqrt{2}\overline{v_{\text{out},v}} = 130 \text{ nV}/\sqrt{\text{Hz}}.$$

9.3 Amplifier noise in microsystems

We are now ready to put together the lessons of this and prior chapters by comparing the noise in piezoresistive and capacitive accelerometers. The mechanical and electrical noise are both included in the analysis and the accelerometer noise floors are estimated.

9.3.1 Case study: A piezoresistive accelerometer

The mechanical and the resistor noise in a piezoresistive accelerometer was covered in Examples 5.4 and 5.7 on pages 81 and 87 respectively. It was found that the noise is dominated by the piezoresistor $1/f$-noise. To include the amplifier noise, we will assume that the accelerometer is sensed with an idealized differential amplifier with a gain of ten. The amplifier noise is assumed to be due to the op amp input transistors and the actual value of gain will not affect our analysis as increasing the gain would increase both the signal and the amplifier noise.

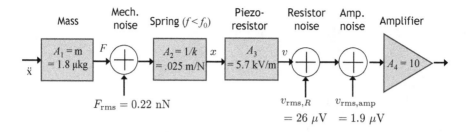

Figure 9.13: Diagram for system level noise analysis of piezoresistive accelerometer.

The op amp input is formed by a differential pair with $W = 20$ μm, $L = 5$ μm, total bias current $2I_D = 90$ μA. The input referred amplifier $1/f$-noise from Equation (9.2) for $0.1 - 100$ Hz bandwidth is

$$v_{\text{rms,amp}} = \sqrt{2\frac{K}{C_{ox}WL}\ln\frac{f_H}{f_L}} = 1.9\ \mu\text{V},$$

where the factor of two comes from the two input transistors.

The complete noise model is shown in Figure 9.13. The total noise at the output is

$$v_{\text{rms,out}} = \sqrt{(A_2A_3A_4F_{\text{rms}})^2 + (A_4v_{\text{rms},R})^2 + (A_4v_{\text{rms,amp}})^2} \approx 0.26\ \text{mV},$$

where F_{rms} is the mechanical noise and $v_{\text{rms},R}$ is the piezoresistor noise. The input referred noise is

$$\ddot{x}_{\text{rms,in}} = \frac{v_{\text{rms,out}}}{A_1A_2A_3A_4} \approx 10\ \text{mG}.$$

This answer is essentially the same in Example 5.7 where the piezoresistor noise is analyzed without the amplifier. We see that in the example piezoresistive accelerometer, the piezoresistors are the main noise source.

As the amplifier noise contribution is insignificant, the design focus should be on mitigating the effect of the piezoresistor noise, which is the dominant noise source. This can be accomplished by reducing the resistor noise or by increasing the "amplification" before the resistor by increasing $A_1 = m$ and/or $A_2 = 1/k$.

9.3.2   Case study: A capacitive accelerometer

The capacitive accelerometer was covered in Example 6.6 on page 102. The capacitive detection results in very good intrinsic noise performance but the small capacitance changes will also result in a small signal. The capacitive detection scheme is therefore easily corrupted by the amplifier noise.

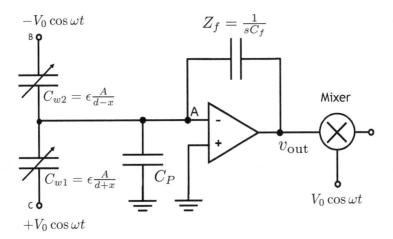

(a) The signal current from differential capacitor is measured with transconductance amplifier. The ac signal is demodulated (=transferred to dc) by the mixer.

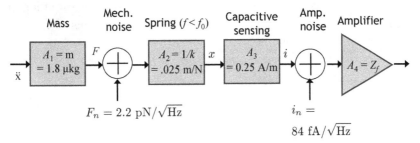

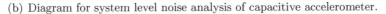

(b) Diagram for system level noise analysis of capacitive accelerometer.

Figure 9.14: Noise in capacitive accelerometer.

Figure 9.14 shows a common measurement circuit for the capacitive accelerometer. The current sensing amplifier of Figure 6.10 is implemented with the transimpedance amplifier of Figure 8.5. As with the piezoresistive accelerometer in the previous section, we will assume that the accelerometer is sensed with an idealized op amp amplifier whose noise is dominated by the input differential pair ($W = 20$ μm, $L = 5$ μm, total bias current $2I_D = 90$ μA). The sense capacitances are $C_w = 5$ pF, the parasitic capacitance is $C_P = 2$ pF, and the feedback capacitance is $C_f = 1$ pF.

The output voltage from the transimpedance amplifier is a sinusoidal wherein the amplitude is proportional to the acceleration. To extract the signal amplitude, the ac-voltage is demodulated by multiplying it with another sinusoidal. From

$$A \cos \omega t \cdot \cos \omega t = \frac{A}{2}(1 + \cos 2\omega t) \tag{9.12}$$

the output of the multiplier (mixer) will have a dc component proportional to the signal amplitude. The demodulation with a mixer will give very good noise performance. The drawback of this method is that the circuit requires fairly large area and consumes a fair amount of power. Here we will not consider the details of the demodulation any further but focus on the signal and noise up to the first amplifier stage.

The $1/f$-corner frequency for the differential pair from Equation (9.4) is $f_c = 5$ kHz. Since the measurement frequency $f = 100$ kHz is significantly higher than the corner frequency: The thermal noise dominates and the $1/f$-noise can be ignored. This is a significant advantage for the modulated detection.

We will now proceed with the noise analysis. From Example 6.6, the signal current for a 1-G acceleration is 120 nA. The output voltage from the transimpedance amplifier at $f = 100$ kHz is

$$v_{\text{out}} = i_{\text{sig}} Z_f = 190 \text{ mV},$$

where we have assumed $C_f = 1$ pF. For the noise analysis, the amplifier topology is identical to the inverting amplifier. From the noise analysis in Examples 9.4 and 9.5, the amplifier output noise voltages are

$$\overline{v_{\text{out,in}-}} = \overline{v_{n,\text{in}-}} \cdot \frac{Z_s + Z_f}{Z_s} \approx 93 \text{ nV}/\sqrt{\text{Hz}}$$

and

$$\overline{v_{\text{out,in}+}} = \overline{v_{n,\text{in}+}} \cdot \frac{Z_s + Z_f}{Z_s} \approx 93 \text{ nV}/\sqrt{\text{Hz}}$$

for the positive and negative differential pair transistors, respectively. Here the noise gain $\frac{Z_s + Z_f}{Z_s}$ is determined by the transimpedance amplifier feedback impedance

$$Z_f = \frac{1}{sC_f} \approx -j1.6 \text{ M}\Omega$$

and the source impedance from a parallel combination of all the capacitances to the ground

$$Z_s = \frac{1}{s(C_{w1} + C_{w2} + C_P)} \approx -j130 \text{ k}\Omega.$$

This shows that the parasitic capacitance C_P increases the noise gain and lowers the noise performance. Thus, although the current measurement configuration eliminates the effect of the parasitic capacitance on signal current, the parasitic capacitance should be minimized for the best noise performance. In the present example, the parasitic capacitance is small in comparison to work capacitances C_{w1} and C_{w2} and the effect of C_P on noise gain is small. But for surface micromachined accelerometers that have small work capacitances, two-chip integration may not be attractive as the parasitic capacitance can result in poor noise performance.

The noise voltages for the differential pair noise currents are

$$\overline{v_{\text{out},i-}} = \overline{i_{n,\text{in}-}} Z_f \approx 2.8 \text{ nV}/\sqrt{\text{Hz}}.$$

and

$$\overline{v_{\text{out},i+}} = 0.$$

The total amplifier noise voltage at the output due to differential pair noise is

$$\overline{v_{\text{out,amp}}} = \sqrt{\overline{v_{\text{out},v+}}^2 + (\overline{v_{\text{out},v-}} + \overline{v_{\text{out},i-}})^2} \approx 0.13 \text{ } \mu\text{V}/\sqrt{\text{Hz}}$$

The input referred amplifier noise current is

$$\overline{i_{\text{in,amp}}} = \frac{\overline{v_{\text{out,amp}}}}{A_4} = \frac{\overline{v_{\text{out}}}}{Z_f} \approx 84 \text{ fA}/\sqrt{\text{Hz}},$$

which can be compared to the noise current of 170 fA/$\sqrt{\text{Hz}}$ due to thermal motion of the proof mass.

Figure 9.14(b) shows the system level schematic for the amplifier noise. The total output noise is

$$\overline{v_{\text{out}}} = \sqrt{(\overline{F_n} A_2 A_3 A_4)^2 + v_{\text{out,amp}}^2} \approx 0.30 \text{ } \mu\text{V}/\sqrt{\text{Hz}}.$$

The input referred acceleration is

$$\overline{\ddot{x}_{\text{in}}} = \frac{\overline{v_{\text{out}}}}{A_1 A_2 A_3 A_4} \approx 1.6 \text{ } \mu\text{G}/\sqrt{\text{Hz}}.$$

We see that the total input referred noise is only slightly larger than the intrinsic mechanical noise of 1.4 μG/$\sqrt{\text{Hz}}$.

The noise over 2 kHz bandwidth is

$$\overline{v_{\text{out,rms}}} = \sqrt{(x_{\text{rms}} A_3 A_4)^2 + \overline{v_{\text{out,amp}}^2} BW} \approx 7.2 \text{ } \mu\text{V}$$

and the input referred rms-acceleration is

$$\ddot{x}_{\text{in,rms}} = \frac{v_{\text{out,rms}}}{A_1 A_2 A_3 A_4} \approx 39 \text{ } \mu\text{G}.$$

This case study showed that it is possible to make accelerometers where the noise performance is limited by the mechanical noise. Yet, while mechanical noise limited accelerometers have been demonstrated [71], the typical commercial capacitive accelerometers analyzed in Chapter 3 offer noise performance that is several times worse than the intrinsic noise limit. Clearly, in the commercial devices, the noise performance has been sacrificed perhaps for increased stability, lower power consumption, or lower total cost.

Key concepts

- The main noise sources in transistors are the thermal channel noise and the $1/f$-gate noise.

- Transistor noise can be modeled with input referred noise sources.

- Input referred channel noise is proportional to $1/g_m$. Thus, increasing g_m improves the amplifier noise performance at the cost of increased power consumption.

- Small $1/f$-noise requires large transistor area WL.

- Noise in operational amplifiers is dominated by the differential pair noise. Focusing on the differential pair allows quantifying of noise/power trade-offs in the amplifier circuits.

Exercises

Exercise 9.1
Calculate the input referred noise voltage densities (both thermal and $1/f$-noise) at $f = 2$ kHz for a common source amplifier if $I_D = 300$ μA, $W = 94$ μm, and $L = 3.5$ μm. For the transistor, use parameters in Table 8.1 and assume $K = 10^{-25}$ V^{2}F.

Exercise 9.2
A low power op amp has a 5-μA current consumption. Assume that of the 5 μA, 3 μA is divided equally between the two input transistors and the 2 μA is used by the subsequent stages. i. What is the input referred thermal noise voltage for the op amp? ii. What is the equivalent resistor value that would generate the same thermal noise? iii. How large is the total rms-noise over 2 kHz bandwidth? iv. Compare your results for commercial op amps with similar current consumption for example from Maxim-IC. Assume that the input noise can be modeled as a differential pair and that $V_{GS} = 2V_T$.

Exercise 9.3
Consider the transimpedance amplifier Figure 8.5 and assume $Z = -j10^6$ Ω for the feedback. The op amp has input referred noise voltages of $\overline{v_n} = 3$ nV/$\sqrt{\text{Hz}}$ and currents of $\overline{i_n} = 150$ fA/$\sqrt{\text{Hz}}$. What is the noise voltage at the output if the bandwidth is 10 kHz? What is the input referred rms-noise current? Assume that transimpedance is connected to a noiseless current source with a high impedance ($Z_S = \infty$ Ω).

Exercise 9.4
A commercial CMOS based op amp is specified to have the input capacitance of 5 pF for the positive and the negative input. Estimate the input transistor size $A = WL$.

Exercise 9.5

Design a transimpedance amplifier for a photodiode measurement that has input referred noise current $\overline{i_n} = 50$ fA/$\sqrt{\text{Hz}}$ at 1 MHz. Assume the source impedance is $|Z_S| = 100$ kΩ and use capacitive feedback so the feedback does not add noise.

10

Switched capacitor circuits

Switched capacitor techniques are used for implementing circuits without using resistors. Thus, the switched capacitor circuits are well suited for modern CMOS processes that have high quality capacitors and switches but are not ideal for making large resistors. In microsystems, the switched capacitor techniques are used for capacitance measurements and signal processing. Compared to the continuous-time circuits covered in Chapters 8 and 9, the switched capacitor sensors are simple to implement and have low power consumption. The noise in switched capacitor circuits, however, is higher than that of continuous-time amplifiers.

In this chapter, we review the operation of switched capacitor circuits. First, we cover the switched capacitor amplifier operation. Next, we analyze the noise in switched capacitor circuits. Finally, a switched capacitor accelerometer is analyzed and the noise performance is evaluated.

10.1 Switched capacitor amplifier

The absolute manufacturing accuracies in IC processes are poor but CMOS technology is well suited for making capacitors that have accurately set ratios. The poor manufacturing tolerances can thus be overcome by relying on the capacitance ratios, for example, in setting the amplifier gain. The monolithic capacitors are also robust against changes in temperature, and in comparison to large resistors, the capacitors can be implemented in a small area. The capacitive feedback is also attractive for signal amplification because unlike resistive feedback it does not introduce thermal noise or dc load to the amplifier.

Given the advantages of capacitors over resistors, the amplifier with capacitive feedback shown in Figure 10.1 is attractive for amplifying ac signals. The

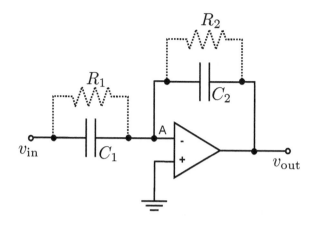

Figure 10.1: An amplifier with capacitive feedback. The resistors are needed to dc bias the point 'A'. Low-frequency operation requires large RC-products.

circuit has a high frequency gain $-C_1/C_2$ set by the capacitance ratio. However, the circuit in Figure 10.1 still requires resistors to provide the dc operation point. Without the resistors, the node 'A' is floating and the circuit will not work. At low frequencies, the resistive feedback dominates and the capacitive feedback is dominant only if $\omega \gg 1/RC$.

To amplify low frequency signals while maintaining the capacitive feedback, large RC-products are needed. The difficulty becomes obvious when we consider the maximum resistance $R = 1$ MΩ and capacitance $C = 10$ pF that are still economical to implement with integrated circuits. For these values, the 3-dB frequency is still $\omega = 1/RC = 100$ kHz. Increasing the RC-product further will quickly result in prohibitively large component values.

An alternative way to obtain capacitive feedback is shown in Figure 10.2. The feedback resistors have been eliminated and biasing is achieved with the switches S_1:s and S_2:s that operate in two phases. In phase one shown in Figure 10.2(b), switches S_1:s are closed and S_2:s are open. The charge over the capacitor C_1 is $q = C_1 v_{in}$. As both sides of capacitor C_2 are grounded, it has no charge.

In the second phase shown in Figure 10.2(c), switches S_1:s are open and S_2:s are closed. The left side of the capacitor C_1 is grounded and the right side is a virtual ground. From the Golden Rule 2, the input of the op amp cannot sink the charge from C_1 and it is therefore transferred to C_2 to maintain op amp input at the ground potential. An alternative way to describe the charge transfer operation is that the op amp feedback drives the negative input to zero and the output voltage required to achieve this is

$$v_{out} = \frac{q}{C_2} = \frac{C_1}{C_2} v_{in}. \tag{10.1}$$

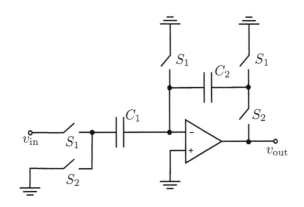

(a) Schematic of the amplifier with ideal switches.

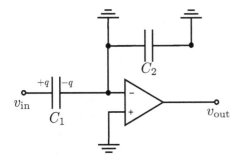

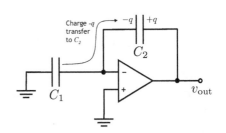

(b) Amplifier connections with S_1:s closed and S_2:s open.

(c) Amplifier connections with S_1:s open and S_2:s closed.

Figure 10.2: Switched capacitor amplifier.

Thus, the circuit amplifies the input signal by a factor $\frac{C_1}{C_2}$. This gain can be set precisely since the capacitance ratios can be accurately controlled in CMOS processes.

Unlike the continuous-time amplifiers, the switched capacitor amplifier does not track the input signal but samples it in C_1. In the second phase, the sampled charge is transferred to C_2 and the signal sampled in the previous stage is amplified as is illustrated in Figure 10.3. If the sampling rate is high compared to the change rate of the input voltage, the switched capacitor amplifier output follows closely to the input. The switched capacitor circuits are best suited for relatively low frequency applications such as sensor and audio amplifiers. For high frequency signals, the required switching frequency becomes impractically high and the continuous-time amplifiers are preferred.

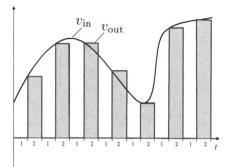

Figure 10.3: Switched capacitor amplifier output is a square wave. In the sampling stage '1', the output is zero. In the amplification stage '2', the output is C_1/C_2 times the input voltage at the end of the sampling stage.

10.2   Noise in switched capacitor amplifiers

The sampling operation causes noise aliasing and shapes the amplifier noise spectrum [72]. The noise spectrum analysis of switched capacitor circuits is beyond the scope of this chapter; however, the rms-noise voltage is easy to evaluate. From Chapter 2, the rms-noise voltage over a capacitor is $\sqrt{k_B T/C}$. The capacitor rms-noise is the main noise source in switched capacitor amplifiers and is analyzed in this section.

Referring to the amplifier in Figure 10.2, the noise is sampled in the two phases. In the first phase, the rms-noise charge on capacitor C_1 is $q_{rms,1} = \sqrt{k_B T C_1}$. This charge is sampled and transferred to capacitor C_2. The capacitor C_2 also has its own noise induced charge $q_{rms,2} = \sqrt{k_B T C_2}$. The total noise induced charge over the capacitance C_2 is therefore

$$q_{rms} = \sqrt{q_{rms,1}^2 + q_{rms,2}^2} = \sqrt{k_B T (C_1 + C_2)}. \tag{10.2}$$

The output noise voltage is

$$v_{rms} = \frac{q_{rms}}{C_2} = \sqrt{\frac{k_B T}{C_2} \frac{C_1 + C_2}{C_2}}. \tag{10.3}$$

To rms-noise given by Equation (10.3) can be significant. To reduce the noise, the switched capacitor systems are often oversampled. The Nyquist criteria for the minimum sampling rate states that signals should be sampled at twice the signal bandwidth. Oversampling means that the signal is sampled at a frequency higher than twice the signal bandwidth. The output of several samples can then be averaged to reduce the noise while maintaining the signal integrity. For example, an accelerometer circuit with a 500-Hz bandwidth would require a sampling rate of 1 kHz. Oversampling by a factor of 100 means that the signal is sampled at 100 kHz, allowing averaging of 100 samples. This is further illustrated in Example 10.1.

In addition to the $k_B T/C$-noise, the amplifier will also contribute to the total sampled noise. The analysis of the amplifier noise is not trivial: The rms-noise depends on the amplifier bandwidth and the amplifier circuit switching frequency f_{CLK} [72]. As a rule of thumb, the op amp bandwidth should be 5× the switching frequency f_{CLK} for the system to properly settle after switching. Assuming single-pole gain roll-off, the effective noise bandwidth is then $BW \approx 8f_{\text{CLK}}$ (see Example 2.2 on page 17). This is further explored in the Example 10.2.

Example 10.1: Switched capacitor amplifier
Problem: Calculate the rms-noise voltage of the switched capacitor amplifier in Figure 10.2 with $C_1 = 12$ pF and $C_2 = 1$ pF. The amplifier switching frequency is 100 kHz and the output is filtered by taking a rolling average of 100 measurements. What is the filtered rms-noise voltage at the amplifier output and what is the input referred noise?
Solution: The noise voltage for one sample given by Equation (10.3) is

$$v_{\text{rms}} = \sqrt{\frac{k_B T}{C_2} \frac{C_1 + C_2}{C_2}} \approx 0.23 \text{ mV}.$$

As shown in Section 2.6, the average of 100 measurements will reduce the noise level to

$$< v_{\text{rms}} >= \frac{1}{\sqrt{100}} v_{\text{rms}} \approx 23 \ \mu\text{V}.$$

Practical circuits will implement "averaging" with digital filters [25].

Since the amplifier voltage gain is $A_v = C_1/C_2 = 12$, the input referred noise is

$$< v_{\text{rms,in}} >= \frac{< v_{\text{rms}} >}{A_v} \approx 1.9 \ \mu\text{V}.$$

Example 10.2: Amplifier noise in switched capacitor circuit
Problem: Show that the contribution of the amplifier noise in Example 10.1 is small if the amplifier circuit from Example 9.5 is used.
Solution: From Example 9.5 on page 150, the total opamp output noise is $\overline{v_{\text{out}}} = 130$ nV/$\sqrt{\text{Hz}}$. The noise bandwidth is $BW \approx 8f_{\text{CLK}}$, where the $f_{\text{CLK}} =$

100 kHz is the switched capacitor clock rate. The rms-noise at the output due to amplifier noise is

$$v_{\text{rms}} = \overline{v_{\text{out}}} \sqrt{BW} \approx 0.12 \text{ mV}$$

which is can be compared to the $k_B T/C$-noise 0.23 mV in Example 10.1. The total noise due to the amplifier noise and $k_B T/C$-noise is $v_{\text{rms}} = \sqrt{(0.11 \text{ mV})^2 + (0.23 \text{ mV})^2} = 0.26$ mV. Thus, the contribution of the amplifier noise to the total noise is small in this example.

10.3   Case study: A switched capacitor accelerometer

Switched capacitor circuits are a compromise between performance, power consumption, and cost. Figure 10.4 shows an accelerometer circuit where the demodulation is achieved with switches. The sensor element values are based on Example 6.6 allowing a direct comparison to the capacitive accelerometer in Section 9.3.2 that was measured with the synchronous demodulation. From Example 6.6, we have $C_0 = 5.1$ pF and $d = 2.5$ μm. In addition, we assume that the capacitors are charged to voltage $v_{\text{in}} = 1$ V.

The circuit operates in two phases. In the first phase, the S_1 switches are closed and the S_2 switches are open. The capacitors $C_{w1} = \epsilon A/(d-x)$ and $C_{w2} = \epsilon A/(d+x)$ are charged to potential $+v_{\text{in}}$ and $-v_{\text{in}}$, respectively. In the second phase, the S_2 switches are closed and the S_1 switches are open forcing the charge stored in the capacitors C_{w1} and C_{w2} to be transferred to C_f. The total charge on C_f is

$$q_f = C_{w1}v_{\text{in}} - C_{w2}v_{\text{in}} = (C_{w1} - C_{w2})v_{\text{in}} \approx \frac{2C_0}{d}v_{\text{in}}x \equiv A_3 x \qquad (10.4)$$

where we have defined $C_0 = \epsilon A/d$. We have $A_3 = \frac{2C_0}{d}v_{\text{in}} = 4.1$ μC/m. The output voltage is

$$v_{\text{out}} = \frac{q_f}{C_f} \equiv A_4 q_f \qquad (10.5)$$

where $A_4 = 1/C_f = 10^{12}$ V/C.

The noise from sampling the capacitors C_{w1} and C_{w2} in the first phase is

$$q_{\text{rms},1} = \sqrt{k_B T(C_{w1} + C_{w2})} \approx 2.1 \cdot 10^{-16} \text{ F} \qquad (10.6)$$

and the sampling noise from the second phase is

$$q_{\text{rms},2} = \sqrt{k_B T C_f} \approx 6.4 \cdot 10^{-17} \text{ F}. \qquad (10.7)$$

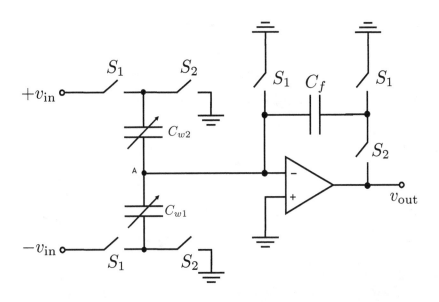

(a) The charge difference in work capacitors is transferred to C_f.

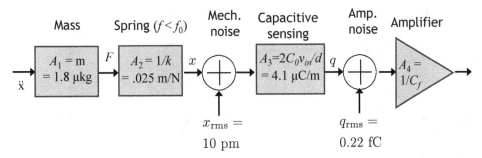

(b) Diagram for system level noise analysis.

Figure 10.4: A switched capacitive accelerometer ($C_f = 1$ pF).

The total noise charge is

$$q_{rms} = \sqrt{q_{rms,1}^2 + q_{rms,2}^2} = 0.22 \text{ fC}. \tag{10.8}$$

and the noise voltage due to charge noise at the switching amplifier output is

$$v_{rms,q} = \frac{q_{rms}}{C_f} = 0.22 \text{ mV}. \tag{10.9}$$

Here we have ignored the amplifier noise but this could be easily included in the analysis following Example 10.2.

Figure 10.4(b) shows the system level noise analysis for the switched capacitor accelerometer. From Example 6.6, the noise displacement due to thermal vibrations is

$$x_{\text{rms}} = \sqrt{\frac{k_B T}{k}} = 10 \text{ pm}. \tag{10.10}$$

The output noise voltage due to thermal vibrations is

$$v_{\text{rms},x} = A_3 A_4 x_{\text{rms}} = 42 \ \mu\text{V}, \tag{10.11}$$

which is small compared to the $k_B T/C$-noise at the output. The total noise voltage at the output is

$$v_{\text{rms,out}} = \sqrt{v_{\text{rms},q}^2 + v_{\text{rms},x}^2} = 0.22 \text{ mV}. \tag{10.12}$$

The input referred noise is

$$\ddot{x}_{\text{rms}} = \frac{v_{\text{rms,out}}}{A_1 A_2 A_3 A_4} = 0.12 \text{ mG}. \tag{10.13}$$

This result is much worse than for the synchronous demodulation in Section 9.3.2 even though the measurement voltage here is $10\times$ larger.

To reduce the amplifier noise, the output can be oversampled by averaging multiple samples. For example, using a sampling frequency of 100 kHz and taking an average of 100 samples to obtain effective signal bandwidth of 500 Hz would give the input referred noise of

$$< \ddot{x}_{\text{rms}} >= \frac{\ddot{x}_{\text{rms}}}{\sqrt{100}} = 12 \ \mu\text{G}. \tag{10.14}$$

Even the averaged noise is four times more than the thermal noise limit, but the result is comparable to the commercial bulk micromachined accelerometer analyzed in Section 3.5.2 on page 44 where the noise of the complete product was much higher than the intrinsic noise limit.

This case study demonstrates the trade-offs in microsensor design. Given enough effort, it is possible to reduce the amplifier noise to a level where the mechanical noise dominates. The end user, however, does not care whether the noise floor is due to mechanical or electrical noise. The customer is happy with good overall performance in terms of noise, power consumption, linearity, and stability at low price. In many cases, the easiest way to reduce the total noise is by making the mechanical sensor element more sensitive by increasing the mass or reducing the spring constant.

Key concepts

- Switched capacitor circuits do not require resistors for dc biasing.

- The main noise source in switched capacitor circuits is the $k_B T/C$-noise.

- Switched capacitor accelerometers are power and area efficient.

Exercises

Exercise 10.1
Calculate the rms-noise voltage of the switched-capacitor amplifier in Figure 10.2 with $C_1 = 2$ pF and $C_2 = 0.1$ pF. The amplifier switching frequency is 200 kHz and the output is filtered by taking a rolling average of 3,000 measurements. What is the filtered rms-noise voltage at the amplifier output and what is the input referred rms-noise? You may ignore the amplifier noise.

Exercise 10.2
Assume that the surface micromachined accelerometer in Figure 3.6 on page 45 is sensed with a switched capacitor circuit as illustrated in Figure 10.4. Assume that the electrode gap is $d = 2$ μm, the feedback capacitance is $C_f = 20$ fF, the switching frequency is 250 kHz, and the switching voltage is 2 V. The amplifier output is filtered by taking a rolling average of 1,000 measurements. What is the input referred rms-noise acceleration.

Exercise 10.3
Design a switched capacitor amplifier that has gain $A = 20$ and input referred noise $v_{rms} = 50$ μV (without averaging).

Exercise 10.4
Calculate the rms-noise voltage v_{rms} and charge q_{rms} on a 100-fF capacitor. How many electrons does this charge corresponds to?

11

Sensor specifications

In the previous chapters, we have focused on the sensor noise performance. The obsession with the noise is a natural result of working with the microstructures that by nature provide small signal levels. The cost pressures are constantly driving further miniaturization and maintaining the signal-to-noise ratio is a challenge. In many systems, the noise specifications set a fundamental limitation for the miniaturization.

The noise specification, however, is not the only criteria for a successful design. We already saw in Chapters 9 and 10 how the power consumption and cost considerations resulted in commercial MEMS accelerometers being much noisier than the intrinsic mechanical noise limit. In this chapter, we explore the other design considerations including selectivity, accuracy, stability, and linearity. The specifications are illustrated by comparing the data sheet values for two commercial accelerometers.

11.1 System specifications

Sensors convert a physical parameter such as pressure or acceleration to an electrical output. The sensors are therefore part of the more general class of devices or transducers, which convert energy from one form to another. A good sensor should be minimally invasive, that is, the sensor should not disturb the system being measured. For example, the inertial sensor should not add significant mass to the system. In this regard, microsensors are often seen as an enabling technology, as the small size makes them easy to integrate within larger systems without changing the system properties.

Choosing the correct sensor for the application requires considering multiple and often conflicting parameters. Several terms have been defined to describe the sensor performance:

Transfer function is the functional relationship between the physical input and the electrical output signal. For example, the pressure sensor transfer function has the form $V = f(p)$ where the function f relates the pressure p to signal voltage V. The sensor sensitivity is obtained from the slope of the transfer function as is illustrated in Figure 11.1(a).

Sensitivity is the ratio of a small change in electrical signal to a small change in physical signal. For example, pressure sensor sensitivity is defined as a change in output voltage for a change in pressure $\Delta V/\Delta p$. As is illustrated in Figure 11.1(a), the sensitivity can be obtained by taking the slope of the transfer function:

$$\text{sensitivity} \equiv \frac{\Delta \text{Output}}{\Delta \text{Input}} = \frac{\partial f(\text{Input})}{\partial \text{Input}} \tag{11.1}$$

Noise in sensors is loosely defined as the random fluctuations at the sensor output with no input signal. All sensors produce some noise in addition to converting the physical input to output signal.

Resolution is the minimum detectable signal change. For sensors with digital output, this can the smallest bit change. The **noise limited resolution** is the minimum detectable signal change limited by the sensor noise.

Dynamic range is the span of physical input which may be converted to electrical signals as is illustrated in Figure 11.1(b). The lower range of the span is determined by the sensor resolution and the upper range is set by the largest physical signal that can be measured before the sensor saturates. In digital sensors, the dynamic range may be set by the analog to digital converter resolution. For example, a pressure sensor with 8-bit digital output has a dynamic range of $1 : 2^8$ or $1 : 256$. If the resolution (one bit change) is $\Delta p = 1$ kPa, the maximum output before the sensor saturates is $p_{\text{MAX}} = 256$ kPa and the dynamic range in the pressure scale is $1 : 256$ kPa.

Nonlinearity is the deviation of the transfer function from the linear transfer function as is illustrated in Figure 11.1(c). The nonlinearity is often described as the percentage error between an ideal linear sensor output and measured nonlinear output:

$$\text{Nonlinearity error} \equiv \frac{\text{Output}_{\text{nonlin}} - \text{Output}_{\text{linear}}}{\text{Output}_{\text{linear}}}. \tag{11.2}$$

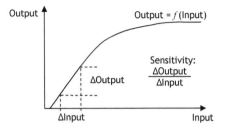

(a) The sensor transfer function relates the physical input to sensor output.

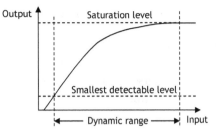

(b) The dynamic range is the range of physical signals that change the sensor output.

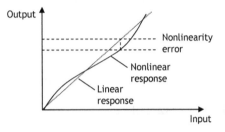

(c) Nonlinearity in the sensor output is the deviation from the linear response.

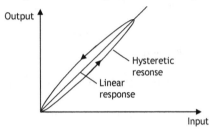

(d) Hysteresis in sensor causes the output to depend on past output. The output depends on whether the prior input was higher or lower than the current input.

Figure 11.1: Sensor specifications illustrated.

Accuracy is the largest expected error between actual and ideal output signals. The sources of inaccuracy include change in sensor characteristics over time or temperature, initial sensors offsets, and nonlinearity.

Selectivity describes sensors sensitivity to other than the desired input signal. Typically, the sensors are somewhat sensitive to other physical parameters such as temperature, stress, and supply voltage variations. The sensitivity to temperature variations can be especially problematic. The temperature dependency is inherent to some sensing mechanisms such as piezoresistive sensing, but it will also affect capacitive sensing due to temperature dependency of spring constants and package deformations due to thermal expansions mismatches.

Stability defines how constant the sensor output is in constant conditions. The change in output is called **drift**. The importance of stability is illustrated in Figure 11.2 wherein the sensor price is compared to the stability. The high-end aerospace market, which demands high stability, can have much higher sales prices than consumer-oriented products. Somewhat confusingly, the change due to temperature fluctuation is often called the **temperature stability**, although this effect does not alter the sensor output in constant conditions.

Hysteresis describes the variations in the sensor output when the input stimulus is cycled up or down as is illustrated in Figure 11.1(d). For mechanical sensors, the hysteresis is often associated with plastic deformation. For example, when a sensor encapsulated in plastic is heated and cooled, the packaging stress changes, which causes changes in the sensor output.

Bandwidth is the range of frequencies that can be measured. As discussed in Appendix B, the bandwidth is also related to sensor response time.

Repeatability describes the sensor's ability to represent the same value under identical conditions.

Power consumption is the amount of electrical power that the sensor uses.

Reliability describes the probability of the sensor failing.

Cost is the most important sensor characteristics for everyone but engineers.

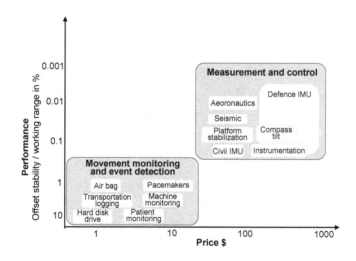

Figure 11.2: Correlation between the sensor stability and price (After [73]).

11.2   Element specifications

MEMS design would be easy if there was only one way to satisfy the system level specifications for a given application. Fortunately, the reality is more interesting and there are endless numbers of possible design variations that will work. The optimal design will meet the specifications at the lowest manufacturing cost.

Exceeding specifications is recommended only if it does not increase the cost. Over-engineered "best possible" designs are costly and have a limited market.

In setting the element specifications, the first thing to note is that a MEMS element is not a perfect component. To meet the system level specifications for the final product, the non-idealities in the mechanical element may need to be corrected with the sensor electronics. A typical adjustment in such an instance is the calibration step that accounts for the manufacturing variations. For example, the accelerometer spring constant may vary by tens of percentages, which is not acceptable for the final product. To calibrate the sensor, it is rotated around its axis to change the acceleration from +1 G to −1 G. This known input is compared to the sensor output and the calibration constants are programmed into electronics. Obviously, this calibration significantly increases the sensor cost.

Other element non-idealities include temperature dependency and nonlinearity. In the best case scenario, the temperature dependency is due to change in silicon Young's modulus which is both a linear and small effect (-60 ppm/K). More often, the temperature dependency is larger and arises from using materials with different thermal expansion coefficients. If the temperature dependency does not vary from sample to sample, it can still be corrected relatively easily. Calibration by measuring the sensor in multiple temperatures is possible but costly.

The element linearity is another area for trade-offs. For example, the capacitive Wheatstone bridge in Figure 6.3 on page 95 has a linear voltage-displacement transfer function but the design is sensitive to parasitic capacitances. The current measurement shown in Figures 6.9 and 6.10 on pages 103 and 104, respectively, is not sensitive to parasitics but the sensor transfer function is nonlinear. For the final product, the transfer function may need to be electronically linearized, which is an easy task if the transfer function is same for all the samples, but is more challenging if the (non)linearity varies from sample to sample for example due to a varying electrode gap d.

In studying amplifier noise in Chapter 8, we noted that the commercial accelerometers are noisier than expected from the mechanical noise alone. This means that the designers have made a decision to use electronics that do not provide the lowest possible noise. If better noise performance is necessary, then either the mechanical element can be improved by increasing the element sensitivity or the electronics can be improved. A simplistic system level analysis can be very valuable in taking away some of the guess work in making component specifications that meet the system specifications in a cost optimal way.

The biggest challenges in MEMS design are related to the stability. For example, charging of dielectrics or change in package pressure can change the sensor output. In addition, external disturbances can cause the sensor characteristics to change over time. Figure 11.3 illustrates how MEMS sensor is deformed when it is soldered onto a circuit board. Compared to most materials, silicon has a low coefficient of thermal expansion and the circuit board will expand more with temperature. The deformations due to thermal expansions mismatches can be tens of nanometers. This may sound small but recalling that the MEMS sensors can measure sub-atomic displacements, the effect is significant. Anchoring the sensor element from a single point has proved effective in mitigating the effect of package deformations [74, 75].

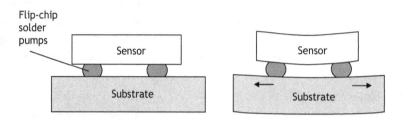

Figure 11.3: The thermal expansions mismatch between plastic substrate and silicon sensor deforms the MEMS device.

11.3   Commercial accelerometer comparison

Table 11.1 compares the data sheet specifications for two commercial accelerometers. The SCA810 is a single-axis bulk micromachined accelerometer for automotive applications from VTI Technologies [76]. The ADXL322 is a dual axis surface micromachined accelerometer for consumer electronics from Analog Devices [77]. When comparing the two devices, the ADXL322 is better in almost all specifications: It is dual axis, has ten times smaller current consumption, slightly smaller noise, higher bandwidth, and lower price! The SCA810, however, is guaranteed over a temperature range of −40 to +125 °C and the offset errors include all errors (calibration, drift over lifetime, and supply voltage). This guaranteed stability over a wide temperature range makes all the difference in targeted automotive applications. In comparison, the ADXL322 is specified over smaller commercial temperature (−20 to +70 °C) range and the specified values are not guaranteed to be the maximum errors over the device life. This example illustrates that although the noise is often emphasized in academic publications, it is only one specification among many. In many applications, noise is not the hardest specification to meet nor the most important for the application.

Table 11.1: Comparison of two commercial accelerometer specifications.

Parameter	SCA810	ADXL322	Units
Range	±2	±2	G
Direction	Y-axis	X/Y-axis (dual)	G
Current consumption	5	0.45	mA
Resolution	1	Analog output	mG
Linearity error	±2	±2	%
Offset	±5	±3	%
Offset temperature dependency	±4	±3	%
Operation temperature	-40 to +125	-20- to +70	°C
Supply voltage	3.3	2.4-6	V
Bandwidth	50	100	Hz
Noise floor	5	3	mG (rms)
Price	> 5	< 5	$

Key concepts

- The most important parameter in characterizing the sensor performance is the price. Other key parameters are the accuracy, the dynamic range, the noise, the resolution, and the stability.

- The system level specifications define the product performance.

- Element specifications define the characteristics for the mechanical component. Systematic element imperfections such as nonlinearity error can be compensated at the system level but unpredictable errors such as drift cannot be compensated.

- The difference between consumer and automotive or aerospace devices can be a single parameter such as guaranteed stability over the specified temperature range.

Exercises

Exercise 11.1

Explain why stability is very important for sensors.

Exercise 11.2

Compare two consumer oriented accelerometers, LIS302DL from ST Microelectronics and MMA7455L from Freescale, for power consumption, size, full scale range, sensitivity, zero-G offset, and self-test capability. Present your comparison in table format.

12

Damping

Understanding damping is critical for microsystem design. Depending on the application, the desired damping may be very small for high quality factor resonators, high for over damped accelerometers and actuators, or something in between for gyroscopes. The damping in microsystems differs from macro-scale, as in micro-scale, the air damping is comparatively much more significant. This is a mixed blessing: By adjusting the air pressure, the quality factor can be tailored over a wide range. On the other hand, low damping typically requires operation in a vacuum. Hermetic packaging plays an important role in controlling the damping characteristics.

In this chapter, the different damping mechanisms are reviewed. We start with the formalism for combining the damping from different sources. Next, we cover the mechanical damping mechanisms such as intrinsic material losses and anchor losses. Finally, the chapter is concluded with an overview of gas damping in microsystems.

12.1 Damping and quality factor

Referring to the harmonic resonator in Figure B.1 on page 410, the equation of motion for a damped mechanical resonator is

$$m\ddot{x} + \gamma_{\text{tot}}\dot{x} + kx = F, \tag{12.1}$$

where x is the displacement of mass m, γ_{tot} is the total damping coefficient, k is the spring constant and F is the excitation force. The damping force $F_\gamma = -\gamma\dot{x}$ is proportional to the mass velocity $\dot{x}$. If there are multiple loss mechanisms affecting the mass, the damping forces sum and the total damping coefficient is

$$\gamma_{\text{tot}} = \gamma_1 + \gamma_2 + \ldots + \gamma_N. \tag{12.2}$$

The quality factor for the resonator is defined as $Q_{\text{tot}} = \frac{\omega_0 m}{\gamma_{\text{tot}}}$, where $\omega_0 = \sqrt{k/m}$ is the resonant frequency. We can also attribute a quality factor for each individual loss mechanisms, for example $Q_1 = \frac{\omega_0 m}{\gamma_1}$, and rewrite Equation (12.2) as

$$\frac{1}{Q_{\text{tot}}} = \frac{1}{Q_1} + \frac{1}{Q_2} + \ldots + \frac{1}{Q_N}. \tag{12.3}$$

Equation (12.3) shows that the total quality factor is smaller than the smallest quality factor in the system. For many systems, one loss mechanism is dominant and the smallest quality factor dominates the total quality factor.

Example 12.1: Summing of the quality factors
Problem: A microresonator has the intrinsic mechanical quality factor $Q_i = 200,000$ because of material losses and the anchor quality factor $Q_a = 10,000$ due to support losses. Calculate the total quality factor.
Solution: From Equation (12.3) we have

$$\frac{1}{Q_{\text{tot}}} = \frac{1}{Q_i} + \frac{1}{Q_a} = 1.05 \cdot 10^{-4}$$

and the total quality factor is $Q_{\text{tot}} \approx 9,500$.

12.2   Damping mechanisms

In microsystems, there are several different damping sources. The intrinsic material losses set the limit for the maximum quality factor for a given material. The mounting related anchor losses are due to energy leaking into supporting device packaging. The anchor losses can be much bigger than the material loss limit especially for flexural structures. Finally, the air damping is significant unless the microstructure is operated in a vacuum. Devices that benefit from a high quality factor, for example resonators and gyroscopes, are often vacuum encapsulated to reduce the air damping losses.

12.2.1   Material losses

Material losses are due to movement at the atomic level that results in heat generation. This "internal friction" sets the fundamental limit for the maximum obtainable quality factor. The material damping depends on the temperature, the vibration frequency, and the vibration mode. This dependency can be complex due to a large number of physical mechanisms that contribute to the damping. At room temperature, however, the simple viscous damping model is adequate for most low loss materials.

Adding the velocity dependent viscous damping term to the constitutive equation $T = ES$ for stress T and strain S, we have

$$T = ES + \mu_\gamma \frac{\partial S}{\partial t} = ES + j\omega\mu_\gamma S, \tag{12.4}$$

where μ_γ and E are the viscosity and the Young's modulus for the material, respectively, and ω is the vibration frequency. The loss term is analogous to damping in the harmonic resonators where force acting on the mass is the sum of elastic and viscous forces $F = kx + \gamma\frac{\partial x}{\partial t}$.

The elastic damping losses given by Equation (12.4) are proportional to the time derivate of the strains, and the damping increases with vibration frequency. The intrinsic quality factor Q_i due to material losses is

$$Q_i = \frac{E}{\mu_\gamma\omega}. \tag{12.5}$$

Equation (12.5) can also be written as a universal scaling law for material losses

$$f \cdot Q = \text{constant}, \tag{12.6}$$

where f is the frequency and Q is the quality factor.

Table 12.1 tabulates the $f \cdot Q$-products for different materials. The $f \cdot Q$-product is a good figure-of-merit for choosing a low-loss material. Typically, single-crystal and/or hard materials such as silicon, AlN, or diamond have low intrinsic losses while softer materials such as metals or plastics have high losses.

Table 12.1: The $f \cdot Q$-products for selected materials.

Material	E[GPa]	ρ[kg/m^3]	μ_γ[Ns/m^2]	$f \cdot Q$[10^9]
Aluminum	11	2700	1.0	1,800
Epoxy	9	1210	2300	0.60
Fused quartz	78	2200	3.1	4,100
Gold	210	19700	78	420
Lithium niobate	390	7300	0.7	91,000
PVC	8	1380	490	2.6
Silicon	160	2320	2.5	10,000

The $f \cdot Q$-product can also be used to compare different resonator designs. Figure 12.1 shows examples of experimental $f \cdot Q$-products for silicon resonators. The highest obtained $f \cdot Q$-product for a silicon resonator is $1 \cdot 10^{13}$ [82]. The

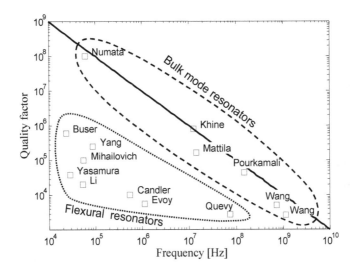

Figure 12.1: The $f \cdot Q$-products for various silicon resonators [78–85].

trend of decreasing the quality factor with increasing frequency is clear but
there are large variations between different resonators. Especially for the low
frequency flexural resonators, the measured quality factors are lower than the
intrinsic limit due to other loss sources such as the anchor loss and thermoele-
lastic dissipation (TED) [86]. The exception in the low frequency range is the
very high quality factor sample at 62.3 kHz by Numata *et al.* that was obtained
with a cylindrical sample that was 6 cm tall, 10 cm in diameter, and vibrating
in bulk torsional mode supported at the nodal points [85]. The anchor losses
will be further discussed in the next section but we can already make a conclusion that the intrinsic quality factor limit is typically approached with the
"bulk mode" resonators. For the "flexural modes", other loss mechanisms are
more significant.

12.2.2 Anchor losses

The microdevices do not float but are anchored to a substrate for mechanical
support. When the device vibrates, a small but non-negligible portion of the
vibration energy leaks into the support. These anchor losses depend on the
location of the support point relative to the vibration mode shape and the type
of anchoring used.

Figure 12.2 qualitatively illustrates the effect of support design on the anchor
losses. The lowest quality factors are obtained with the clamped-clamped beam
resonators, shown in Figure 12.2(a). The resonator exerts significant stress at
the anchor, which results in the anchor vibrating as well. Quantitative analysis
of the anchor loss is surprisingly difficult. One approach is to model the support

as an infinitely large elastic body. The stress and strain at the anchor launches longitudinal and transverse sound waves that propagate into the support. Based on this analytical approach, an estimate for the anchor support losses for a cantilever is obtained as

$$Q_{\text{anchor}} \approx 2.17 L^3/t^3, \tag{12.7}$$

where L is the cantilever length and t is the cantilever thickness in the direction of vibration [87]. Similar analysis approach can be extended to modeling losses in bulk mode resonators [88, 89].

Many applications including accelerometers and actuators do not benefit from a high quality factor. In this case, the simple support shown in Figure 12.2(a) for the beam is more than adequate. If higher quality factor is needed, the anchor losses in the clamped-clamped beam resonator can be significantly reduced by making the device symmetric as shown in Figure 12.2(b). The two beams vibrate out-of-phase to form a double-ended tuning fork (= a tuning work anchored on both beam ends). The tuning fork structure minimizes the anchor losses as the stresses from the two arms of the resonator cancel at the mounting location. The tuning fork structures are used extensively in gyroscopes and low frequency time references.

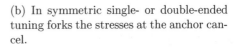

(a) Clamped-clamped beam resonator exerts net stress on the anchor with results in high anchor losses.

(b) In symmetric single- or double-ended tuning forks the stresses at the anchor cancel.

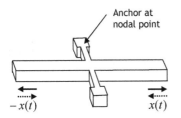

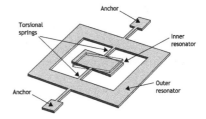

(c) Longitudinal mode beam resonator. The anchor is at the nodal point and there is no net stress exerted on the anchors.

(d) Torsional mode resonator is decoupled from the substrate with another resonator that has a lower resonant frequency.

Figure 12.2: Anchoring of microresonators to the substrate.

The high frequency "bulk mode" resonators can maintain quality factors close to the intrinsic material quality factor if they are anchored at the nodal point. This is illustrated for a longitudinal beam resonator in Figure 12.2(c). The beam resonator has a nodal point at the beam center and there is no net stress exerted on the anchor [90]. Lamé-mode silicon plate resonators anchored at the nodal point have achieved the record $f \cdot Q$-product of 10^{13} [82].

The previous examples have illustrated how the anchor losses are minimized by ensuring that the net stress at the anchor is zero. An alternative approach is to trap the energy in the resonator by a suspension design is shown in Figure 12.2(d). The energy loss to the substrate is avoided by a filtering suspension between the resonator and the substrate. The torsion resonator is mounted on a frame that has a much lower resonant frequency than the primary resonator [84, 91]. This ensures that frame motion is not significantly excited and very little energy leaks to the substrate. The energy trapping is extensively used for quartz resonators operating at MHz frequencies [92].

12.2.3   Surface related losses

In small cantilevers, surface related losses have been observed to reduce the resonator quality factor [78, 79, 93]. The surface roughness, formation of the oxide layer, and surface contamination can all reduce the resonator quality factor. In typical experiments, the resonator quality factor is measured before and after a heat treatment at an elevated temperature. If the quality factor is changed, the effect is attributed to the surface effects.

While the surface losses can be significant for devices that approach nanoscale, the surface loss effects have not been found to be significant for clean silicon microresonators.

12.2.4   Mode conversion losses

Ideally for a resonating device, only one high-Q resonator mode should be excited. If other modes have the resonance frequency close to the main resonance, they are also excited. Assuming that two modes with a quality factor Q_1 and Q_2 are excited, the total quality factor is

$$\frac{1}{Q_{\text{tot}}} = \frac{a_1}{Q_1} + \frac{a_2}{Q_2}, \tag{12.8}$$

where a_1 and a_2 are the fraction of the total energy in the modes 1 and 2, respectively. As Example 12.2 shows, the overall quality factor can be significantly degraded by nearby modes that have a low-Q even if the vibration amplitude for the parasitic mode is small.

Example 12.2: Two mode quality factor
Problem: Calculate the overall quality factor for a resonator that has two vibrations modes; a main mode with $Q_1 = 100{,}000$ and a parasitic mode with $Q_2 = 100$. Assume that 99.9% of the vibration energy is in the main mode.
Solution: Using $Q_1 = 100{,}000$, $Q_2 = 100$, $a_1 = 0.999$, and $a_2 = 0.001$, Equation (12.8) gives

$$Q_{\text{tot}} = \left(\frac{a_1}{Q_1} + \frac{a_2}{Q_2} \right)^{-1} = 50{,}000.$$

The mode conversion losses explain the relatively low quality factors observed in some solidly mounted film bulk acoustic resonators (FBARs) that rely on acoustic mirrors to prevent vibrations from leaking to the substrate [94]. Best membrane type FBAR resonators have quality factors greater than 2,000 at 1 GHz. The quality factors for solidly mounted resonators, however, can be much lower ($Q \sim 500$) even though the mirrors are effective in preventing the longitudinal mode from leaking to the substrate. This is explained by the excitation of shear modes that have a much lower quality factor. Even though the vibration amplitudes of the shear modes are small, they can significantly degrade the total quality factor.

Another example where mode conversion losses can play a role are the silicon bulk-mode plate resonators. A solid plate with no holes has shown the record $f \cdot Q$-product of $1 \cdot 10^{13}$ [82] but an identical structure with etch holes shows quality factors that are 5× smaller [95]. A likely explanation is that the etch holes act as local perturbations that excite flexural modes with a lower quality factor. This energy leakage to lower-Q modes lowers the overall quality factor as shown by Equation (12.8).

12.2.5 Air damping

Viscous losses due to fluid flow (gas or liquid) are often the dominant loss mechanisms in micro-scale. The relatively large importance of air damping in micro-scale is obvious when we consider how small dust particles can "float" in air as the air drag limits the free fall velocity to mere millimeters per hour.

As the air damping is significant in micro-scale, many MEMS devices are operated in lowered pressure to reduce the viscous effects. Figure 12.3 illustrates how the pressure affects a microresonator quality factor [96]. At pressures close to the atmospheric pressure, the air viscosity is independent of pressure. In this *viscous regime*, the air acts as a viscous fluid and the air damping is independent of pressure.

The transition from viscous flow to molecular flow is characterized by the mean free path in which an air molecule travels before colliding with another molecule. In the air at atmospheric pressure, the mean free path is 65 nm, which is small compared to the typical device dimensions. As pressure is lowered, there are fewer molecules with which to collide, and the mean free path increases. In low enough pressures, the mean free path starts to be comparable to the characteristic device dimension d_c. At this point, the air stops behaving as a viscous fluid with a constant viscosity but individual gas molecule-device interactions become the dominant loss mechanism. Reducing the air pressure in this *molecular regime* reduces the air damping losses as the number of individual gas molecules interacting with the structure decreases. Finally, at very low pressures, the gas damping becomes insignificant and other loss mechanisms dominate. In this instance *intrinsic regime* the air damping contributes little to the total losses and the total quality factor Q_{tot} is again independent of pressure.

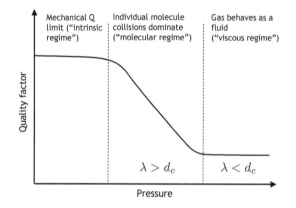

Figure 12.3: The effect of pressure on microresonator quality factor (after [96]).

While the air damping and its pressure dependency is easily understood qualitatively, modeling of the air damping is one of the most difficult problems in MEMS design. We therefore devote the entire next section for the quantitative analysis of air damping.

12.3   Models for the air damping

The gas flow analysis is a challenging problem even at macro-scale.[1] Even simple geometries that are analytically solvable can require several pages of analysis. Complex problems such as turbulent flow over an airplane wing requires hours of supercomputer time. The design of macroscopic systems such as airplanes that are sensitive to fluid flow often involves a combination of simplified analyt-

---

[1]For illustration purposes, the Reynold's equation that governs the compressible air flow is $\rho \left(\frac{\partial \mathbf{v}}{\partial t} + \mathbf{v} \cdot \nabla \mathbf{v} \right) = -\nabla p + \mu \nabla^2 \mathbf{v} + \mathbf{f} + (4\mu/3 + \mu^v) \nabla (\nabla \cdot \mathbf{v})$.

ical solutions to give design intuition, expensive and time consuming computer analysis, and very expensive experimentation in wind tunnels.

In micro-scale, the analysis is complicated by the fact that the continuum model approximation breaks down when the mean free path of the gas molecules approaches the device dimensions. At this point, the gas should be treated as individual molecules that interact with the device. Although simulation tools for this exist, they are neither easy to use nor standardized.

In this section, the models for air damping are reviewed. We will first cover the calculation of the mean free path of gas molecules as a function of pressure. Knudsen number is introduced to characterize the transition from the viscous to the molecular flow regime. The effective viscosity is adopted to analyze the gas flow in micro-scale, where the classical continuum models are no longer valid. The viscous damping is reviewed by analyzing the damping in two important geometries. The squeeze film effect governs damping in parallel plate geometries found in many devices such as accelerometers and electrostatic actuators. The Couette flow models damping between two plates sliding parallel. This geometry is common for laterally moving surface micromachines. These two simple geometries allow us to understand the scaling of gas damping with device dimension and pressure. Finally, this section is concluded with ideas on analyzing the air damping in more complex geometries.

12.3.1 Mean free path and Knudsen number

In normal pressures and temperatures, the gas viscosity is essentially independent of the pressure. This continuum model breaks down at rarefied pressures where gas stops behaving as a continuous fluid and individual device/gas molecule interactions become important. The transition pressure is characterized by the mean free path of gas molecules λ that is the average distance that the gas molecules travel before colliding with another molecule. The mean free path is given by

$$\lambda = \frac{1}{\sqrt{2}\pi d_g^2 n_V}, \tag{12.9}$$

where d_g is the effective diameter of the gas molecules and n_V is the number of gas molecules per unit volume. Using the ideal gas law $n_V = N_A p / RT$, Equation (12.9) can be written as

$$\lambda = \frac{RT}{\sqrt{2}\pi d_g^2 N_A p}, \tag{12.10}$$

where $N_A = 6.022 \cdot 10^{23}$ 1/mol is the Avogadro's number, $R = 8.3145$ J/mol·K is the universal gas constant, and T is the gas temperature. According to Equation (12.10), the mean free path decreases as the inverse of the pressure.

Example 12.3: Mean free path of gas molecules in air
Problem: Calculate the mean free path for air ($d_g = 3.65$ Å) as a function of pressure.
Solution: The mean free path is obtained using Equation (12.10). Table 12.2 tabulates the mean free path for several pressure values. For the benefit of the US readers, the pressure is also tabulated in units of Torr (1 Pa = $7.5 \cdot 10^{-3}$ Torr). In atmospheric pressures ($p = 10^5$ Pa), the mean free path is only about 70 nm and the gas behaves as a viscous fluid. As the pressure approaches 100-1000 Pa (0.75-7.5 Torr), the mean free path becomes comparable to the microdevice dimensions.

Table 12.2: Mean free path of molecules in air.

p[Pa]	10^5	10^4	10^3	10^2	10^1
p[Torr]	750	75	7.5	0.75	0.075
$\lambda[\mu m]$	0.07	0.7	7	70	700

The ratio of mean free path λ to critical device dimensions d_c occurs frequently in fluid dynamics and is defined as the Knudsen number

$$K_n = \frac{\lambda}{d_c}, \tag{12.11}$$

where the critical dimension d_c is the characteristic length for the problem. Typically, the characteristic length is the smallest dimension; for example the microchannel diameter or electrode gap. If the Knudsen number is greater than 1, the gas collisions with the device are more frequent than collisions with other gas molecules and the gas starts to lose in cohesive nature.

The gas flow is typically divided to three regimes:

1. In the *continuum region* $K_n < 0.1$ and the continuum flow models are valid.

2. In the *transition region* $0.1 < K_n < 10$, the molecular effects start to be significant and the continuum models need to be modified.

3. In the *molecular region* $K_n > 10$, the gas molecule interactions are no longer significant and continuum models cease to be valid.

Many MEMS devices fall into the transition region. In the next section, we cover how gas flow in this region can be analyzed by adopting the effective viscosity.

12.3.2   Squeeze film damping

The squeeze film effect illustrated in Figure 12.4 governs the gas flow between two surfaces moving toward each other [97, 98]. As a gap between the plate and substrate is reduced, the volume between the two surfaces is reduced. If the plate movement is slow, the gas is squeezed out, resulting in dissipation losses. Fast movement compresses the air, which results in spring forces. Thus, we expect the squeeze film effect to result in both damping and spring forces. In addition, the moving air will have inertia, which increases the effective mass of the plate. This effect is typically small but can be noticeable for resonators operating at high frequencies.

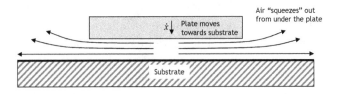

Figure 12.4: Schematic of the squeeze film damping. As the plate moves toward the substrate, the air flows from under the plate.

An effective way to model the squeeze flow at large Knudsen numbers is to introduce effective viscosity that depends on gas pressure. Following reference [99], the effective viscosity for the squeeze film damping in narrow gaps is

$$\mu_{\text{eff}} = \frac{\mu}{1 + 9.638 K_n^{1.159}}, \tag{12.12}$$

where μ is the viscosity in normal conditions ($K_n \ll 0.1$). The effective viscosity model explains the observed microstructure quality factor behavior as a function of pressure shown in Figure 12.3. When $K_n \ll 0.1$, Equation (12.12) simplifies to $\mu_{\text{eff}} = \mu$ and damping is independent of pressure. At low pressures, $K_n > 0.1$ and viscosity decrease with decreasing pressure.

The functional form of Equation (12.12) has been shown to agree with experiments for squeeze film damping in narrow gaps. The formula is semi-empirical and care should be taken when applying it to other geometries. The approach of using effective viscosity can be applied to other problems as well but the functional form and/or constant may have to be adjusted.

Using the effective viscosity, the squeeze film damping coefficient and spring constant for a solid plate are

$$\gamma_{\text{gas}} = \frac{64 \sigma p A}{\pi^6 d \omega} \sum_{m,n \text{ odd}} \frac{m^2 + c^2 n^2}{(mn)^2 \left[(m^2 + c^2 n^2)^2 + \sigma^2/\pi^4 \right]} \tag{12.13}$$

and

$$k_{\text{gas}} = \frac{64\sigma^2 pA}{\pi^8 d} \sum_{m,n \text{ odd}} \frac{1}{(mn)^2 \left[(m^2 + c^2 n^2)^2 + \sigma^2/\pi^4\right]}, \quad (12.14)$$

respectively, where m and n are odd integers, p is the pressure, $c = W/L$ is the ratio of plate width W and length L ($W < L$), and the squeeze number σ is

$$\sigma = \frac{12\mu_{\text{eff}} W^2}{pd^2}\omega. \quad (12.15)$$

Here μ_{eff} is the effective viscosity given by Equation (12.12) and ω is the frequency of vibration [98, 99].

At low squeeze numbers, the air squeezes out without compression and the fluid stiffness given by Equation (12.14) can be neglected. A high squeeze number means that the gas does not have sufficient time to squeeze out and air acts like a compression spring. The model assumes that the Reynold's number is small ($\rho\omega d^2/\mu \ll 1$) which allows neglecting the gas inertia. If this condition is not satisfied, the squeeze film damping will add mass to the plate.

If $\sigma/\pi^2 \ll 1$ and the plate is square ($c = 1$), Equations (12.13) and (12.14) can be simplified significantly to

$$\gamma_{\text{gas}} = 0.42 \frac{\mu_{\text{eff}} A^2}{d^3} \quad (12.16)$$

and

$$k_{\text{gas}} = 0.25 \frac{\mu_{\text{eff}}^2 A^3}{pd^5}\omega^2. \quad (12.17)$$

Equations (12.16) and (12.17) show that the squeeze film damping is very sensitive to gap d and also to plate area A.

Example 12.4 illustrates the squeeze film damping in a bulk micromachined accelerometer. The air damping in atmospheric pressures is excessive and operation in a lowered pressure is required. This is typical for structures that have large parallel plates that move toward each other.

Example 12.4: Squeeze film damping
Problem: Consider the capacitive accelerometer in Example 6.6 on page 102. Calculate the squeeze film spring constant, damping coefficient, and the accelerometer quality factor as a function of pressure in the pressure range of $10\,\text{Pa}$ to $10^5\,\text{Pa}$ at $f = 1\,\text{Hz}$. Assume the mechanical quality factor $Q_{\text{mech}} = 10$.
Solution: Using Equations (12.16) and (12.17), $\mu = 1.8 \cdot 10^{-5}\,\text{Pa·s}$ for air, and Example 12.3, the squeeze film spring constant and damping coefficient can

be readily calculated as a function of pressure. The quality factor for the gas damping is obtained from

$$Q_{\text{air}} = \frac{m\omega_0}{\gamma},$$

where $\omega_0 = \sqrt{(k + k_\mu)/m} \approx \sqrt{k/m}$ and $k = 40$ N/m is the mechanical spring constant. The total quality factor from Equation (12.3) is

$$\frac{1}{Q_{\text{tot}}} = \frac{1}{Q_{\text{air}}} + \frac{1}{Q_{\text{mech}}}. \tag{12.18}$$

At atmospheric pressure, the effective viscosity from Equation (12.12) with $d_c = 2.5$ μm is

$$\mu_{\text{eff}} = \frac{\mu}{1 + 9.638 K_n^{1.159}} \approx 1.56 \cdot 10^{-5} \text{ Pas},$$

which gives $\gamma = 0.87$ Ns/m and $Q = 0.01$.

Figure 12.5 shows the calculated squeeze film spring constant and quality factor as a function of pressure. As expected, the damping coefficient decreases with decreasing pressure. This increases the quality factor until the mechanical Q_{mech} is approached. In atmospheric pressure, the design is heavily over damped and has $Q \sim 0.01$. To improve the step response characteristics, operation at $p \sim 300$ Pa is needed to obtain $Q \approx 0.5$.

If low damping is needed but operation in a lower pressure is not an option, the design could be adjusted for example by changing the proof mass aspect ratio from square to thin and long to obtain a low W/L-ratio. Alternatively, the plate can perforated with holes to reduce the distance that the air has to travel in the narrow gap.

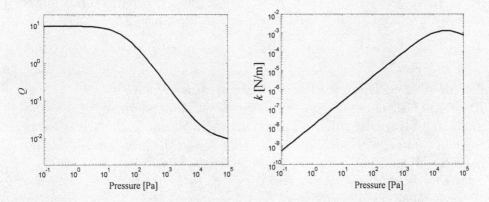

Figure 12.5: Squeeze film damping as a function of pressure for the accelerometer in Example 6.6.

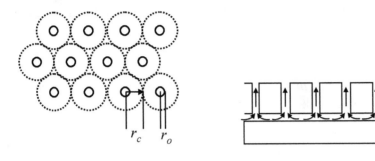

(a) Top view of the holed plate. Each hole occupies a cell with radius r_c and the hole diameter is r_o.

(b) Side view of the holed plate. The air flows through the plate with thickness H.

Figure 12.6: Squeeze film flow in an infinite plate with holes. Each hole has radius r_o and it occupies a cell with radius r_c.

To reduce the air damping in plates without using vacuum packaging, the plate can be perforated as is illustrated in Figure 12.6. The air only has to move to the nearest hole, which for large plates significantly reduces the damping in comparison to the solid plate. A simple model for the perforated plate assumes an infinite plate with holes. Referring to Figure 12.6, each hole occupies a cell with radius r_c and the hole radius is r_o. Assuming the plate is thin so that the flow resistance through the hole can be ignored, the damping coefficient is

$$\gamma_{\text{gas}} = \frac{3\mu_{\text{eff}} r_c^2}{2d^3} A K(\beta),\tag{12.19}$$

where $K(\beta) = 4\beta^2 - \beta^4 - 4\ln\beta - 3$, $\beta = r_o/r_c$, and $A = N\pi r_c^2$ is the area for a plate with N holes [100]. This is illustrated in Example 12.5.

Example 12.5: Damping for a thin perforated plate
Problem: Calculate the damping coefficient in atmospheric pressure for a thin perforated plate with lateral dimensions of 1.2 mm × 1.2 mm and electrode gap $d = 2.5$ μm. The hole radius is $r_o = 2$ μm and the holes are laid out in the pattern of Figure 12.6 with the cell radius of $r_c = 6$ μm.
Solution: Using the effective viscosity $\mu_{eff} \approx 1.56 \cdot 10^{-5}$ Pa·s from Example 12.4, Equation (12.19) gives

$$\gamma_{\text{gas}} = \frac{3\mu_{\text{eff}} r_c^2}{2d^3} A K(\beta) \approx 1.3 \cdot 10^{-4} \text{ Ns/m},$$

where we have used $\beta = r_o/r_c = 1/3$ and $K(\beta) = 4\beta^2 - \beta^4 - 4\ln\beta - 3 \approx 1.83$. The damping coefficient is 10,000 times smaller than for a solid plate of Example 12.4 with identical dimensions.

Equation (12.19) assumes that the plate is thin and does not account for the flow resistance through the holes. In the other extreme, the plate thickness H is so large that the flow resistance through the channel dominates. The flow resistance through a channel is

$$R_{gas} = \frac{\Delta p}{\dot{V}} = \frac{8\mu_{eff}H}{\pi r_h^4}, \qquad (12.20)$$

where Δp is the pressure difference over the channel lenght and $\dot{V}$ is volume flow rate through the channel. For a plate moving at velocity $\dot{x}$, the cell volume changes with the rate

$$\dot{V} = A_c \dot{x} = \pi r_c^2 \dot{x}, \qquad (12.21)$$

where $A_c = \pi r_c^2$ is the cell area. This must equal the flow rate through the channel. The force acting on the plate surface over the cell area A_c is

$$F_\mu = \Delta p A_c = \Delta p \pi r_c^2. \qquad (12.22)$$

By combining Equations (12.20) to (12.22), the damping coefficient $\gamma = F_\mu / \dot{x}$ for a plate with N holes is

$$\gamma = N\Delta p A_c \frac{\dot{V}}{\dot{x}} = \frac{8\pi N \mu_{eff} r_c^4 H}{r_h^4}. \qquad (12.23)$$

For a long channel, the effective viscosity is

$$\mu_{eff} = \frac{D\mu}{4B(D)} \qquad (12.24)$$

where

$$B(D) = \frac{D}{4} + 1.485\frac{1.78D+1}{2.625D+1} \qquad (12.25)$$

and $D = r_h\sqrt{\pi}/2\lambda$ [101]. The use of Equation (12.23) is illustrated in Example 12.6.

Equations (12.19) and (12.23) give two opposing extremes for the damping of a holed plate: Either the flow resistance through the hole is taken to be a negligible or it is the dominant loss mechanism. Most designs fall somewhere in between with air flow to the hole and through the hole, both contributing to the total losses. Approaches for modeling complex flow problems are discussed in Section 12.3.4

Example 12.6: Damping for a thick perforated plate
Problem: Calculate the damping coefficient for a 20 μm thick perforated plate
with lateral dimensions of 1.2 mm × 1.2 mm and electrode gap $d = 2.5$ μm.
The hole radius is 2 μm and the holes are laid out in the pattern of Figure 12.6
with the cell radius of $r_c = 6$ μm.
Solution: The number of holes is the approximate ratio of plate area to cell area
$N \approx A/A_c = A/\pi r_c^2 \approx 12,732$. From Equation (12.24), the effective viscosity is

$$\mu_{\text{eff}} = \frac{D\mu}{4B(D)} \approx 1.3 \cdot 10^{-5} \text{ Pas},$$

where we used where $A = \frac{D}{4} + 1.485\frac{1.78D+1}{2.625D+1} \approx 610$ and $D = r_h\sqrt{\pi}/2\lambda \approx 17.1$.
Equation (12.23) gives

$$\gamma = \frac{8\pi N \mu_{\text{eff}} r_c^4 H}{r_h^4} \approx 0.0066 \text{ Ns/m},$$

Comparing to Example 12.5, this 40 times larger the damping for a thin perforated plate and the channel flow resistance is the dominant damping term.

12.3.3 Lateral damping

Laterally moving structures are common in surface micromachining. The commercial applications include accelerometers and gyroscopes. Since the structure
moves parallel to the substrate, the squeeze film damping is effectively eliminated [102, 103]. The main damping mechanism is the shear flow between the
plate and substrate as is illustrated in Figure 12.7. This effect is smaller than
the squeeze film damping, and the laterally moving structures can have quality
factors greater than one in air, as exemplified by the surface micromachined
accelerometer in Figure 3.6 on page 45.

    Following reference [104], the damping coefficient for a plate with the area
A and the plate-substrate gap d is

$$\gamma = \frac{\mu_{\text{eff}} A}{d}, \tag{12.26}$$

where the effective viscosity is

$$\mu_{\text{eff}} = \frac{\mu}{1 + 2K_n}. \tag{12.27}$$

Equation (12.26) assumes that the gas velocity distribution is linear. This is
not valid at high plate velocities. Also, the effect of the moving air mass is

neglected. These effects can be included in the model with significant increase in complexity [104].

The lateral damping model omits the air drag on the top and sides of the plate. For structures with a large number of fingers, the model only gives qualitative results, as the squeeze film type damping between the fingers is also significant. Unfortunately, modeling this type of structure is not easy. Some modeling strategies for analyzing complex geometries are provided in Section 12.3.4.

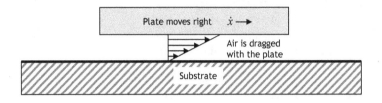

Figure 12.7: Air damping in a laterally moving structure. The shear force between the plate and the substrate is the biggest damping mechanism.

12.3.4 Air damping in complex geometries

The simple squeeze film and lateral damping model provided in Section 12.3 are useful for understanding the effect of geometry and pressure on damping, and as test cases for more complex simulation tools. The analytical models, however, are highly developed simplifications of fundamentally very complex fluid flow phenomena. As such, they cannot be directly applied to more complex geometries such as those shown in Figure 12.8. The perforated plate in Figure 12.8(a) has elements of the squeeze film models of Section 12.3.2 but the none of the models apply directly as the transition region at the plate edge effectively increases the plate area and the flow resistance from under the plate to hole and through the hole are both significant, and the pressure transition region at the hole end effectively makes the channel longer. The comb-finger in Figure 12.8(b) combines elements of squeeze damping between the fingers and lateral shear damping between the finger and the substrate. The assumptions of large areas relative to the gap, however, are clearly not valid and our simple models cannot be expected to give accurate results.

In this section, strategies for modeling complex structures such as those shown in Figure 12.8 are discussed. The section is by no means an extensive "how to guide" but will provide insights into trade-offs between different modeling approaches.

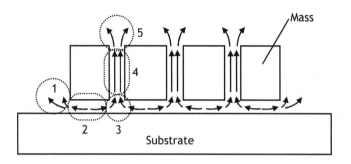

(a) Air flow in a holed plate moving toward substrate. Region 1 is the transition from from squeeze film to free space (edge effect), region 2 is the squeeze film flow under plate, region 3 is transition from squeeze film flow to pipe flow, 4 is the flow through the pipe, and region 5 is the transition from pipe flow to free space.

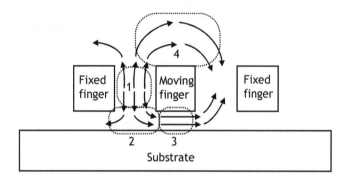

(b) Air flow around a comb-drive finger. Region 1 is the squeeze film flow between fingers, region 2 models transition from squeeze film to shear flow, region 3 is the shear flow under finger, and region 4 is the flow over the finger.

Figure 12.8: Examples of microstructures that cannot be modeled with simple models.

Analytical modeling of approximate nature

The easiest approach is to model the complex damping using the simple analytical models. The analytical models have the advantage that they are easy to implement and give valuable information about scaling effects. For example, it is easy to see what happens if a gap is increased by 2× or pressure dropped by 10×. Depending on how far removed the real structure is from the ideal model, the results can match fairly well. Even if the answer is off by a large factor, the estimation can be a starting point for a feasibility study, for example, by indicating whether the expected quality factor for an accelerometer is $Q \approx 1$

(good) or $Q \approx 0.001$ (bad).

The drawback of applying analytical models to structures where they don't apply is that it may not be possible to know in advance how far off the analytical results are. For example, the simple squeeze film model suggests that electrostatically actuated resonators cannot have high quality factors in atmospheric pressures. To much surprise, the resonators operating at high frequency ($f_0 \sim 1$ GHz) actually show quality factors $Q > 500$ in normal air pressures where the high vibration velocity $\dot{x}$ makes squeeze film models invalid [105].

Semi-analytical simulation

In circuit analysis, the analytical models have been successfully combined with simulation algorithms to make very powerful simulation tools such as SPICE or Aplac. A single transistor is modeled by a set of equations and to model a complex circuit with several transistors, the circuit designer just needs to code the transistor connections using a text or graphical interface. The designer does not need to worry about the model details other than specifying the transistor dimensions and connections.

This circuit design inspired approach can be applied to the air damping as illustrated in Figure 12.9. The gas flow is modeled as a combination of simple structures for which a model can be easily developed. To model the air flow in perforated plates, the flow resistance can be viewed as a series combination of flow from under the plate to the channel and flow through the channel. The plate edge is modeled with an appropriate element. The device dimensions can be easily adjusted by changing the model parameters and large number of design candidates can be quickly evaluated.

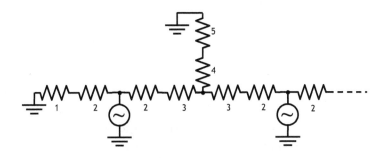

Figure 12.9: A complex shape can be broken into simpler structures that are then combined with a system level simulation tool such as a circuit simulator. Resistors 1-5 model the the flow resistances in the different regions in Figure 12.8(a)

This approach has been successfully applied for modeling air damping in atmospheric pressures [106] but currently the experimental data is not available for evaluating the validity of the proposed models in rarefied pressures when the mean free path is much longer that the structure dimensions.

A limitation of this approach is that unlike in circuit design, a large number of models is needed to cover different geometries. For example, most circuits are implemented with just two transistors (NMOS and PMOS) but even the relatively simple perforated plate requires several element types. Also, implementation of realistic model for truly 3D shapes is a challenge.

Computational flow simulation

Given the difficulties in analytical modeling of the air flow, computation using flow simulation tools is an attractive complement to the analytical approaches. Finite element modeling (FEM) of the sensor element is almost always done anyway to verify and fine tune the design. At this point, carrying out a simulation of the gas damping would appear straightforward. An ideal simulation tool would:

1. Solve the gas flow for small and large Knudsen numbers.

2. Model large displacements where gas damping is nonlinear.

3. Calculate the mechanical displacements and gas flow simultaneously as these two are likely to interact through the gas spring forces.

4. Be easy to use and fast to execute.

Unfortunately, no software currently meets these four criteria. The finite element packages are typically based on standard flow models where viscosity is independent of pressure. To use these tools, the user must change to viscosity to "effective viscosity" based on analytical estimates from device geometry. Models for large displacement may be available at significant computation expense, but most likely, the analysis of the expected vibration mode shapes does not include large displacement effects. Some tools are better implementing the gas/structure interactions but many simulation tools require first analyzing the expected mechanical displacement that is then passed to the fluid simulator. Obviously, this approach does not correctly predict how the gas damping affects the resonant mode shapes. The problem is especially important for parallel plate structures where air spring can cause tilting behavior instead of desired parallel plate movement. Finally, while the software is constantly improving, setting up a new simulation may take several days and is a task that must be repeated if the design changes.

In summary, all analysis approaches have merits and all three approaches are often used depending on design stage. The analytical approaches offer fast feedback and can be made into adaptable "blocks" for simulating more complex structures. The full-blown finite element simulation still requires manually accounting the effect of lowered effective viscosity at large Knudsen numbers. The simulations are also very time consuming. Thus, the FEM is typically carried out at later stage in development when the basic design topology has already been selected based on the rough analytical analysis.

Key concepts

- The desired damping depends on the application. Sensors and actuators are typically close to the critical damping. Resonating devices benefit from a high quality factor.

- Material losses set the fundamental limit for the attainable quality factor. The $f \cdot Q$-product is approximately constant for given material and is a figure of merit for high-Q resonator design.

- The anchor losses can be significant especially for flexural structures. Bulk mode resonators anchored at the nodal point have demonstrated the highest $f \cdot Q$-products.

- Air damping is significant at micro-scale. By adjusting the package pressure and device design, the damping characteristics can be tailored for the application.

- The air flow is characterized by the Knudsen number $K_n = \lambda/d_c$ that relates the mean free path of the gas molecules to the device dimensions. At $K_n < 0.1$, continuum models are valid. For $K_n > 10$, gas molecule interactions are rare and the molecule-structure interactions are dominant.

- Continuum models can be adopted for rarefied gases by using the effective viscosity that decreases with pressure.

- The air damping is typically independent of pressure in normal pressures but decreases for lower Knudsen numbers.

- Analyzing air damping for non-trivial shapes requires either combinations of analytical analysis and computation or full-blown simulation. In either case, the analysis accuracy depends on the validity of the assumptions in the model.

Exercises

Exercise 12.1
Calculate the mean-free path for argon atoms ($d_g = 360$ pm) and helium atoms ($d_g = 190$ pm) in atmospheric pressure.

Exercise 12.2

A surface micromachined accelerometer has the intrinsic mechanical quality factor $Q_{\text{intrinsic}} = 50,000$, the anchor quality factor $Q_{\text{anchor}} = 1,000$, and the air damping quality factor $Q_{\text{air}} = 40$. Calculate the total quality factor.

Exercise 12.3

Based on Table 12.1, estimate the maximum quality factor for a silicon resonator and a fused quartz resonator at 2 GHz.

Exercise 12.4

Estimate the anchor loss limited quality factor for a silicon cantilever with the length $L = 200$ μm and thickness $h = 4$ μm in the direction of vibration.

Exercise 12.5

Plot the effective viscosity for air as a function of pressure for squeeze film damping with the gap $d = 2.5$ μm.

Exercise 12.6

Consider the squeeze film damping in Example 12.4. Assuming that the accelerometer is operated in atmospheric pressure and there is no perforation to reduce damping, how large the electrode gap d must be to obtain quality factors greater than $Q > 0.1$?

Exercise 12.7

A square micromirror that moves up and down is made of SOI wafer. The mirror thickness is $t = 5$ μm and the gap under the mirror is $d = 10$ μm. Using Equation (12.16), estimate the maximum mirror size if the desired quality factor is $Q = 1$ and the desired resonant frequency is $f_0 = 500$ Hz.

Exercise 12.8

Derive Equation (12.8) for mode conversion losses. Assume that the energies stored in the modes 1 and 2 are $E_1 = a_1 E_{\text{tot}}$ and $E_2 = a_2 E_{\text{tot}}$, respectively. Hint: you can start with dissipation loss power $P = \omega E/Q$ and calculate the total dissipation loss.

13

Pressure sensors

The silicon-based pressure sensors were developed and commercialized in the 1970s, a long time before the word MEMS was even coined. Pressure sensing remains a large market with total revenue of over $500M/year. The number of sold pressure sensors keeps increasing but due to price erosion, the total market size has remained flat in the recent years. The emergence of consumer-oriented applications such as altimeters in sport computers and swatches may generate another growth spurt in the pressure sensor market.

Despite the large market, the number of pressure sensor publications in recent years is small compared to, for example, RF MEMS publications. The lack of academic interest is due to the maturity of the field: The basic pressure sensor structure, a micromachined diaphragm has remained unchanged since the 70's although the fabrication processes have evolved.

In this chapter, we will cover the piezoresistive and capacitive pressure sensors. First, the theory of diaphragm deflection due to static pressure is reviewed. Next, the operation of piezoresistive and capacitive pressure sensors is analyzed. The performance of the two transduction mechanisms is exemplified and compared. Finally, the packaging and other commercial device issues are discussed.

13.1 Pressure sensing with micromechnical diaphragms

Essentially all micromechanical pressure sensors are based on thin diaphragms that deflect due to a pressure difference [107]. The diaphragm displacement is translated to an electrical signal either directly with capacitive sensing or indirectly with piezoresistive sensing.

Figure 13.1 illustrates different ways to achieve differential pressure across the diaphragm. The absolute pressure sensor in Figure 13.1(a) has a cavity that is in vacuum pressure and the net pressure acting on the diaphragm is p_{input}. The pressure gauge in Figure 13.1(b) has the cavity in a reference pressure and the net pressure acting on the diaphragm is $p_{\text{input}} - p_{\text{reference}}$. The pressure gauge can offer a higher sensitivity around the reference pressure. For example, to measure the atmospheric pressure, the reference pressure is chosen to be around the normal pressure of 101.3 kPa. The differential pressure sensor in Figure 13.1(c) has both sides of the diaphragm open to external pressures.

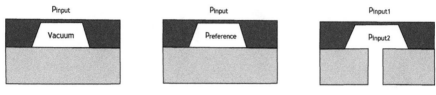

(a) Absolute pressure sensor has the cavity in vacuum.

(b) Pressure gauge has the cavity in a reference pressure, typically atmospheric pressure.

(c) Differential pressure sensor does not have a closed cavity but measures the pressure difference between two inputs.

Figure 13.1: Classification of pressure sensors based on the reference pressure (After [9]).

The shape of the diaphragm can be almost arbitrary but typically it is either circular or square. In next sections, we will cover the membrane deflection theory for these two common shapes. In the analysis, it is assumed that the deflection is small in comparison to the diaphragm thickness and that the diaphragm is clamped from the edges.

13.1.1   Circular diaphragm

The circular diaphragm is illustrated in Figure 13.2(a). The shape is popular for surface micromachined structures where the membrane is typically made of polycrystalline silicon or silicon nitride. Following reference [108], the deflection w at a radial location r from the diaphragm center is

$$w(r) = \frac{pa^4}{64D} \left(1 - \frac{r^2}{a^2}\right)^2, \tag{13.1}$$

where p is the pressure difference acting on the diaphragm, a is the diaphragm radius, and D is the flexural rigidity given by

$$D = \frac{Eh^3}{12(1 - \nu^2)}. \tag{13.2}$$

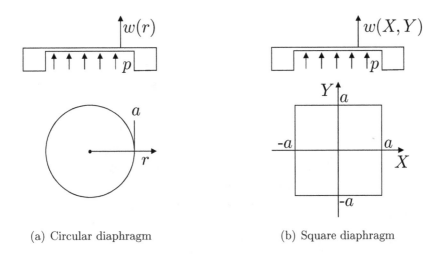

(a) Circular diaphragm   (b) Square diaphragm

Figure 13.2: Common pressure sensor geometries.

Here ν is the Poisson's ratio, E is the Young's modulus, and h is the diaphragm height. From Equation (13.1), the maximum deflection at the diaphragm center $(r = 0)$ is

$$w(0) = \frac{3(1 - \nu^2)pa^4}{16Eh^3}. \tag{13.3}$$

The membrane stresses are important for the piezoresistive sensor design. As with the beam analysis, the net membrane stress is zero and the stress on the opposing membrane surfaces have equal magnitude but opposite sign. The radial stress on the diaphragm surfaces is

$$T_r(r) = \pm\frac{3}{8}\frac{pa^2}{h^2}\left[(3 + \nu)\frac{r^2}{a^2} - (1 + \nu)\right], \tag{13.4}$$

where the positive sign is for the lower surface and the negative sign is for the top surface in Figure 13.2(a).

The transverse stress on the diaphragm surface is

$$T_t(r) = \pm\frac{3}{8}\frac{pa^2}{h^2}\left[(3\nu + 1)\frac{r^2}{a^2} - (1 + \nu)\right], \tag{13.5}$$

where the positive sign is for the upper surface and the negative sign is for the lower surface.

The maximum stresses are obtained at the diaphragm edge $(r = a)$ where

$$T_r(a) = \pm\frac{3}{4}\frac{pa^2}{h^2} \tag{13.6}$$

and

$$T_t(a) = \pm\frac{3}{4}\nu\frac{pa^2}{h^2}. \tag{13.7}$$

At the diaphragm center $(r = 0)$ we have

$$T_r(0) = T_t(0) = \mp\frac{3}{8}\frac{pa^2}{h^2}(1 + \nu). \qquad (13.8)$$

Example 13.1: Surface micromachined circular diaphragm
Problem: A needle mounted arterial blood pressure sensors are used to measure the patient blood pressure waveform during heartbeats. A surface micromachined pressure sensor with the Young's modulus $E = 130$ GPa, Poisson's ratio $\nu = 0.3$, the radius $a = 150$ μm, and the thickness $h = 2.5$ μm is used for the measurement. Calculate the diaphragm center displacement and the radial stress at the diaphragm edge with $\Delta p = 16$ kPa across the diaphragm.
Solution: From Equation (13.3) the diaphragm center displacement is

$$w(0) = \frac{3(1 - \nu^2)pa^4}{16Eh^3} \approx 0.68\ \mu\text{m}.$$

From Equation (13.6), the radial stress at the diaphragm edge is

$$T_r(a) = \frac{3}{4}\frac{pa^2}{h^2} \approx \pm 43\ \text{MPa}.$$

13.1.2 Square diaphragm

The square shaped diaphragm illustrated in Figure 13.2(b) results from wet etching of single crystalline silicon with anisotropic etchants such as KOH. Unlike for the circular geometry, the displacement and stresses for the square membrane can only be analyzed approximately. A simple expression for the diaphragm displacement at location X, Y is

$$w(X, Y) \approx \frac{1}{47}p\frac{a^4}{D}\left(1 - \frac{X^2}{a^2}\right)^2\left(1 - \frac{Y^2}{a^2}\right)^2, \qquad (13.9)$$

where $2a$ is the square edge length, p is the net pressure acting on the diaphragm and and D is the flexural rigidity given by Equation (13.2) [108]. The maximum displacement occurs at diaphragm center $(X = Y = 0)$ where

$$w(0, 0) \approx \frac{pa^4}{47D}. \qquad (13.10)$$

The approximate normal stress in X-direction is

$$T_{XX} \approx \mp \frac{24}{47} p \frac{a^2}{h^2} \left[\left(1 - \frac{Y^2}{a^2} \right)^2 \left(1 - 3\frac{X^2}{a^2} \right) + \nu \left(1 - \frac{X^2}{a^2} \right)^2 \left(1 - 3\frac{Y^2}{a^2} \right) \right],$$
(13.11)

where the minus sign is for the lower surface and positive sign is for the upper surface. The approximate normal stress in Y-direction is

$$T_{YY}(x,y) \approx \mp \frac{24}{47} p \frac{a^2}{h^2} \left[\left(1 - \frac{X^2}{a^2} \right)^2 \left(1 - 3\frac{Y^2}{a^2} \right) + \nu \left(1 - \frac{Y^2}{a^2} \right)^2 \left(1 - 3\frac{X^2}{a^2} \right) \right].$$
(13.12)

The stresses along the X-axis on the diaphragm edge ($X = a$, $Y = 0$) are

$$T_{XX}(a,0) \approx \pm \frac{48}{47} p \frac{a^2}{h^2}$$
(13.13)

and $T_{YY} = \nu T_{XX}$. The stresses at the diaphragm center ($X = Y = 0$) are

$$T_{XX}(0,0) = T_{YY}(0,0) \approx \mp \frac{24}{47} (1 + \nu) p \frac{a^2}{h^2}.$$
(13.14)

Example 13.2: Square diaphragm displacement and stresses
Problem: Consider a square single-crystal silicon diaphragm with the edge length $2a = 1$ mm, the thickness $h = 10$ μm, Young's modulus $E = 170$ GPa, and Poisson's ratio $\nu = 0.06$. Calculate the diaphragm center displacement, the average displacement, the stresses T_{XX} and T_{YY} at diaphragm edge ($X = a$ and $Y = 0$) for a pressure difference $\Delta p = 10$ kPa accross the diaphragm.
Solution: The flexural rigidity is $D = \frac{Eh^3}{12(1-\nu^2)} \approx 13.4$ μNm and the center displacement from Equation (13.10) is

$$w(0,0) \approx \frac{pa^4}{47D} \approx 0.93 \ \mu\text{m}.$$

From Equation (13.13), the stresses at $(a,0)$ are

$$T_{XX}(a,0) \approx \frac{48}{47} p \frac{a^2}{h^2} \approx 26 \ \text{MPa}$$

and

$$T_{YY} = \nu T_{XX} \approx 15 \ \text{MPa}.$$

The average displacement is obtained by integrating the Equation (13.9) and is given by

$$w_a = \frac{1}{4a^2} \int_{-a}^{a} \int_{-a}^{a} \mathrm{d}X \mathrm{d}Y \, w(X,Y) = \frac{64a^4}{10575D} p \approx 0.27 \ \mu\text{m}.$$

13.2   Electromechanical transduction

The first step in the micromechanical pressure sensing is to convert the pressure into mechanical displacement using a diaphragm. The next step is to translate the displacement into an electronic signal. Drawing from our experience in accelerometers, we have essentially two options: The piezoresistive sensor readout and the capacitive sensor readout. The piezoelectric conversion is not an option as the displacement is in dc and the signal from the piezoelectric readout is proportional to the rate-of-change. In this section, the piezoresistive and capacitive sensing methods are analyzed and compared.

13.2.1   Piezoresistive pressure sensors

Figure 13.3(a) shows a schematic for the piezoresistive pressure sensor. The displacement-induced stress in the diaphragm is translated into an electrical signal by incorporating piezoresistors into or onto the diaphragm. For bulk micromachined single-crystal silicon diaphragms, the resistors are usually made by diffusion. In case of surface micromachining, the resistors can be made of deposited polycrystalline silicon. In either case, it is beneficial to integrate the resistors near the diaphragm edge which is the location of the maximum stress.

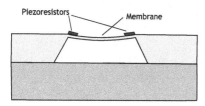
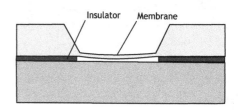

(a) Piezoresistive sensor has piezoresistors integrated near the location of maximum stress.

(b) Capacitive pressure sensor sense the membrane displacement by detecting the capacitance over a narrow gap.

Figure 13.3: Piezoresistive and capacitive pressure sensors.

The piezoresistive sensing for accelerometers was discussed in detail in Chapter 5 and the results are directly applicable to the pressure sensors. The piezoresistive sensing has the advantages of simple measuring circuitry, fairly linear response, and inherently shielded structure of the piezoresistors. The disadvantages are the large power consumption, large temperature dependency, and the noise introduced by the piezoresistors. The piezoresistive sensing is further illustrated in Example 13.3.

The resistive sensing has historically been the dominant sensing principle for the pressure sensors but, as with the accelerometers, the capacitive sensors have been gaining market share. This is especially true for battery powered devices such as sport computers or tire pressure sensors where the low power consumption is critical.

Example 13.3: Piezoresistive pressure sensor

Problem: Consider the square diaphragm in Example 13.2. The membrane displacement is measured with two piezoresistors aligned along X-axis at the diaphragm edges (the resistor locations are $X = \pm a$ and $Y = 0$) as shown in Figure 13.4. The diaphragm edge is aligned to [110]-crystal axis. Two reference resistors located outside the diaphragm where the stress is zero. The resistors are connected as a Wheatstone bridge. Calculate:

(1) the sensor transfer function,
(2) signal voltage for $\Delta p = \pm 50$ kPa pressure,
(3) Wheatstone bridge power consumption, and
(4) the smallest measurable pressure change limited by resistor noise.

Assume that the noise is dominated by the resistor $1/f$-noise and the noise bandwidth is $0.1 - 10$ Hz. The piezoresistor has dimensions of $l_r = 20$ μm, $h_r = 0.5$ μm, and $w_r = 10$ μm, the resistor carrier concentration is $n = 5 \cdot 10^{16}$ cm^{-3}, the resistivity is $\rho = 0.34$ Ωcm, and $\alpha = 10^{-5}$, the bias voltage is $V_{\text{bias}} = 2.5$ V, and the piezoresistive coefficients are $\pi_l = 70 \cdot 10^{-11}$ Pa^{-1} and $\pi_t = -66 \cdot 10^{-11}$ Pa^{-1}. Use $E = 170$ GPa and $\nu = 0.06$ for [110]-silicon.

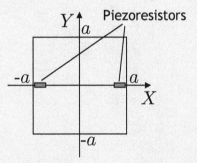

Figure 13.4: Piezoresistive pressure sensor.

Solution: (1) From Equation (13.13), the relationship between the pressure and longitudinal stress acting on the piezoresistors is

$$T_l = T_{XX}(a, 0) \approx \frac{48}{47} \frac{a^2}{h^2} p = 2550p,$$

and the transverse stress is $T_t = T_{YY}(a, 0) = \nu T_{XX}(a, 0) \approx 153p$. The fractional change in resistance for a pressure change from Equation (5.7) is

$$\frac{\Delta R}{R} = \pi_l T_l + \pi_t T_t \approx 1.6840 \cdot 10^{-6} \text{Pa}^{-1} p,$$

which gives the transfer function between the resistance change and pressure

$$A_1 = \frac{\Delta R/R}{p} \approx 1.6840 \cdot 10^{-6} \text{Pa}^{-1}.$$

The Wheatstone bridge transfer function between the resistance change $\Delta R/R$ and voltage change ΔV is

$$A_2 = \frac{\Delta V}{\Delta R/R} = \frac{1}{2} V_{\text{bias}} = 1.25 \text{ V}.$$

The total transfer function between the output voltage and pressure is

$$A = A_1 A_2 \approx 2.11 \ \mu\text{V}/\text{Pa}.$$

(2) The output voltage for the pressure $p = \pm 50$ kPa is $v_s = Ap \approx \pm 105$ mV.
(3) The piezoresistor resistances are

$$R = \rho \frac{l_r}{h_r w_r} = 13.6 \text{ k}\Omega$$

and the total power consumed by the Wheatstone bridge is $p = V_{\text{bias}}^2/R \approx$ 460 μW
(4) To calculate the piezoresistor noise, we will first obtain the $1/f$-noise corner frequency. The number of carriers in the piezoresistor is $N = nl_r w_r h_r$. From Equations (5.17) and (2.25), the Flicker noise corner frequency is obtained as

$$f_c = \frac{\alpha V^2}{N 4 k_B T R} \approx 55.5 \text{ kHz}.$$

The thermal resistor noise is $\overline{v_n} = \sqrt{4 k_B T R} \approx 15$ nV/$\sqrt{\text{Hz}}$. From Equation (2.13), the rms-noise due to Flicker-noise is

$$v_{\text{rms}} = \sqrt{\overline{v_n^2} f_c \ln \frac{f_H}{f_L}} \approx 16 \ \mu\text{V}.$$

The input referred pressure noise is

$$p_{\text{rms}} = \frac{v_{\text{rms}}}{A} \approx 7.7 \text{ Pa}.$$

This is roughly equal to the smallest measurable pressure change.

13.2.2 Capacitive pressure sensors

Figure 13.3(b) shows a schematic for the capacitive pressure sensor. The diaphragm displacement is measured directly by measuring the capacitance between the movable membrane and a fixed electrode. The capacitive sensing was covered in Chapter 6 and the associated measurement electronics was analyzed in Chapters 8-10. In this section, we will apply the results from these chapters to the pressure sensors.

The main benefits of capacitive sensing are the low power consumption and excellent noise performance. The disadvantages are the nonlinear capacitance-displacement relationship, the complex measurement electronics in comparison to piezoresistive sensors, and the sensitivity to parasitic capacitances. An added challenge for capacitive pressure sensors is the difficulty of implementing the differential measurement as there is only one movable electrode. In addition, the integration of the reference capacitor on to the pressure-sensing element is costly due to large size of the typical diaphragm. Implementing the reference capacitor with the IC is straightforward but introduces additional temperature dependency as the sense and reference capacitor will not have identical temperature dependencies.

Example 13.4 further illustrates the capacitive pressure sensors. Compared to the piezoresistive sensor, the noise-limited resolution is significantly better. This performance advantage is confirmed by the commercially available capacitive pressure sensor SCP1000 by VTI Technologies [109]. The device has resolution of 1.5 Pa ($\sim$ 10 cm height change in air at sea level) with $< 10~\mu$A current consumption illustrating excellent power/noise performance.

Example 13.4: Capacitive pressure sensor
Problem: Consider the square diaphragm in Example 13.2. The diaphragm displacement is measured capacitively by measuring the capacitance change between the diaphragm and fixed electrode. The electrode gap is $d_0 = 3~\mu$m. For the capacitive readout, the switched capacitor amplifier shown in Figure 10.4(a) on page 165 is used. In Figure 10.4(a), C_{w1} is the pressure sensor capacitance, C_{w2} is a constant reference capacitance with a value of C_{w1} at zero displacement, the sense voltage is $v_{\text{in}} = 1$ V, and $C_f = 1$ pF. Assume that the sensor output is an average of hundred measurements and calculate
i. the sensor transfer function,
ii. the smallest measurable pressure change limited by thermal noise, and
iii. the output voltage for pressures of ±50 kPa.
Solution: The diaphragm capacitance at zero displacement is

$$C_0 = \epsilon_0 \frac{A}{d_0} \approx 3 \text{ pF}.$$

Due to the pressure, the diaphragm shape changes and the gap is no longer uniform. The gap at location (X, Y) is $d(X, Y) = d_0 - w(X, Y)$ where $w(X, Y)$ is the diaphragm displacement given by Equation (13.9). The displaced diaphragm capacitance is obtained from integrating

$$C = \int_{-a}^{a} \int_{-a}^{a} \epsilon_0 \frac{\mathrm{d}X \mathrm{d}Y}{d_0 - w(X, Y)}$$

for which there is no closed form solution. However, if the displacement w is small compared to the gap, we can use the Taylor series expansion

$$\frac{1}{d_0 - w(X, Y)} \approx \frac{1}{d_0} \left(1 + \frac{w(X, Y)}{d_0} \right)$$

to approximate the capacitance as

$$C \approx \int_{-a}^{a} \int_{-a}^{a} \epsilon_0 \frac{\mathrm{d}X \mathrm{d}Y}{d_0} \left(1 + \frac{w(X, Y)}{d_0} \right) = C_0 \left(1 + \frac{w_a}{d_0} \right)$$

where w_a is the average displacement given by

$$w_a = \frac{1}{4a^2} \int_{-a}^{a} \int_{-a}^{a} \mathrm{d}X \mathrm{d}Y w(X, Y) = \frac{64a^4}{10575D} p.$$

The switched capacitor amplifier will translate the capacitance change into output voltage as given by Equations (10.4) and (10.5). The output voltage is

$$v_{\text{out}} = \frac{C_0 v_{\text{in}}}{C_f d_0} w_a$$

which is a factor of two smaller than for a differential measurement with both C_{w1} and C_{w2} changing. The complete sensor transfer function is

$$A = \frac{v_{\text{out}}}{p} = \frac{C_0 v_{\text{in}}}{C_f d_0} \frac{64a^4}{10575D} \approx 26.2 \ \mu\text{V/Pa}.$$

and the output voltage for $p = \pm 50$ kPa is $v_{\text{out}} = \pm 1.4$ V.

The noise at the amplifier output due to the $k_B T / C$-noise is given by Equations (10.6) to (10.9) and is

$$v_{\text{rms}} = \frac{\sqrt{k_B T (C_{w1} + C_{w2} + C_f)}}{C_f} \approx 0.17 \ \text{mV}.$$

and the averaged noise from 100 measurement is

$$< v_{\text{rms}} > = \frac{v_{\text{rms}}}{\sqrt{100}} \approx 17 \ \mu\text{V}.$$

The input referred pressure noise

$$p_{\mathrm{rms}} = \frac{< v_{\mathrm{rms}} >}{A} \approx 0.6 \text{ Pa}$$

which corresponds to average diaphragm displacement of 17 pm.

As with the accelerometers, we see that the capacitive sensor noise performance is significantly better than for the piezoresistive sensor.

13.3   Large deformation effects

The diaphragm deflection theory reviewed in Section 13.1 assumes that the diaphragm displacement is small in comparison to the diaphragm thickness. At large deflections, however, the restoring forces result from a combination of the bending stiffness and the membrane stretching. This stretching results in nonlinear restoring force that is not covered by the small deflection theory in Section 13.1.

The large deflection analysis requires either numerical simulations, for example with finite element analysis, or a combination of analytical analysis and numerical calculations. Reference [110] is a useful starting point for analyzing large deflection induced stresses in a single-crystal silicon diaphragm.

13.4   Packaging and specifications

The focus of this chapter has been on the diaphragm design and the electromechanical transduction. The noise limited resolution was evaluated for the piezoresistive design in Example 13.3 and for the capacitive design in Example 13.4. In addition, commercial pressure sensors have to meet other considerations such as linearity, temperature stability, drift, and robustness to environment. This section will focus on the stability and protective packaging that are profound challenges in pressure sensor manufacturing.

The challenge of achieving stable operation over temperature is evident when we note that the diaphragm displacement due to pressure change can be nanometers or less. In comparison, the thermal expansion of a 1 mm silicon membrane for -40 to $+85$ °C temperature change is 2.6 ppm/K·(85 K+40 K)·1 mm$\approx$ 300 nm which is already significant. An even bigger problem is that the sensor is mounted in a package that has a different thermal expansion coefficient than the diaphragm. This thermal expansion mismatch results in large bending stresses that can result in huge temperature dependency. Even with a careful design, the temperature dependency of

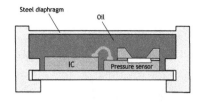 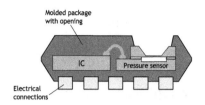

(a) A micromachined pressure sensor and measuring circuit packaged inside an oil filled steel package. The thin steel diaphragm transmits the outside pressure inside the package.

(b) A low-cost molded plastic package. The thin micromachined diaphragm is exposed to the elements.

Figure 13.5: MEMS pressure sensor package examples.

packaged pressure sensor can be unacceptably high.

Temperature dependency will also vary from sample to sample due to variations in the element mounting which makes the compensation difficult. To compensate the unpredictable temperature dependency, the part can be calibrated in two or more temperatures but this will add to the testing cost. Also, the sensor characteristic may change due to stress of mounting the sensor to the final product, for example by soldering it to the circuit board, or due to creep of the sensor package itself. These changes cannot usually be compensated and result in drift.

The second challenge in pressure sensor design is the protection of the sensor and the electronics while exposing the diaphragm to external pressure variations. A robust but expensive solution is illustrated in Figure 13.5(a): The entire sensor is inside an oil filled metal package. The package has a metal diaphragm that allows the external pressure to be transmitted to the sensor element. For less aggressive environments, it may be sufficient to just encapsulate the sensor IC and the electrical connections in a low cost plastic package as is illustrated in Figure 13.5(b). The electrically active parts are encapsulated in epoxy and just the diaphragm is exposed.

These challenges demonstrate that it is not meaningful to design a pressure sensor without considering the final package. This lesson can be extended to other MEMS components and has been relearned for example in the RF switch and resonator development.

Key concepts

- Pressure sensors are the oldest MEMS product and the yearly revenue is over $500M.

- All commercial micromachined pressure sensors are based on measuring diaphragm deflection.

- Most pressure sensors require a cavity for the reference pressure. The pressure inside the reference cavity should be stable which requires good hermetic sealing process.

- Piezoresistive pressure sensors measure the diaphragm displacement induced strain.

- Capacitive pressure sensors measure the diaphragm displacement induced capacitance changes.

- Packaging of pressure sensors cheaply while protecting the thin diaphragm without introducing temperature or time dependent drift is a challenge.

Exercises

Exercise 13.1
A surface micromachined circular silicon nitride membrane is used as an absolute pressure sensor. How thick does the membrane have to be if membrane diameter is 50 μm, the cavity height under the membrane is 1.5 μm, and the operating pressure range is $0 - 6$ bar? Use $E = 210$ GPa and $nu = 0.3$ for the silicon nitride.

Exercise 13.2
Consider the piezoresistive pressure sensor in Example 13.3. Repeat the example but assume the piezoresistors located at $X = \pm a$ and $Y = 0$ are rotated by $90°$ to be transverse to the X-axis.

Exercise 13.3
Design an absolute capacitive pressure sensor capable of measuring pressures up to 5 MPa and having a noise limited resolution better than 2 kPa.

Exercise 13.4
Design a capacitive pressure sensor for 0.5 bar to 1.5 bar pressure range. The noise limited resolution should be better than 0.1 mbar.

Exercise 13.5
Explain why the pressure gauge is a better design than the absolute pressure sensor for sensors operating atmospheric pressure range.

Exercise 13.6
Figure 13.6 shows a piezoresistive pressure sensor with lateral piezoresistors (current along Y-axis). Derive an expression for the resistance change for resistors 1 and 2. How should the four resistors be connected in a Wheatstone bridge to maximize the signal?

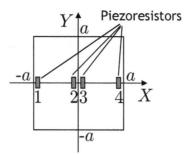

Figure 13.6: Piezoresistive pressure sensor with transverse piezoresistors.

14

Actuation

Actuators are essential for many microsystems: For example, accelerometers employ electrostatic actuation for self-test, micromirror arrays use mechanical motion to modulate the light intensity, and gyroscopes are based on detecting the Coriolis force on electrically excited resonators. Microactuators have been under intensive research and several different actuation methods have been proposed. The demonstrated microactuators usually do not resemble the macroscopic motors. This is partially because the micromanufacturing methods are not well suited for defining complex 3D shapes required by the conventional engines. More fundamentally, the physics in micro-scale is different, and good microactuators are based on different operational principles than the macroscopic actuators.

To understand the challenges in micro-scale actuation, we start this chapter by studying the scaling laws for the fundamental forces. This will build "microintuition" as we will learn what effects become significant at small scale. Next, the different actuation methods are reviewed. The scaling laws for electrostatic, magnetic, thermal, and piezoelectric forces are developed. The focus here is on how the different effects scale to micro-scale. The scaling analysis helps us to focus on the most promising microactuation schemes. More detailed analysis of different actuation methods is carried out in the subsequent chapters: The capacitive actuation is analyzed in Chapter 15, the piezoelectric actuation is covered in Chapter 16, and the thermal actuation is covered in Chapter 17.

14.1 Scaling laws

The physics in micro-scale is not intuitive to humans that are used to macro-objects. For example, as we noted in Chapter 2, the mechanical noise that is inconsequential for macro-sized objects, becomes important as the devices are miniaturized. Scaling laws allow us to study how physics scales to different sizes. The scaling laws are obtained by scaling objects isometrically meaning that all dimensions are changed by an equal factor. For example, the area is proportional to the dimensions squared and the scaling law for the area is $A \sim l^2$ where l is the object dimension. Similarly, the scaling law for the volume is $V \sim l^3$.

Taking the ratio of two scaling laws allows comparison of relative influence of the different physical effects at different size scales. For example, the surface-to-volume ratio scales as $A/V \sim l^{-1}$, which shows that the surface-to-volume ratio increases as the device dimension l decreases. Consequently, surface effects such as adhesion forces become more important in micro- and nano-scale. The effect of large surface-to-volume ratio is demonstrated by considering how sugar in different forms interacts when dropped onto for example a shirt. A sugar cube will roll down to the floor as body forces (gravity) are more important than the surface forces. But powdered sugar will stick to the shirt, demonstrating that the surface forces are larger than the gravity forces.

Table 14.1 shows common scaling laws [10, 111]. The power of scaling laws becomes apparent when we compare them to natural phenomena at small scale:

1. Gravity force is less important in micro-scale. Insects can walk on water and land on ceilings as surface forces are enough to counter the gravity.

2. Small things are fast. Humming bird wings beat 50–200 times per second and ants take dozens of steps per second. The resonant frequency scales as l^{-1}, and the resonators used in high frequency communication systems have critical dimensions in micrometer range. Due to the high resonant frequency, the small devices have a fast response time.

3. Viscous forces scale with area and become significant in micro-scale. Macroscopic objects fall down due to gravity force but dust particles seemingly float in air. Air damping is significant for microstructures. For example, accelerometers may have to be packaged at reduced pressures to reduce the air damping.

4. Surface tension is significant in micro-scale. The effect of water surface tension in a drinking glass is barely noticeable, but a 10-μm diameter capillary will raise water three meters against the gravity force. Porous surfaces will suck liquid and a simple cloth or sponge can be used to hold water.

5. Flying objects or animals require a certain lift-to-weight ratio to stay in the air. Using Table 14.1, the lift-to-weight ratio scales as $F_L/m \sim \dot{x}^2/l$. In order to maintain a constant lift-to-weight ratio, the velocity of flying objects must scale as $\dot{x} \sim l^{1/2}$. Small birds can fly more slowly than large birds.

Table 14.1: Collection of scaling laws

Parameter	Formula	Scaling	Notes
Distance	d	l	
Area	A	l^2	
Volume	V	l^3	
Mass	$m = \rho V$	l^3	ρ = the density
Spring constant	k	l	
Capillary force	$F_\gamma = 2\pi r \gamma$	l	γ = the surface tension
Resonance frequency	$\omega_0 = \sqrt{\frac{k}{m}}$	l^{-1}	
Heat conduction	$P = \kappa \frac{A}{d} \Delta T$	l	κ = the thermal conductance
Lift force	$F_L = \frac{1}{2} C_L \rho A \dot{x}^2$	$l^2 \dot{x}^2$	C_L = the lift coefficient
			ρ = the air density
			A = the wing area
Viscous drag force	$F_D = \frac{1}{2} C_D \rho A \dot{x}^n$	$l^2 \dot{x}^n$	C_D = the drag coefficient
			A = the projected area
			$n \approx 1$ for low speed
			$n \approx 2$ at high speed
Pipe flow resistance	$R = \frac{\pi r^4}{8 \mu L}$	l^3	μ = the viscosity
			r = the pipe radius
			L = the pipe length
Reynold's number	$\mathrm{Re} = \frac{\rho \dot{x} D}{\mu}$	$l \dot{x}$	μ = the viscosity
			D = the critical dimension

6. Drag force is important at small scale. The lift-to-drag ratio scales as $F_L/F_D \sim \dot{x}$. The drag forces are dominant for small creatures such as insects that fly at low speeds, and the insect flight can be compared to swimming as drag of the wings is more significant than lift.

7. Sample volume is lost quickly due to evaporation in small volumes. Some insects cope with this by having waxed outer shells to reduce loss of fluids. Similarly in microfluidics, the small sample volumes cannot be exposed to air, as it would quickly disappear due to evaporation.

Further information on scaling laws in nature can be found in the beautiful book "On Growth and Form" by D'Arcy Thompson [112].

Example 14.1: Scaling of surface tension
Problem: Using Table 14.1, analyze how surface tension force scales in com-

parison to body forces such as gravity.

Solution: The surface tension scales as $F_\gamma \sim l$ and the body forces scale as $F_V \sim l^3$. The surface tension to body force ratio scales as

$$\frac{F_\gamma}{F_V} \sim l^{-2}$$

demonstrating that in micro-scale, the surface tension is comparatively much more important.

Example 14.2: Jump height

Problem: Using the scaling laws, estimate how the height that an animal can jump varies with the animal size.

Solution: The work done in leaping up is proportional to the mass m and height of the jump h

$$W = m\mathrm{G}h \sim l^4.$$

Energy available for the jump is proportional to the mass of the muscle $m \sim l^3$, which is proportional to the mass of the animal. Thus h tends to be constant irrespective of the animal size. Grasshoppers and horses whose mass differs by several orders of magnitude can jump about the same height.

Example 14.3: Accelerometer displacement

Problem: Derive a scaling law for accelerometer proof-mass displacement.

Solution: Using Table 14.1, the spring constant scales as $k \sim l$ and the mass scales as $m \sim l^3$. The accelerometer displacement scales as

$$x \sim \frac{m}{k} \sim l^2.$$

Reducing all dimensions by 10×, will reduce the displacement by 100×.

14.2   Scaling of actuation forces

Human sized actuators are usually based on combustion engines or magnetic forces (electric motors). Neither actuation principle is popular for microactuators where electrostatic, piezoelectric, and thermal actuators are dominant [113–117]. To understand why some actuators scale better than others, we will review the scaling laws for different actuation principles [111].

14.2.1 Electrostatic forces (capacitive actuation)

Capacitive or electrostatic transduction is based on the attractive forces between positive and negative charges. At macro-scale, the electrostatic forces are weak in comparison to body forces such as gravity. But as devices are scaled to micro-scale, electrostatic attraction becomes significant. The main advantages of the capacitive actuators are the relative ease of construction and low power consumption. The disadvantage of the electrostatic actuation is that large voltages may be needed to generate significant forces.

The capacitive actuation is covered in detail in Chapter 15. Looking ahead, the attractive force between two parallel plates is

$$F = \epsilon \frac{A}{2d^2} V^2 \sim l^0, \qquad \text{(constant } V) \qquad (14.1)$$

where ϵ is the permittivity, A is the plate area, d is the plate separation, and V is the voltage over the plates. Equation (14.1) shows that for a constant voltage, the actuation force is size independent. As other forces are usually reduced with size, the electrostatic force become comparatively significant in micro-scale.

The constant voltage scaling may not be realistic at small gaps or high voltages. The maximum voltage that can be applied between the plates depends on the breakdown field $\mathcal{E}_{\text{MAX}}$ that is the maximum field before the medium between the plates becomes conductive. In gas environment, the breakdown is due to runaway ionization of the gas and $\mathcal{E}_{\text{MAX}}$ is a pressure and gas dependent constant. Using the relationship $\mathcal{E} = V/d$ between the voltage and electric field, the scaling law for the electrostatic force with a constant electric field is

$$F = \epsilon \frac{A}{2} \mathcal{E}^2 \sim l^2, \qquad \text{(constant } \mathcal{E}) \qquad (14.2)$$

which decreases when the device size is reduced, but the rate of decrease is slower than for body forces.

The estimation of the maximum electric field in electrostatic actuators is complicated as the continuum breakdown model breaks down for micrometer-sized gaps and the breakdown field $\mathcal{E}_{\text{MAX}}$ becomes gap dependent: the short distance between the gaps is not sufficient for the gas molecules to gain sufficient velocity for the breakdown ionization and the maximum electric field $\mathcal{E}_{\text{MAX}}$ increases for small gaps. This effect was first observed by Friedrich Paschen in 1889, and the plot of breakdown voltage vs. gap shown in Figure 14.1 is called the "Paschen curve" [118, 119]. In athmosperic pressure, the minimum breakdown voltage is about 200 V for 5 μm gap but the breakdown voltage

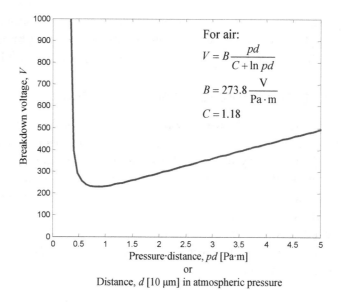

Figure 14.1: Paschen curve in air. For small gap d and/or pressure p, the breakdown voltage increases as the gap is reduced.

increases for smaller gaps. In vacuum or for very small gaps, the Paschen curve will lose validity and field emission will in turn limit the maximum electric field.

It is therefore impossible to obtain a universal scaling law for the electrostatic actuation. The electrostatic force scales as l^0 to l^2 depending on weather the voltage or electric field is held constant. In both cases, the force-mass ratio increases as the device is scaled to smaller dimensions and the electrostatic force therefore scales well for microactuation.

The capacitive actuation is the most used actuation method in commercial MEMS devices. Due to its importance, the capacitive actuation is analyzed in detail in Chapter 15.

14.2.2   Magnetic forces

Magnetic transduction is based on the Lorentz force between moving charged particles. Magnetic actuation is the most common method for electromechanical transduction at macro-scale. For example, modern kitchens can house more than a dozen electric motors that are based on magnetic forces. Unfortunately, good magnetic actuators require coils that are hard to miniaturize and fabricate with lithographic techniques. Despite the challenges, several elegant magnetic actuators have been proposed [120, 121], but commercial success has been elusive.

To gain insights into why magnetic actuators do not scale well to microscale, we'll analyze the scaling law for the force between two conductors carrying current I. The force is

$$F = \frac{\mu_0 I^2 L}{2\pi d} \sim l^0, \qquad \text{(constant } I) \qquad (14.3)$$

where μ_0 is the permeability of space which equals $4\pi \cdot 10^{-7}$ N/A^{-2}, L is the length of the conductors, and d is the distance between the conductors. Equation (14.3) suggests that the force for a given current is independent of the dimensions. The current, however, is limited by the maximum current density J_{MAX} that depends on the conductor material. The maximum current for the given conductor cross sectional area A is $I = J_{\text{MAX}} A$. Keeping the current density constant, the force scales as

$$F = \frac{\mu_0 I_1 I_2 L}{2\pi d} = \frac{\mu_0 J_{\text{MAX}} A_1 J_{\text{MAX}} A_2 L}{2\pi d} \sim l^4. \qquad \text{(constant } J) \qquad (14.4)$$

Equation (14.4) shows that the magnetic force does not scale well to small dimensions but decreases faster than the body forces.

Because of the unfavorable scaling and due to the difficulty at fabricating 3D coils, the magnetic actuation has not found commercial success in microdevices. Due to the lack of commercial interest, we have chosen not cover the magnetic actuation in detail in this book. For a more detailed review of magnetic actuation in microdevices, the reader is referred to references [5,9].

14.2.3   Thermal forces

Thermal actuation is based on the thermal expansion of materials. The actuator is heated, usually by passing current through, and the thermal expansion is used to generate displacements. The thermal expansion is characterized by large forces but small displacements. Often leverage is used to trade the force for a larger displacement.

Heat expansion of a rod with the cross sectional area A and the Young's modulus E generates a force

$$F = EA\alpha\Delta T \sim l^2, \qquad (14.5)$$

where α is the coefficient of thermal expansion and ΔT is the temperature change. Equation (14.5) shows that the scaling is similar to electrostatic actuators and thermal actuation scales well to small dimensions.

The actuation speed is limited by the cooling time of the actuator. This too scales well. The thermal energy is proportional to the mass and the heat conduction scales linearly with the dimensions. Thus, the cooling time scales as

$$t \sim l^2. \qquad (14.6)$$

Micro-scale thermal actuators can be fast and thermally actuated resonators operating at MHz frequency have been demonstrated [122].

As with the capacitive actuators, the thermal actuators are easy to construct. Additionally, the heating can be achieved with low voltages whereas the electrostatic actuation can require large voltages. The main disadvantage of thermal actuation is that the power consumption for heating is significant especially when compared to electrostatic or piezoelectric actuators that have zero dc power consumption.

The thermal actuation is covered in detail in Chapter 17 where we will quantitatively analyze the thermal actuator power consumption and response speed.

14.2.4   Piezoelectric forces

Piezoelectric transduction is based on materials that deform due to electric field. The piezoelectric materials also work in the opposite direction: Deformation due to external force results in electric field. Thus, the piezoelectric materials can be used both as sensors and as actuators.

The maximum dimension change in piezoelectric materials is usually small, only about 0.1% of the original dimension. But this small change can be controlled to a high precision, and piezoelectric macroscopic transducers are used, for example, in optical devices where the small displacement is sufficient. In MEMS, the piezoelectric effect has been used for example in micromotors [123], microfluidic pumps, and surgical tools [124, 125].

A piezoelectric plate with the area A, the thickness h, and the Young's modulus E generates a force

$$F = EAd_{33}\frac{V}{h} \sim l, \qquad \text{(constant } V) \qquad (14.7)$$

where d_{33} the piezoelectric constant. As with the electrostatic actuators, the maximum voltage is limited by the breakdown field of the piezoelectric material. For constant electric field $\mathcal{E} = V/h$, the force scales as

$$F = EAd_{33}\mathcal{E} \sim l^2. \qquad \text{(constant } \mathcal{E}) \qquad (14.8)$$

In practice, the breakdown field is usually high for the piezoelectric materials, about 100 kV/mm, and the available voltage limits the actuation force.

The piezoelectric actuation force is shown to scale well (l to l^2), perhaps even better than the electrostatic force. This strong coupling is the basis for many RF resonators that use piezoelectric materials to convert mechanical vibrations to electrical signals and vice versa. It is, however, difficult to obtain large displacements and the requirement for special piezoelectric materials increases the processing costs. For these reasons, the piezoelectric transducers are seldom used in commercial microactuators that require displacements larger than a few hundred nanometers.

The piezoelectric actuation is covered in detail in Chapter 16 where we will quantify the actuator force and displacement.

Key concepts

- Scaling laws provide an intuitive way to quantify the relative importance of different physical effects in micro-scale.

- Body forces become less important in micro-scale and surface forces dominate.

- Small devices are fast.

- Electrostatic, thermal, and piezoelectric actuation scale well to small dimensions. Magnetic forces diminish faster than the body forces, which makes the magnetic actuation unattractive in micro-scale.

Exercises

Exercise 14.1
Verify that the $k \sim l^1$ scaling law for the spring constant applies for both rod extension and cantilever springs.

Exercise 14.2
Explain why small devices are fast.

Exercise 14.3
How does the microresonator resonant frequency change if all dimension are reduced by $10\times$?

Exercise 14.4
A capacitive microactuator is reduced by $2\times$. How does the actuator displacement $x = F/k$ scale? Account both the actuator scaling and spring constant scaling. Assume that the actuation voltage is constant.

Exercise 14.5
Is the scaling of *practical* capacitive actuators limited by the breakdown field or can constant voltage scaling be used?

Exercise 14.6
If Reynold's number is less than 1,000, a flow is laminar. Reynold's number greater than 1,000 indicates turbulent flow. How does the maximum laminar flow velocity and flow rate (volume per second) scale with the pipe diameter?

Exercise 14.7
How does a pressure sensor diaphragm displacement scale if all dimensions are scaled by an equal factor?

Exercise 14.8
Derive a scaling law for a capacitive accelerometer electrode gap if all other dimensions are scaled by an equal factor but the sensitivity $\Delta C/C_0/\ddot{x}$ is to remain constant.

Exercise 14.9
Derive Equation (14.6).

Exercise 14.10
An average 180 cm tall human can carry approximately his/her own weight. Derive a scaling law for the weight a human can carry. Note that the weight scales as l^3 but the muscle force scales as l^2. Based on the scaling law, how tall can a human be before the body mass is too much to carry?

15

Capacitive actuation

Electrostatic or capacitive actuation is based on the attraction of electric charges. At macro-scale, the electrostatic forces are usually too small for actuation but as the device size is reduced to the micron-scale, the electrostatic forces become significant. The capacitive actuators are easy to manufacture using lithographic techniques, do not require special materials, and consume no dc power. Due to these advantages, the capacitive actuation is currently the most popular method to move microdevices. Successful commercial products include micromirror devices and gyroscope sensors. More recently, electrostatically actuated resonators, switches, and varactors for RF applications have been introduced to the market.

In this chapter, the principles of electrostatic actuation are reviewed. First, a general expression for electrostatic force is derived. Next, steady-state solutions of spring balanced electrostatic actuators are reviewed. For parallel plate actuators, the actuation force is found to be a nonlinear function of displacement. To characterize the effect of the nonlinearity, the stability of capacitive transducers is investigated and expressions for the maximum stable displacement, the "pull-in" point", are derived. After the dc analysis, we will develop models for the ac actuation used in vibrating structures such as gyroscopes and RF resonators.

15.1 Force acting on a capacitor

The electrostatic force depends on the actuation voltage and the device geometry. Before analyzing different actuator geometries, we wish to obtain a general expression for the capacitive force that is valid for any actuator shape. To derive the general actuation force law, we will first calculate the total energy stored in the system comprising of the capacitor and the voltage source. Once the total energy is known, the electrostatic force is given by the gradient of the stored energy: $F_e = -\frac{dW}{dx}$, where W is the stored potential energy and x is the displacement. Example 15.1 illustrates the approach of calculating force from the stored energy in familiar examples of a loaded spring and a mass at height x above ground.

Example 15.1: Force as a gradient of potential energy
Problem: Consider (1) a mass that is at height x above ground and (2) a spring k stretched by x. Derive the force acting on the mass and the spring by taking the derivate of the potential energy with respect to displacement.
Solution: (1) The potential energy for a mass m at height x is $W = mGx$ where $G = 9.81$ m/s^2 is the gravity constant. The gravity force is $F = -\frac{dW}{dx} = -mG$ where the negative sign shows that the force is in the direction of decreasing height (=toward ground).
(2) The potential energy for a spring k stretched by x is $W = \frac{1}{2}kx^2$. The spring force is $F = -\frac{dW}{dx} = -kx$ where the negative sign shows that the force works to reduce the displacement x.

The potential energy for the capacitive transducer can be calculated with the aid of Figure 15.1 that shows a schematic view of a capacitive transducer connected to a voltage supply. The voltage V is held constant, but the capacitance C and the charge stored in the capacitor Q change as the distance between the capacitor electrodes changes. The energy stored in the capacitor is

$$W_C = \frac{1}{2}CV^2, \tag{15.1}$$

where C is the capacitance and V is the voltage over the capacitor. The charge stored in the capacitor is

$$Q = VC. \tag{15.2}$$

Differentiating Equation (15.1) with respect to capacitance C gives the capacitor energy change due to the capacitance change dC

$$dW_C = \frac{1}{2}V^2 dC. \tag{15.3}$$

As the capacitance changes, the charge stored in the capacitor changes by

$$dQ_C = V dC. \tag{15.4}$$

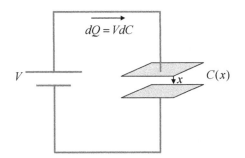

Figure 15.1: Schematic of capacitor C connected to voltage supply V.

This charge is provided by the voltage supply so the supply charge changes by $dQ_V = -dQ_C$ and the energy stored in the supply is reduced by

$$dW_V = V dQ_V = -V dQ_C. \tag{15.5}$$

Combining Equations (15.3) to (15.5) gives the total change in the stored electrical energy

$$dW_e = dW_C + dW_V = -\frac{1}{2}V^2 dC. \tag{15.6}$$

The force $F_e = -\frac{dW_e}{dx}$ acting on the capacitor is obtained by dividing Equation (15.6) with dx leading to

$$F_e = -\frac{dW_e}{dx} = \frac{1}{2}V^2\frac{dC}{dx}. \tag{15.7}$$

Equation (15.7) shows that the capacitive force is positive for positive $\frac{dC}{dx}$, which means that the force is in the direction of *increasing* capacitor capacitance. The second observation is that the force is proportional to the voltage squared V^2, and therefore does not depend on the sign of the voltage.

The gradient of capacitance in Equation (15.7) depends on the capacitor geometry. The two most common capacitors are the parallel plate capacitor and the comb capacitor that are analyzed in the next section.

15.2   Parallel plate transducer

The parallel plate transducer in Figure 15.2 is the most common capacitor geometry for MEMS devices and is frequently used to actuate optical and tunable electrical components. For example, the parallel plate capacitor can be used as a tunable RF capacitor.

The parallel plate capacitance is given by

$$C = \epsilon\frac{A}{d-x}, \tag{15.8}$$

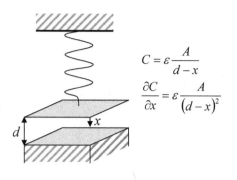

Figure 15.2: Schematic parallel plate transducer connected to a spring k. The capacitor capacitance C depends on the plate area A, the initial gap d, and the displacement x.

where $\epsilon = \epsilon_R \epsilon_0$ is the permittivity, A is the electrode area, d is the initial electrode gap, and x is the movement of the capacitor plate from the initial position. From Equation (15.7), the force acting on the plate is

$$F_e = -\frac{\mathrm{d}W_e}{\mathrm{d}x} = \frac{1}{2}V^2\frac{\mathrm{d}C}{\mathrm{d}x} = \frac{1}{2}\frac{\epsilon A}{(d-x)^2}V^2. \tag{15.9}$$

Equation (15.9) is complicated as $1/(d-x)^2$ term is very nonlinear: The force F increases toward infinity as the displacement x approaches the gap d. In what follows, the capacitive force given by Equation (15.9) is analyzed in more detail and expressions for the equilibrium point, unstable pull-in point, and capacitive spring forces are developed.

15.2.1 Equilibrium and pull-in point

For the equilibrium analysis of the transducer in Figure 15.2, we assume that the actuation voltage V is changing slowly in comparison to mechanical resonance frequency. This allows for ignoring the inertial effects in the analysis. Combining the electrostatic force F_e given by Equation (15.9) and the mechanical spring force $F_m = -kx$, the total force acting on the plate is

$$F = F_e + F_m = \frac{1}{2}\frac{\epsilon A}{(d-x)^2}V^2 - kx. \tag{15.10}$$

At small voltages, the electrostatic force is countered by the spring force, and the equilibrium displacement is obtained by solving Equation (15.10) for zero. As the voltage increases, the electrostatic force increases and the plates will eventually snap together. Estimating this pull-in voltage V_P and the plate travel distance x_P before the plates snap together is required for a successful design of electrostatic actuators, switches, varactors, and sensors.

Our analysis of the pull-in will proceed as follows: First, we will derive an expression for the equilibrium point where the mechanical and electrical forces are in balance. Next, we will investigate the stability of the equilibrium point to find out the point where the system becomes unstable.

At the equilibrium point, the electrostatic force is balanced by the mechanical spring force. Solving Equation (15.10) for $F = F_e + F_m = 0$ gives the equilibrium displacement x_0

$$\frac{1}{2}\frac{\epsilon A}{(d - x_0)^2}V^2 - kx_0 = 0. \tag{15.11}$$

Equation (15.11) can be solved analytically for the displacement x_0 but the result is not intuitive to interpret [5]. If the equilibrium displacement is small, the solution to Equation (15.11) can be approximated by

$$x_0 \approx \frac{d}{4}\left(1 - \sqrt{1 - \frac{8V^2}{V_C^2}}\right) \text{ for } x_0 < 0.15d \text{ and } V < 0.3V_C, \tag{15.12}$$

where we have defined the characteristic voltage $V_C = \sqrt{\frac{2kd^3}{\epsilon A}}$.

Instead of solving for x_0, a simpler way to proceed is to write the actuation voltage V as a function of plate displacement x_0. From Equation (15.11) we have

$$V^2 = \frac{2kx_0(d - x_0)^2}{\epsilon A} = V_C^2\frac{x_0}{d}\left(1 - \frac{x_0}{d}\right)^2, \tag{15.13}$$

which has been plotted in Figure 15.3(a). As expected, starting from the zero displacement, the actuation voltage grows with the displacement x_0. With the increasing displacement, however, the slope of the $V - x_0$-curve decreases indicating that at larger displacements, a smaller change in voltage is needed to increase the displacement by a fixed amount. At $x_0 = d/3$, the $V - x_0$-curve reaches the maximum. In what follows, we will show that the solutions $x_0 \geq d/3$ of Equation (15.13) are unstable.

The stability of the equilibrium point can be analyzed by deriving Equation (15.10) to obtain the *stiffness* of the system:

$$\frac{\partial F}{\partial x} = \frac{\epsilon A}{(d - x)^3}V^2 - k. \tag{15.14}$$

The stiffness characterizes how the system behaves around the equilibrium point. Negative stiffness means that a small displacement away from the equilibrium point will lead to restoring force toward the equilibrium point and the system is stable. A positive stiffness means that a small displacement away from the equilibrium point will lead to positive force away from equilibrium point. This positive feedback makes the system unstable. A macroscopic example of an unstable equilibrium point is a pencil balanced on its tip. If the pencil is

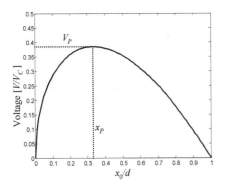

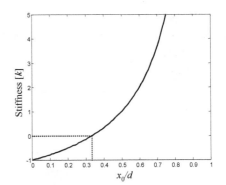

(a) The relationship between the equilibrium plate displacement x_0 and the voltage V.

(b) The stiffness as a function of equilibrium point x_0. Positive stiffness indicates unstable equilibrium point.

Figure 15.3: Pull-in effect in the parallel plate capacitor. No equilibrium solutions are obtained above the pull-in voltage (horizontal line). Above the pull-in displacement (vertical line), the actuator is unstable.

perfectly upright, all forces will balance and the pencil will stay upright. However, even a small displacement will cause the pencil to fall as the gravity will act in the direction of initial displacement and there are no restoring forces.

Coming back to the capacitive actuator, we substitute Equation (15.13) to (15.14) and evaluate the expression at $x = x_0$ to obtain an expression for the stiffness at the equilibrium point

$$\left.\frac{\partial F}{\partial x}\right|_{x=x_0} = \frac{2kx_0}{(d - x_0)} - k. \tag{15.15}$$

With no applied voltage, Equation (15.15) is simply $\frac{\partial F}{\partial x} = -k$; a small positive movement δx results in a negative restoring force $F = \frac{\partial F}{\partial x}\delta x = -k\delta x$. This is the familiar Hooke's law for a loaded spring.

Increasing the equilibrium displacement x_0 lowers the stiffness given by Equation (15.15) until the restoring force (stiffness) becomes zero and the pull-in point is reached. Solving Equation (15.15) for $\frac{\partial F}{\partial x} = 0$ gives the pull-in displacement

$$x_P = \frac{1}{3}d \tag{15.16}$$

at which point the stiffness vanishes. Beyond this point, the stiffness becomes positive as shown in Figure 15.3(b) and the system is unstable: A small positive movement δx result in a positive force that further increases x.

One way to look at the stiffness is that it provides the effective spring constant for the system. As the stiffness of the system decreases with increasing bias voltage, its resonance frequency $\omega = \sqrt{k/m}$ decreases. At the unstable point, the resonance frequency becomes zero. As a historical side note, this has been used for monitoring the loading of construction scaffolds. As more weight is put onto a scaffold, its resonance frequency is decreased. By hitting the scaffold bars and listening to the resonance frequency, it is possible to determine when the resonant frequency is dangerously low and the scaffold is in danger of collapsing due to buckling of the beams.

We will now obtain an expression for the voltage needed to obtain the pull-in displacement x_P. This voltage is called the pull-in voltage and it is the maximum dc voltage that can be applied to the parallel plate transducer without entering into the unstable region. Substituting Equation (15.16) to (15.13) gives the pull-in voltage at which the spring supported parallel transducer becomes unstable

$$V_P = \sqrt{\frac{8}{27}\frac{kd^3}{\epsilon A}} = \sqrt{\frac{4}{27}}V_C. \qquad (15.17)$$

Physically, when the pull-in point is reached, the mechanical spring force can no longer counter the electrostatic force and the plates snap together. Consequently, at voltages above the pull-in voltage V_P, Equations (15.10) and (15.13) have no stable solutions. Equation (15.17) for the pull-in voltage is a weak function of spring constant k and area A, but is strongly dependent on the gap d. To obtain low actuation voltages, a small gap d is desired, but this means that the maximum travel distance x_P is also reduced.

Example 15.2: Tuning of a mechanical RF varactor
Problem: Figure 15.4 shows a top view and a schematic model for a RF varactor that can be used to tune communication systems. The varactor is actuated with dc voltage that pulls the top electrode closer to the bottom electrode thus increasing the capacitance. The plate area A is 400 μm$\times$400 μm, the nominal gap d is 2 μm, and the total spring constant k for the four beams in parallel is 22 Nm. Calculate the capacitance tuning range and the tuning voltage range.

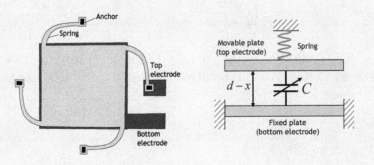

Figure 15.4: Top view and schematic model for RF varactor

Solution: The pull-in voltage from Equation (15.17) is

$$V_P = \sqrt{\frac{8}{27}\frac{kd^3}{\epsilon_0 A}} = 6.1 \text{ V}$$

so the tuning voltage range is approximately 0-6 V. The capacitance at zero bias voltage is

$$C = \epsilon_0 \frac{A}{d} = 0.7 \text{ pF}.$$

Increasing the bias voltage decreases the gap and increases the capacitance. The maximum plate displacement from Equation (15.16) is $x_P = d/3$ and the capacitance at the pull-in point is

$$C = \epsilon_0 \frac{A}{d - x_P} = \epsilon_0 \frac{3A}{2d} = 1.1 \text{ pF}.$$

Thus, the theoretical maximum tuning range is 0.7-1.1 pF or 50% increase of the nominal value of 0.7 pF. Practical tuning range without closed loop control is lower, typically less than 30% to avoid the pull-in.

15.2.2 Capacitive spring forces

As was shown in Section 15.2.1, the voltage biased parallel plate capacitor reduces the system stiffness effectively reducing the restoring spring force kx. This effect can be modeled by calculating the capacitive spring forces around the equilibrium point x_0.

To derive an expression for the capacitive spring, we start by summing the electrical and mechanical forces. The force acting on the parallel plate capacitor is

$$F = F_m + F_e = -k(x + x_0) + \frac{1}{2}\frac{\epsilon A}{(d - x_0 - x)^2}V^2, \qquad (15.18)$$

where x_0 is the equilibrium point given by Equation (15.10) and x is the displacement from the equilibrium point. By taking a series expansion of Equation (15.18) gives

$$F = -kx_0 - kx + \frac{V^2 \epsilon A}{2(d - x_0)^2}\left(1 + \frac{2x}{d - x_0} + \frac{3x^2}{(d - x_0)^2} + \frac{4x^3}{(d - x_0)^2} + \cdots\right)$$
$$(15.19)$$

$$= -kx + \frac{V^2 \epsilon A}{2(d - x_0)^2}\left(\frac{2x}{d - x_0} + \frac{3x^2}{(d - x_0)^2} + \frac{4x^3}{(d - x_0)^2} + \cdots\right), \qquad (15.20)$$

where in Equation (15.19), the second and the third terms cancel out at the equilibrium point (Equation (15.11)) to give Equation (15.20).

Equation (15.20) can be written as $F = -kx + k_e(x)x$ where the electrostatic spring around the equilibrium point is

$$k_e(x) = k_{0e}(1 + k_{1e}x + k_{2e}x^2)$$

$$k_{0e} = -\frac{V^2 C_0}{(d-x_0)^2}, \ k_{1e} = \frac{3}{2(d-x_0)}, \ \text{and} \ k_{2e} = \frac{2}{(d-x_0)^2}, \qquad (15.21)$$

In Equation (15.21), the capacitance is defined as $C_0 = \epsilon \frac{A}{d-x_0}$. The linear electrostatic spring k_{0e} is negative and reduces the mechanical spring constant. The spring constant k_{0e} gives an alternative approach to the pull-in phenomenon: it is easy to show that at the pull-in point x_P, the electrostatic spring and the mechanical spring have equal magnitudes and the net spring constant, the "stiffness", is zero as expected.

The higher order nonlinear spring terms in Equation (15.21) are significant for microresonators as they limit the linear vibration range and reduce the maximum power handling capacity of micro-oscillators and RF filters. This is covered in more detail in Section 20.4.

15.3   Longitudinal capacitor (comb drive)

In the longitudinal capacitor geometry, the electrodes move parallel to each other keeping the electrode gap constant but changing capacitor area. The main benefits of the longitudinal geometry over the parallel plate capacitor are that the stroke distance is not limited by the pull-in and the capacitance changes linearly with the displacement. Figure 15.5 shows a schematic view of the longitudinal comb finger capacitor where the displacement is in the direction of the fingers. In the MEMS community, this structure is commonly referred to as the "comb drive".

The capacitance for one finger to finger electrode overlap is

$$C = \epsilon \frac{h(l+x)}{d} + C_f, \qquad (15.22)$$

where h is the height of the finger, l is the initial overlap between the fingers, x is the displacement, d is the gap between the fingers and C_f is the fringe capacitance that accounts for the non-overlapping portion of the fingers. As is illustrated in Figure 15.5(a), the fringe capacitance does not change with displacement and the gradient of capacitance is

$$\frac{dC}{dx} = \epsilon \frac{h}{d}. \qquad (15.23)$$

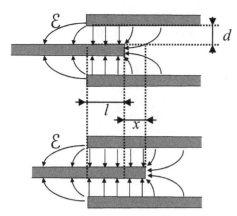
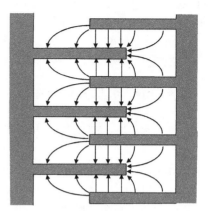

(a) A schematic view of the comb capacitor. The fringe fields are independent of displacement x as long as there is sufficient space between the comb finger tip finger supports.

(b) Larger capacitances and actuation forces can be obtained by putting several fingers in parallel. If fingers are not sufficiently long, the finger support contributes to the fringe fields.

Figure 15.5: Schematic view of the longitudinal comb finger capacitor where the displacement is in the direction of the fingers. Arrows show the field lines, l is the initial overlap between the fingers, x is the displacement, and d is the gap between the fingers.

The force acting on the capacitor from Equations (15.7) and (15.23) is

$$F_e = -\frac{\mathrm{d}W_e}{\mathrm{d}x} = \frac{1}{2}V^2\frac{\mathrm{d}C}{\mathrm{d}x} = \epsilon\frac{h}{2d}V^2. \tag{15.24}$$

The force given by Equation (15.24) does not depend on displacement x, and the comb drive capacitance does not introduce spring forces (linear or nonlinear).

The force from the one finger-to-finger overlap is small, but higher forces are obtained by putting a large number of fingers in parallel as shown in Figure 15.5(b) where there are six finger overlaps. From Equation (15.24), the force from N finger electrode overlaps is

$$F_e = N\epsilon\frac{h}{2d}V^2. \tag{15.25}$$

Compared to the parallel plate geometry, the comb drive offers linearity and large travel range but has two disadvantages: First, it is difficult to fabricate comb capacitance in direction perpendicular to the wafer surface [126, 127]. Thus, the structure is mainly suited for laterally moving devices. Second, the force generated by the comb drive is smaller than for a parallel plate capacitor with an equal volume and capacitor gap. Thus, large actuation forces require large actuators or high voltages.

Example 15.3: Comparison of comb drive and parallel plate actuator
Problem: Consider a comb drive with $N = 100$ finger overlaps, the finger height $h = 5$ μm, the finger gap $d = 2$ μm, and the actuation voltage $V = 20$ V. (1) Calculate the electrostatic actuation force for the comb drive. (2) What is the total area of the actuator if the finger length is $L = 10$ μm, finger overlap is $l = 5$ μm, and finger width is $w = 2$ μm? (3) Compare the force to a parallel plate transducer with the same total area and electrode gap.
Solution: (1) The electrostatic force from Equation (15.25) is

$$F = N\epsilon\frac{h}{2d}V^2 \approx 0.44 \ \mu\text{N}.$$

(2) One hundred finger overlaps means that the transducer will have 50 fingers over one side and 51 fingers on the other side (see Figure 15.5(b)). The total transducer area is

$$A = (50(2w + 2d) + w)(2L - l) \approx 33{,}300 \ \mu\text{m}^2.$$

(3) The force from a parallel plate actuator with the electrode area A and the gap d is

$$F = \epsilon\frac{A}{2d^2}V^2 \approx 15 \ \mu\text{N}.$$

This is about 30× higher than for the comb drive.

15.4   Capacitive actuation with ac voltages

For the equilibrium and pull-in analysis, we assumed that the actuation voltage V was slowly changing and could be considered constant. If this assumption does not prove true, we need to consider the effect of the square force law $F \sim V^2$ on the frequency characteristics of the actuation force. Most situations can be analyzed by breaking the actuation voltage into ac signal v and dc bias V_{dc}. Using $V = V_{\text{dc}} + v$, the actuation force is

$$F = -\frac{\mathrm{d}W_e}{\mathrm{d}x} = \frac{1}{2}(V_{\text{dc}} + v)^2\frac{\mathrm{d}C}{\mathrm{d}x} = \frac{1}{2}(V_{\text{dc}}^2 + 2vV_{\text{dc}} + v^2)\frac{\mathrm{d}C}{\mathrm{d}x}. \tag{15.26}$$

Substituting $v = v_0 \cos\omega t$ into Equation (15.26) gives

$$F = \frac{1}{2}\left(V_{\text{dc}}^2 + \frac{1}{2}v_0^2 + 2V_{\text{dc}}v_0\cos\omega t + \frac{1}{2}v_0^2\cos 2\omega t\right)\frac{\mathrm{d}C}{\mathrm{d}x} \tag{15.27}$$

where v_0 is the amplitude and ω is the frequency of the ac signal.

The two dc force terms $V_{\mathrm{dc}}^2 + \frac{1}{2}v_0^2$ in Equation (15.27) set the equilibrium point for the transducer. The fact that the ac voltage v results also in dc force is significant, for example, in case of capacitive switches where large RF signals can cause self-actuation. The third term gives time harmonic force at the excitation frequency and is utilized, for example, in micromechanical filters and oscillators. The last term gives actuation force at twice the excitation frequency. If the frequency of the last term is much higher than the mechanical resonant frequency $(2\omega \gg \omega_0)$ it can be ignored as it will generate only small displacements. However, actuation is possible when $\omega \leq \omega_0/2$.

In the next sections, the typical cases of ac actuation are analyzed in more detail.

15.4.1   Time harmonic actuation with dc bias

When the effect of 2ω term in Equation (15.27) can be considered small ($V_{\mathrm{dc}} \gg v_0$ or $2\omega \gg \omega_0$), Equation (15.27) simplifies to

$$F = \frac{1}{2}\left(V_{\mathrm{dc}}^2 + \frac{1}{2}v_0^2 + 2V_{\mathrm{dc}}v_0\cos\omega t\right)\frac{\mathrm{d}C}{\mathrm{d}x}. \tag{15.28}$$

The first two terms in Equation (15.28) sets the dc operation point of the transducer. The last term is seen to generate ac force at the drive frequency ω.

For resonators, it is helpful to write this ac force as

$$F_{\mathrm{ac}} = V_{\mathrm{dc}}\frac{\mathrm{d}C}{\mathrm{d}x}v_0\cos\omega t = \eta v, \tag{15.29}$$

where we have defined the electromechanical transduction factor $\eta = V_{\mathrm{dc}}\frac{\mathrm{d}C}{\mathrm{d}x}$ and used $v = v_0\cos\omega t$. For parallel plate capacitors the electromechanical transduction factor is

$$\eta_P = V_{\mathrm{dc}}\frac{\epsilon A}{(d-x_0)^2} = V_{\mathrm{dc}}\frac{C_0}{(d-x_0)}, \tag{15.30}$$

where $C_0 = \epsilon A/(d-x_0)$ and x_0 is the equilibrium displacement at the operating point. For the comb drive we have

$$\eta_C = NV_{\mathrm{dc}}\frac{\epsilon h}{d}. \tag{15.31}$$

Example 15.4: Electromechanical transduction factor for comb drive
Problem: Calculate the electromechanical transduction factor for the comb drive in Example 15.3 with bias voltage $V_{\mathrm{dc}} = 50$ V. What is the ac force if the ac voltage is $v_0 = 2$ V?

Solution: The electromechanical transduction factor from Equation (15.31) is

$$\eta_C = N V_{\text{dc}} \frac{\epsilon h}{d} \approx 0.11 \ \mu\text{N/V}.$$

The magnitude of the ac force is $F_{\text{ac}} = \eta v_0 \approx 0.22 \ \mu\text{N}$.

Example 15.5: Electromechanical transduction factor for parallel plate capacitor

Problem: Calculate the electromechanical transduction factor for a parallel plate capacitor with an area $A = (100 \ \mu\text{m})^2$, an initial gap $d = 2 \ \mu\text{m}$, and bias voltage $V_{\text{dc}} = 50$ V. Assume that the plate is connected to a spring with a spring constant $k = 400$ Nm.

Solution: The characteristic voltage is $V_C = \sqrt{\frac{2kd^3}{\epsilon A}} \approx 269$ V. From Equation (15.12), the equilibrium displacement is

$$x_0 \approx \frac{d}{4} \left(1 - \sqrt{1 - \frac{8V^2}{V_C^2}} \right) \approx 75 \ \text{nm},$$

which is less than $0.15d$ allowing the use of the approximate Equation (15.12). The electromechanical transduction factor from Equation (15.30) is

$$\eta_C = V_{\text{dc}} \frac{\epsilon A}{(d - x_0)^2} \approx 1.2 \ \mu\text{N/V}.$$

15.4.2 $\omega_0/2$-actuation

When $V_{\text{dc}} = 0$, Equation (15.27) simplifies to

$$F = \frac{1}{2} \left(\frac{1}{2} v_0^2 + \frac{1}{2} v_0^2 \cos 2\omega t \right) \frac{\text{d}C}{\text{d}x}, \tag{15.32}$$

where the first term is at dc and sets the operation point of the transducer. The second term is seen to generate ac force at a frequency 2ω. In another words, to obtain ac force at frequency ω_0, we need to choose the actuation voltage frequency to be $\omega = \omega_0/2$ or half of the operation frequency.

The $\omega_0/2$-actuation has been utilized in gyroscopes where small capacitance changes due to the Coriolis force at the MEMS vibration frequency ω_0 are to be measured. If these structures are actuated at the operation frequency using the dc bias, the measurement of the small Coriolis signal can be difficult due to the presence of the large actuation voltage at the same frequency. When the actuation voltage is at frequency $\omega_0/2$, it can be filtered to make the detection of the signal at ω_0 easier.

Example 15.6: $\omega_0/2$-actuation of a comb drive
Problem: Consider the comb drive in Examples 15.4 and 15.3. The comb drive is excited with ac signal $v_0 = 5$ V at $\omega = \omega_0/2$ without dc bias voltage $V_{\rm dc} = 0$ V. How large ac force is generated at frequency ω_0?
Solution: The ac force at ω_0 from Equation (15.32) is

$$F = \frac{1}{2}v_0^2\frac{dC}{dx} = \frac{1}{2}v_0^2 N\frac{\epsilon h}{d} \approx 28 \text{ nN}.$$

15.4.3   High frequency actuation ($\omega \gg \omega_0$)

When the ac voltage frequency is much higher than the mechanical resonance frequency, the ac force does not generate significant mechanical displacement. From Equation (15.27), however, shows that the ac voltage generates an additional dc force term $v_0^2/2$. The effective dc voltage is

$$V^2 = V_{\rm dc}^2 + \frac{1}{2}v_0^2. \tag{15.33}$$

Equation (15.33) is significant as just ac voltage is sufficient to change the operating point of a capacitive transducer. This limits the magnitude of the ac voltages, for example, that MEMS capacitors can handle or that can be used to measure accelerometers. The effective dc voltage us illustrated in Examples 15.7, 15.8, and 15.9.

Example 15.7: Self-actuation of varactor
Problem: The pull-in voltage V_P for the varactor in Example 15.2 was 6.1 V. If the ac signal has maximum amplitude of $v_0 = 4$ V, what is the maximum dc tuning voltage?
Solution: From Equation (15.33), the effective dc actuation voltage from the ac and dc voltages is

$$V^2 = V_{\rm dc}^2 + \frac{1}{2}v_0^2.$$

Requiring that this is below the pull-in voltage, gives the maximum dc voltage of

$$V_{\text{dc}} = \sqrt{V_P^2 - \frac{1}{2}v_0^2} = 5.4 \ V.$$

Example 15.8: Self-actuation in accelerometer

Problem: Consider the capacitive accelerometer in Example 6.6. If the proof mass displacement is too large, the electrode gap becomes so small that the ac voltage used for sensing the proof mass displacement is enough to close the remaining gap. Estimate this maximum displacement before the pull-in effect pulls the mass to the sense electrode.

Solution: From Equation (15.17), the minimum gap given the sense ac voltage $v_0 = 100$ mV is

$$d_{\text{min}} = \sqrt[3]{\frac{27\epsilon_0 A v_0^2}{2 \cdot 8k}} \approx 220 \text{ nm},$$

where the additional factor of two in the denominator is due to ac instead of dc voltage over the capacitor ($V^2 = \frac{1}{2}v_0^2$ from Equation (15.33)). If the gap becomes smaller than this, the ac sense voltage will pull the mass and electrode together.

is from happening, there can be physical stoppers that limit the splacement or the pull-in may be eliminated electronically for moving or lowering the sense voltage if the electrode gap becomes

9: Capacitive spring in capacitive accelerometer

Problem: Calculate the electrical spring force for the capacitive accelerometer in Example 6.6 when the proof mass is in the rest position.

Solution: There are two measurement capacitance each having capactitance $C_0 = 5.1$ pF. The ac measurement voltage is $v_0 = 100$ mV and the effective voltage for the spring constant calculation is $v_0^2/2$. From Equation (15.21), the capacitive spring for one electrode is

$$k_{0e} = -\frac{v_0^2 C_0}{2d^2} \approx -0.0041 \text{ N/m}.$$

The total spring constant for two measurement capacitors is -0.0082 N/m which is about 2% of the mechanical spring.

As the capacitive spring forces grow with the displacement, the effect of the capacitive springs can cause significant deviation from the linear mechanical Hooke's law for large displacements. At large accelerations, the gap can even become so small that the pull-in effect occurs. The prevent this, capacitive accelerometers can either deploy mechanical stoppers that prevent the gap from fully closing or the capacitive force is adjusted to counter the mass movement using closed loop control.

Key concepts

- Capacitive forces are proportional to the actuation voltage squared and directly proportional to the gradient of the capacitance $(F_e = \frac{1}{2}V^2\frac{dC}{dx})$.

- The capacitive force is in the direction of increasing capacitance.

- Parallel plate transducers snap together when the voltage reaches the pull-in voltage $V_P = \sqrt{\frac{8}{27}\frac{kd^3}{\epsilon A}}$. The maximum transducer displacement before the pull-in point is $x_P = d/3$.

- For the comb drive actuators, the force is independent of the displacement. Thus, the comb actuators do not suffer from nonlinear spring effects and the travel distance is not limited by the pull-in voltage.

- The force from the comb drive actuators is smaller than for a parallel plate actuator with similar size.

- A time harmonic force is obtained with a combination of dc bias and ac excitation.

- Due to the square voltage law, ac signal results in dc force and force at twice the ac signal frequency.

Exercises

Exercise 15.1
Figure 15.6 shows a schematic of a MEMS device that has capacitors on both sides of the movable plate. Derive the expression for the pull-in voltage. Assume that equal voltage is applied over both electrodes surrounding the center plate and note that the equilibrium point is always at $x_0 = 0$. Hint: the correct answer is $V_P = \sqrt{kd^3/2\epsilon A}$.

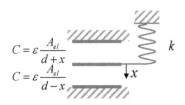

Figure 15.6: Two sided capacitive actuator.

Exercise 15.2

Design a capacitive microactuator for optical switching. The actuator should be capable of a 100-μm linear displacement at actuation voltage $V = 50$ V. You have an access to SOI process where you can make actuators that are vertically etched of single-crystal silicon that is 100 μm deep. The smallest features that you can make are 4 μm wide. The actuator should have a square foot print. Your answer should include a top view schematic (mask layout) of the actuator and calculations for the spring constant and actuation force.

Exercise 15.3

Calculate the one sided (voltage over one electrode) and two sided (voltage over both electrodes as in Exercise 15.1) pull-in voltages for the accelerometer in Figure 3.6 on page 45 if the electrode gap is $d = 1.3$ μm.

Exercise 15.4

Consider the varactor in Example 15.2. How large ac voltage will cause the pull in with no applied dc voltage?

Exercise 15.5

Calculate the capacitive spring constant for the accelerometer in Figure 3.6 on page 45 if the electrode gap is $d = 1.3$ μm and the ac measurement voltage is $v = 100$ mV. How does this compare to the mechanical spring constant?

Exercise 15.6

Capacitive actuation is used to self-test a capacitive micromechanical pressure sensor. How large voltage needs to be applied over $d = 2$ μm gap to obtain a force equivalent of $\Delta p = 0.1$ kPa of pressure.

Exercise 15.7

Show that at the pull-in point, the capacitive and mechanical spring constants cancel.

Exercise 15.8

Plot the capacitive spring constant as a function of displacement for the accelerometer in Example 6.6.

Exercise 15.9

Redo Examples 15.8 and 15.9 but assume that the measurement voltage is $v_0 = 1$ V.

Exercise 15.10

For a low voltages $V < V_C/5$, the linear electrostatic spring constant is approximately $k_{0e} \approx \frac{\epsilon A}{2d^3} V^2$. Noting that the resonator resonance frequency is $f = \sqrt{(k + k_{0e})/m} = f_0 \sqrt{1 + k_{0e}/k}$, derive an expression for the resonance frequency change $(f - f_0)/f_0$ as a function of bias voltage and characteristic voltage V_C only. How large should the bias voltage be to change the resonance frequency by 1%?

16

Piezoelectric actuation

Piezoelectric actuators are based on piezoelectric materials that deform when electric fields are applied over them. This effect is small. Typical strains are only 10^{-5} for electric fields of 100 V/mm. Piezoelectric actuators therefore require large voltages to generate significant displacements. While the displacement is small, the piezoelectric actuation does not consume dc power, provides instantaneous force in response to the electric field, and generates large forces. These advantages are utilized, for example, in adaptive optics and inkjet printers where a small displacement is sufficient. Piezoelectric actuation is also used in resonators that are used as time references, RF filters, or gyroscopes.

The piezoelectric sensing was covered in Chapter 7 and this chapter covers the piezoelectric actuation. First, we will review the relationship between the stress, the strain, and the electric field in piezoelectric materials. Expressions for the actuation force are derived for the two common cases: Longitudinal actuators, where the force is in the direction of the electric field and transverse actuator where the force is perpendicular to the electric field. The focus of this chapter is on dc actuation. The resonant actuation is covered later in Chapter 20 where microresonators are introduced.

16.1 Actuation force

The stress in piezoelectric materials is governed by the constitutive equation

$$T = ES - e\mathcal{E}. \tag{16.1}$$

Here E is the Young's modulus, S is the strain, e is the piezoelectric coefficient, and $\mathcal{E}$ is the electric field. By comparing Equation (16.1) to $T = ES$

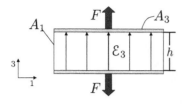
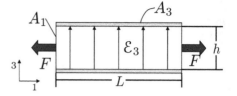

(a) Longitudinal configuration. The force is in the direction of electric field.

(b) Transverse configuration. The force is perpendicular to the electric field.

Figure 16.1: The two common configurations for piezoelectric actuators. The piezoelectric material is sandwiched between two electrodes.

for non-piezoelectric materials, we see that the piezoelectric effect generates an additional stress term $-e\mathcal{E}$.

As with the piezoelectric sensing, a three dimensional analysis of piezoelectric actuators requires matrix formalism to account for the anisotropic nature of piezoelectric materials [59]. Luckily, an approximate one-dimensional analysis is an excellent starting point for the majority of practical piezoelectric actuators.

The two common transducer configurations are shown in Figure 16.1. Figure 16.1(a) shows the longitudinal actuator where the electric field is in the direction of the generated force. Figure 16.1(b) shows the transverse actuator where the electric field is perpendicular to the generated force. As is custom for the piezoelectric materials, the axis directions X, Y and Z are indicated with numbers 1, 2, and 3, respectively.

16.1.1   Longitudinal actuator

The longitudinal configuration in Figure 16.1(a) has electric field in the direction of stress. To calculate the stress, we will start by rewriting Equation (16.1) to explicitly show the material directions:

$$T_3 = ES_3 - e_{33}\mathcal{E}_3. \tag{16.2}$$

The piezoelectric coefficient e_{33} couples the parallel stress and the electric field. The first term in Equation (16.2) is the regular stress due to material deformation and the second term is the stress due to the piezoelectric effect. In Equation (16.2), we have ignored the effect of lateral stresses S_1 and S_2 in piezoelectric coupling. But as shown in Appendix G, this effect is usually small.

If the piezoelectric material is free to deform, the total stress is zero ($T_3 = 0$) and the material stress ES and piezoelectric stress $-e_{33}\mathcal{E}_3$ in Equation (16.2) will cancel. By setting $T_3 = 0$ in Equation (16.2), the strain is

$$S_3 = \frac{e_{33}}{E}\mathcal{E}_3 = \frac{e_{33}}{Eh}v, \tag{16.3}$$

where we have used the voltage over the electrodes $v = \mathcal{E}_3 h$ in the last step. In the case of uniform strain, we have $S_3 = \Delta h / h$ and Equation (16.3) gives

$$\Delta h = \frac{e_{33} h}{E} \mathcal{E}_3 = \frac{e_{33}}{E} v. \tag{16.4}$$

Equation (16.4) shows that the obtained displacement depends only on the piezoelectric coefficient e_{33}, Young's modulus E, and voltage v – not on transducer dimensions. To obtain large displacement with the longitudinal transducer, large voltages must be used. Alternatively, multiple transducers can be stacked to increase the displacement as illustrated in Figure 16.2. In the stacked actuator, the direction of the electric field is in opposite directions in the adjacent actuators. Thus, the adjacent actuators need to have opposite orientations so that each element generates displacement in the same direction.

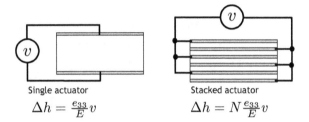

Figure 16.2: The actuator displacement for given voltage can be increased by stacking several actuators.

Example 16.1: Piezoelectric pipette actuator
Problem: Figure 16.3 shows a micromechanical pipette that combines silicon micromachining with milled metal parts [128]. Even with the thin silicon membrane, large actuation forces are needed. Initially, the capacitive actuation was considered but generation of sufficient force was a challenge and the piezoelectric actuation was eventually adopted. The PZT actuator presses against the silicon membrane. When voltage is applied over the PZT, it forces the membrane to deflect which pumps fluid through the needle. Assuming that the PZT actuator is stiff in comparison to the silicon membrane, how large is the voltage necessary to obtain 50 nm membrane displacement?

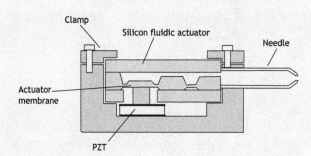

Figure 16.3: Micromachined pipettor made of two micromachined silicon wafers that are bonded together and actuated with a PZT4 plate [128].

Solution: The actuator displacement is given by Equation (16.4). Using, $\Delta h = 50$ nm, $E = 110$ GPa, and $e_{33} = 15.1$ C/m^2 from Table 7.1 on page 111, we have

$$v = \Delta h \frac{E}{e_{33}} \approx 364 \text{ V.}$$

This voltage is high even for laboratory use. To lower the voltage, a stacked actuator shown in Figure 16.2 could be used. For example, a stack of twenty actuators would require voltage of only 18 volts.

It is often useful to model the piezoelectric effect as a force acting on the boundaries of the transducers as shown in Figure 16.4. In this book, we use the sign notation that the positive force at the material boundary stretches the material and the resulting tensile stress has positive sign. With this sign notation, the equivalent force at the boundary that would cause the same stress as the piezoelectric effect is

$$F_{\text{piezo}} = e_{33} A_3 \mathcal{E}_3 \tag{16.5}$$

$$= \frac{e_{33} A_3}{h} v, \tag{16.6}$$

where A_3 is the electrode area and $v = \mathcal{E}_3 h$ is the voltage over the electrodes.

The force analog shown in Figure 16.4 can greatly simplify the analysis of piezoelectric actuators as lumped forces are often easier to analyze than distributed stresses. The lumped force model is exemplified in Examples 16.2 and 16.3. In Chapter 20, we will utilize the lumped force model to analyze piezoelectrically actuated microresonators.

It is interesting to compare the piezoelectric force to the capacitive actuation. According to the Equation (16.5), the force generated can be positive or negative depending on the sign of the actuation voltage v. This is a major difference to capacitive actuation where the force is proportional to voltage squared, and only attractive forces can be generated.

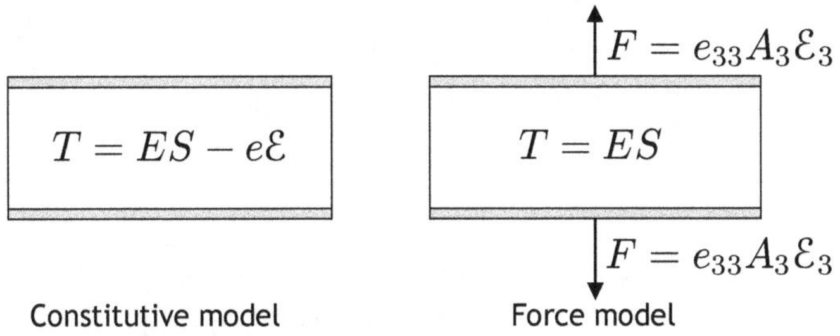

Figure 16.4: Force model for piezoelectric materials. Mathematically equivalent results are obtained by modeling the stress-strain relationship as for a non-piezoelectric material and applying a "piezoelectric" force to the electrodes.

Example 16.2: Equivalent actuator force

Problem: Recalculate the actuator displacement in Example 16.1 using the force analog approach shown in Figure 16.4 assuming that the actuation voltage is $v = 364$ V, the actuator area is $A_3 = 16$ mm^2, and the actuator height is $h = 1$ mm.

Solution: The force from Equation (16.5) is

$$F_{\text{piezo}} = \frac{e_{33} A_3}{h} v \approx 88 \text{ N}.$$

From Equation (4.19), the longitudinal spring constant for the PZT plate is

$$k = \frac{EA}{L} \approx 1.8 \text{ GN/m}.$$

The displacement is

$$\Delta h = \frac{F}{k} \approx 50 \text{ nm},$$

which agrees with Example 16.1.

Example 16.3: Clamped piezoelectric actuator

Problem: Figure 16.5 shows a piezoelectric PZT4 actuator clamped between two metal plates with four screws. Calculate the actuator displacement for an actuation voltage of $V = 200$ V assuming that the metal plates will not deform but screws holding the plates will. The PZT4 actuator has the area

$A = 3$ mm$\times 3$ mm and thickness $h = 4$ mm. The four screws are made of steel ($E = 200$ GPa) and have the length $L = 8$ mm and the diameter $D = 1$ mm.

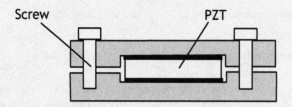

Figure 16.5: Clamped piezoelectric actuator. Two of the four screws are shown.

Solution: The net spring constant for the four screws is

$$k_s = 4E\frac{A}{L} \approx 79 \text{ MN/m}.$$

From Table 7.1, we have $E = 110$ GPa and $e_{33} = 15.1$ C/m^2 for the PZT4. The spring constant for the PZT actuator is

$$k_{\text{PZT}} = E\frac{A}{h} \approx 248 \text{ MN/m}.$$

The total spring constant is $k_{\text{tot}} = k_s + k_{\text{PZT}} \approx 326$ MN/m. The actuation force is

$$F = \frac{e_{33}A_3}{h}V \approx 6.8 \text{ N}$$

The actuator displacement is $x = F/k_{\text{tot}} \approx 21$ nm.

The calculated displacement can be compared to the freely deforming PZT without the screws. From Equation (16.4), the displacement is

$$\Delta h = \frac{e_{33}}{E}V \approx 27 \text{ nm}.$$

The same result would have been obtained from $\Delta h = F/k_{\text{PZT}} \approx 27$ nm.

16.1.2   Transverse actuator

The transverse actuator in Figure 16.1(b) has an electric field orthogonal to the stress. To calculate the stress, we write Equation (16.1) to explicitly show the material directions:

$$T_1 = ES_1 - e_{31}\mathcal{E}_3. \tag{16.7}$$

The first term ES_1 in Equation (16.7) is the regular stress due to the material deformation and the second term $e_{31}\mathcal{E}_3$ is due to the piezoelectric effect. The piezoelectric coefficient e_{31} couples the electric field and stress that are independent in normal materials.

If the piezoelectric material is free to deform, the total stress is zero, and the material stress and piezoelectric stress in Equation (16.7) will cancel. By setting $T_1 = 0$ in Equation (16.7), the strain is

$$S_1 = \frac{e_{31}}{E}\mathcal{E}_3 = \frac{e_{31}}{Eh}v, \tag{16.8}$$

where we have used $v = \mathcal{E}_3 h$. In the case of uniform strain, we have $S_1 = \Delta L/L$ and Equation (16.8) gives

$$\Delta L = \frac{e_{31}L}{E}\mathcal{E}_1 = \frac{e_{31}L}{Eh}v. \tag{16.9}$$

Unlike the longitudinal actuator, the dimensions affect the displacement. To obtain a large stroke, the length-to-thickness ratio L/h should be maximized.

The force analog also applies to the transverse actuators as shown in Figure 16.1(b). The piezoelectric actuator is modeled as a non-piezoelectric material ($T_1 = ES_1$) with the equivalent "piezoelectric force" applied to the boundaries. The analogous force is

$$F_{\text{piezo}} = e_{31}A_1\mathcal{E}_3 \tag{16.10}$$

$$= \frac{e_{31}A_1}{h}v. \tag{16.11}$$

where A_1 is the cross sectional transducer area and $v = \mathcal{E}_3 h$ is the voltage over the electrodes.

Example 16.4: Ultrasonic surgical tool
Problem: Figure 16.6 shows an ultrasonic surgical tool that is designed for cataract surgery [129]. The horn is excited to vibrate in the longitudinal direction so that the vibrating tip will cut hard tissues. To obtain the large forces necessary to stretch the silicon horn, a PZT actuator is used. Estimate the force from the PZT5 plate with length of 3 cm, width of 2 cm, and height of 2 mm if the actuation voltage is 30 V.

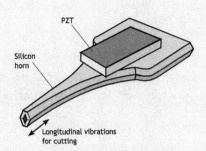

Figure 16.6: Micromachined ultrasonic surgical tool made of two bulk micromachined silicon wafers that are bonded together [129].

Solution: The force from Equation (16.10) is

$$F = \frac{e_{31}A_1}{h}v = e_{31}wv = -3.2 \text{ N},$$

where we have used $e_{31} = -5.4$ C/m^2, $v = 30$ V, $A_1 = wh$, and $w = 2$ cm.

Key concepts

- Piezoelectric actuation generates large forces and does not consume dc power.

- Piezoelectric actuators can be modeled by applying equivalent force at the actuator boundary.

- Unlike capacitive actuators that generate only attractive forces, piezoelectric force can be positive or negative.

- Piezoelectrically induced strains and displacements are small. Large actuation voltages may be required to obtain a sufficient stroke.

- Larger displacement for the given voltage is obtained by stacking the piezoelectric actuators.

Exercises

Exercise 16.1
Calculate the longitudinal and transverse forces for a PZT4 plate with length of 2 cm, width of 0.4 cm, and height of 1 mm if the actuation voltage is 20 V in the height direction.

Exercise 16.2

Design a piezoelectric stack actuators for optical systems that will have a 0.5-μm displacement at 80 V actuation voltage.

Exercise 16.3

Piezoelectric actuators are capable of generating high forces. Design an actuator capable of lifting an object that weights 100 kg a distance of 10 μm. Assume that the maximum available voltage is $V_{max} = 500$ V.

Exercise 16.4

Both capacitive and piezoelectric actuators generator instantaneous force in response to a excitation voltage. What limits the actuator response time? If a piezoelectric actuator spring constant is larger than that of a otherwise equal capacitive actuator, which actuator is faster?

17

Thermal actuation

The thermal actuators are simple to implement, operate at low voltages, and provide large forces. While commercial success has thus far been elusive, the robustness of thermal actuators has made them popular in academic research. The actuation is based on the thermal expansion of the transducer. Usually the heating is achieved by passing current through the device to cause resistive heating. The dimensional change is small but the displacement can be mechanically amplified, for example, by using a bi-morph that is based on two materials with different thermal expansion coefficients.

The drawbacks of thermal actuation are the large power consumption, typically measured in milliwatts, and the slow response time, limited by the heating and cooling time constants. These drawbacks are inherent to the actuation scheme and explain why thermal actuators have had only limited success outside academia.

In this chapter, the thermal actuation is analyzed in the context of microactuators. First, a simple model for the thermal expansion of a rod is given. Next, we cover methods to amplify the small thermal expansion induced displacement. To gain insight in power consumption and the response time trade-offs, we will also analyze the transient response of a thermal actuator. Finally, the chapter is concluded with notes on nonlinear effects and bi-stable actuators.

17.1 Principle of operation

The operation principle of a thermal actuator is illustrated in Figure 17.1. A rod is heated to induce thermal expansion. The obtained displacement can be estimated from the strain due to the temperature change ΔT. The thermal

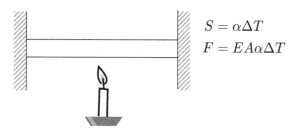

Figure 17.1: Thermal expansion results in stress S. A clamped rod will exert force F on boundaries (After [6]).

strain is

$$S_{XX} = \alpha \Delta T, \tag{17.1}$$

where α is the linear thermal expansion coefficient. For single-crystal silicon at room temperature, the thermal expansion coefficient is $\alpha = 2.6 \cdot 10^{-6}/\text{K}$. For metals, the thermal expansion coefficients are higher, typically $\alpha = 10 - 30 \cdot 10^{-6}/\text{K}$.

If the material is not allowed to expand, the total strain must be zero. Adding the thermal strain and stress induced strain we have

$$S_{XX} = \alpha \Delta T + \frac{T_{XX}}{E} = 0 \tag{17.2}$$

or

$$T_{XX} = -E\alpha \Delta T. \tag{17.3}$$

The negative stress indicates that the rod is compressed by the boundaries. The reactive force acting on the boundaries is

$$F = A T_{XX} = E A \alpha \Delta T. \tag{17.4}$$

Equation (17.4) and Example 17.1 show that the thermal expansion can generate large forces but the absolute displacement from linear expansion alone is small.

The energy required for heating the sample is

$$\Delta W = c_P \rho V \Delta T, \tag{17.5}$$

where ΔW is the thermal energy, c_P is the heat capacity, ρ is the density, V is the sample volume, and ΔT is the temperature change. Practical actuators lose significant energy to boundaries and Equation (17.5) provides the minimum energy required.

Example 17.1: Linear thermal actuator
Problem: Calculate the thermal expansion induced displacement for a silicon rod with $A = 8~\mu m^2$ and $L = 200~\mu m$ assuming temperature change $\Delta T = 100$ K. What is the reactive force pushing the boundaries if it is not allowed to expand? What is the thermal energy change in the beam due to the temperature change? Use $E = 170$ GPa for silicon.
Solution: From Equation (17.1), the strain is

$$S = \alpha \Delta T$$

and from Equation (4.8) the beam displacement is

$$\Delta L = SL = \alpha \Delta T L = 52 \text{ nm.}$$

The force that the rod extends on the boundaries from Equation (17.4) is

$$F = EA\alpha\Delta T = 0.35 \text{ mN.}$$

Compared to the μN:s from capacitive actuators, this force is high. Thermal energy from Equation (17.5) is

$$\Delta W = c_P \rho V \Delta T = 0.26~\mu\text{J.}$$

17.2   Leverage for large displacement

As Example 17.1 showed, the thermal actuators can generate large forces but small displacements. To obtain larger displacements, bi-morphs or geometrical amplification can be used as is illustrated in Figure 17.2. For example, large displacement is obtained if the actuator has a hot arm and cold arm as shown in Figure 17.2(a) [115]. When current is passed through the actuator, the thin arm will heat more than the thick arm. This asymmetry causes large lateral movement.

The second example illustrated in Figure 17.2(b) uses a chevron shaped actuator to geometrically amplify the small length change [130]. This actuator is further investigated in Example 17.2. The third example shown in Figure 17.2(c) is a bi-morph that is made of two dissimilar materials [131]. When the device is heated, the difference in thermal expansion causes the beam to curl. Again, the obtained displacement can be much larger than the length change of the cantilever.

(b) Chevron shaped actuator. Two arms meet in a small angle to amplify the motion.

(a) Asymmetric actuator. Narrow arm heats and expands more than the thick arm.

(c) Bi-morph actuator. Two materials have different thermal expansion coefficients and the actuator bends when heated.

Figure 17.2: Methods to leverage the small thermal expansion into larger displacements [115, 130, 131].

Example 17.2: Chevron shaped actuator

Problem: Figure 17.3 shows a chevron shaped actuator that amplifies the small displacement ΔL. Assuming a silicon transducer is heated by $\Delta T = 300$ K, $y_0 = 10$ μm and $L_0 = 200$ μm, what is the displacement Δy?

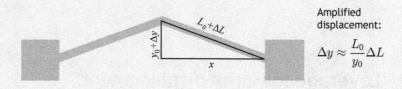

Amplified displacement:

$$\Delta y \approx \frac{L_0}{y_0}\Delta L$$

Figure 17.3: Chevron shaped actuator amplifies the thermal expansion induced displacement ΔL.

Solution: The strain due to thermal expansion from Equation (17.1) is

$$S = \alpha \Delta T \approx 0.00078$$

and the extension is

$$\Delta L = SL \approx 0.16 \ \mu\text{m}.$$

The amplified displacement Δy is obtained from

$$x^2 + y^2 = L^2.$$

Using $x^2 = L_0^2 - y_0^2$, $y = y_0 + \Delta y$, and $L = L_0 + \Delta L$ and neglecting small terms Δy^2 and ΔL^2 gives

$$\Delta y \approx \frac{L_0}{y_0}\Delta L \approx 3.1 \ \mu\text{m}.$$

The Chevron actuator amplifies the linear displacement by a factor of $3.1/0.16 \approx 20$.

17.3 Transient analysis

We will now turn our attention to the actuator response time and power consumption. The thermal actuator is inherently slow as there is a lag between change in excitation voltage and the temperature change. The heating can be sped up by increasing the heating power but the cooling time is set by the heat capacity and thermal conductivity of the sample.

To explore the power consumption vs. response time trade-offs, we'll analyze a simplified model for the chevron actuator shown in Figure 17.4. The actuation consists of three cycles. In the first cycle, current is passed through the actuator to heat the device. As the actuator becomes warmer, it will thermally expand. In the second cycle, the actuator has reached a steady state temperature where the heat transferring out of the device equals the heating power. In the third actuation cycle, the heating is turned off and the actuator will cool down.

As the actuator is symmetrical around the middle point, only half of the actuator needs to be modeled. We will first obtain the steady state displacement for given input power. Next, we will analyze the transient turn-on and turn-off responses. In the analysis, only the conductive heat transfer is considered and the convection and radiation are ignored. The anchors act as a heat sink and will maintain a constant temperature at the right boundary $(x = L)$. Due to symmetry in the half model, no heat will flow through the left boundary.

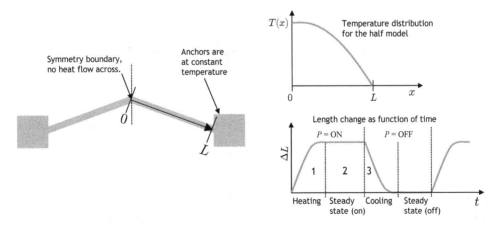

Figure 17.4: Utilizing symmetry, we need to model only half of the chevron actuator. The heating and cooling times limit the actuator frequency response.

17.3.1 Steady state

In the steady state, the heat flow from the actuator balances heat influx due to heating. We assume that the sample is heated with uniform heat power per unit length $P_L = P/L$ where P is the total heating power in the half of the actuator and L is the half actuator length. Referring to Figure 17.4, the heating power to the left of location x is

$$\dot{W} = P_L x = \frac{P}{L} x. \tag{17.6}$$

In the steady state, this heat power must flow out of the actuator. The heat flow through at location x must therefore be equal to the heating power $\dot{W}$ to the left of location x.

The heat flow is governed by the thermal diffusion equation

$$\dot{W} = -\kappa A \frac{\partial T}{\partial x}, \tag{17.7}$$

where κ is the thermal conductance which for silicon is $\kappa = 130$ W/mK and A is the cross sectional area. Equating (17.6) and (17.7) to balance out the heat input and output and solving for the temperature gives

$$T(x) = \frac{P}{2\kappa A L} \left(L^2 - x^2 \right) = T_{MAX} \frac{L^2 - x^2}{L^2}, \tag{17.8}$$

where $T(x)$ is the temperature relative to the anchor temperature. The maximum temperature change at the actuator center is

$$T_{MAX} = \frac{PL}{2\kappa A}. \tag{17.9}$$

Integrating the strain $S(x) = \alpha \Delta T(x)$ over the half length L gives the length change

$$\Delta L = \int_0^L \alpha T(x) \mathrm{d}x = \alpha T_{MAX} \frac{L^3 - L^3/3}{L^2} = \alpha T_{MAX} \frac{2L}{3}. \tag{17.10}$$

17.3.2 Heating

Transient analysis of heating and cooling requires solving the time dependent heat diffusion equation

$$\frac{\partial T}{\partial t} = \frac{\kappa}{\rho c_P} \frac{\partial^2 T}{\partial x^2} + \frac{P_V}{\rho c_P}, \tag{17.11}$$

where P_V is the heating power per unit volume. Solving Equation (17.11) analytically is possible but involves infinite Fourier series [132]. Here we will take an approximate approach to obtain closed form expression for the heating and cooling times. This approach is satisfactory as any accurate analysis would require modeling of the temperature dependencies of α, κ and c_P which makes the problem analytically intractable. The approximations that we will make in the transient analysis are smaller than the errors from neglecting the temperature dependencies of these material parameters.

A simple approximate solution to the heating problem is obtained by assuming that the shape of the temperature distribution is given by Equation (17.8). We can then write the temperature as

$$T(x,t) = T(x)u(t) = T_{MAX}\frac{L^2 - x^2}{L^2}u(t) \qquad (17.12)$$

where $T(x)$ is the shape of the temperature distribution given by Equation (17.8) and $u(t)$ is a still unknown time dependency. We will solve the time dependency in two steps: First, we will calculate the net heat flux entering the system

$$P_{net} = P - P_{out} \qquad (17.13)$$

where P is the heating power and P_{out} is the heat exiting the system through the right boundary. In the second step, we will solve the thermal energy rate of change $\dot{W} = P_{net}$ based on the heat capacity and the net heat flux.

The heat power entering into the system is P. The heat exiting the system at the anchor location $x = L$ from Equations (17.7) and (17.12) is

$$\dot{W}\Big|_{x=L} = -\kappa A\frac{\partial T}{\partial x}\Big|_{x=L} = -T_{MAX}\frac{2\kappa A}{L}u(t). \qquad (17.14)$$

The net heat flux from combining the heat input and outflow is

$$\dot{P}_{net} = P - T_{MAX}\frac{2\kappa A}{L}u(t). \qquad (17.15)$$

The heat energy is related to the temperature by Equation (17.5). The thermal energy per length dL is $dW(x) = c_P\rho T(x)AdL$. Integrating this heat energy over the sample length gives

$$W = \frac{2}{3}T_{MAX}c_P\rho ALu(t). \qquad (17.16)$$

The heat energy rate of change is the time derivative of Equation (17.16):

$$\dot{W} = \frac{2}{3}T_{MAX}c_P\rho AL\dot{u}(t). \qquad (17.17)$$

Equating (17.15) and (17.17) gives a differential equation for the heat energy rate of change $\dot{W} = P_{net}$:

$$P - T_{MAX}\frac{2\kappa A}{L}u(t) = \frac{2}{3}T_{MAX}c_P\rho AL\dot{u}(t). \qquad (17.18)$$

A solution satisfying Equation (17.18) is

$$u(t) = 1 - e^{-\lambda_h t}, \tag{17.19}$$

where

$$\lambda_h = \frac{3\kappa}{\rho c_P L^2}. \tag{17.20}$$

In summary, the approximate sample temperature during the heating cycle is

$$T(x,t) = T(x)u(t) = T_{MAX}\frac{L^2 - x^2}{L^2}\left(1 - e^{-\lambda_h t}\right) \tag{17.21}$$

and

$$\Delta L(t) = \Delta L\left(1 - e^{-\lambda_h t}\right), \tag{17.22}$$

where ΔL is given by Equation (17.10).

17.3.3 Cooling

The cooling is also characterized by the heat diffusion. With no heating, Equation (17.11) simplifies to

$$\frac{\partial T}{\partial t} = \frac{\kappa}{\rho c_P}\frac{\partial^2 T}{\partial x^2}, \tag{17.23}$$

and the initial conditions are given by Equation (17.8). The exact solution requires the use of Fourier series but a simple approximate solution is obtained by approximating the initial conditions by

$$T(x) = T_{MAX}\cos\frac{\pi x}{2L}. \tag{17.24}$$

In Equation (17.24), the $\cos\frac{\pi x}{2L}$-term is in fact the basis function for the first Fourier series term. In this case, the $\cos\frac{\pi x}{2L}$ models the initial conditions quite accurately and including the higher order Fourier terms would not change the results significantly.

Substituting $T(x,t) = T(x)u(t)$ to Equation (17.23) and solving for $u(t)$ gives

$$u(t) = e^{-\lambda_c t}, \tag{17.25}$$

where

$$\lambda_c = \frac{\pi^2 \kappa}{4\rho c_P L^2}. \tag{17.26}$$

The time constant for cooling is almost equal to time constant for heating. In summary, the sample temperature during the cooling cycle is

$$T(x,t) = T(x)u(t) \approx T_{MAX}\frac{L^2 - x^2}{L^2}e^{-\lambda_c t} \tag{17.27}$$

$$\approx T_{MAX}\cos\frac{\pi x}{2L}e^{-\lambda_c t} \tag{17.28}$$

and the length change as a function time is

$$\Delta L(t) = \Delta L e^{-\lambda_c t}, \tag{17.29}$$

where ΔL is given by Equation (17.10).

Example 17.3: Thermal actuator

Problem: Consider the thermal actuator of Figure 17.4 and assume the actuator length $L = 300$ μm, the height $h = 5$ μm, and the width $w = 3$ μm, and the heating power $P = 2$ mW for the half actuator (the total heating power is $P = 4$ mW). Calculate the maximum temperature, the length change, and the time constants for the heating and cooling cycles. Assume $\alpha = 2.6 \cdot 10^{-6}$, $\kappa = 130$ W/m·K, and $c_P = 700$ J/kg·K for single-crystal silicon. What is the amplified displacement if $y_0 = 30$ μm?

Solution: From Equation (17.8), we have the maximum temperature at the center of the actuator

$$T_{MAX} = \frac{PL}{2\kappa wh} \approx 154 \text{ K}.$$

The total length change from Equation (17.10) is

$$\Delta L = \alpha T_{MAX} \frac{2L}{3} \approx 80 \text{ nm}.$$

The amplified displacement is $\Delta y \approx \frac{L_0}{y_0} \Delta L \approx 0.8$ μm.

The time constants for the heating and cooling given by Equations (17.20) and (17.26) are

$$\tau_h = \frac{1}{\lambda_h} = \left(\frac{3\kappa}{\rho c_P L^2} \right)^{-1} \approx 0.38 \text{ ms}$$

and

$$\tau_c = \frac{1}{\lambda_c} = \left(\frac{\pi^2 \kappa}{4\rho c_P L^2} \right)^{-1} \approx 0.46 \text{ ms}.$$

17.4   Higher order models

When the materials are heated by several hundred degrees Kelvin, the temperature dependency of the material parameters is significant. For example, the thermal conductivity of polycrystalline silicon reduces to half when it is heated from 300 K to 520 K [133]. Accounting for these nonlinear effects analytically is very cumbersome and simulations are preferred if the large temperature effects need to modeled. This partially justifies the approximate treatment of the transient analysis in the previous section that provides approximate but usable closed form expressions for the thermal actuator. For further information about modeling the nonlinear effects, the reader is referred to reference [133].

17.5   Bi-stable actuators

The large power consumption is a major drawback for the thermal actuators. One actuator can consume more than 10 mW of power and often several actuators are combined in parallel to obtain a larger force and to guide the motion. The resulting power consumption is not compatible with low power applications that demand μW power consumption. One way to reduce the total power consumption is to make the thermal actuator bi-stable so that power is consumed only during the switching. Figure 17.5 shows a popular implementation for a bi-stable actuator. The actuator has bend springs so that when the actuator is moved, the springs will "snap" between two stable position [134, 135]. In addition to saving power, the bi-stable switch has the advantage of maintaining its position even if power is lost. This can be a design requirement for optical switches and electrical relays. The bi-stable spring design can also be utilized in other, for example, capacitive actuators.

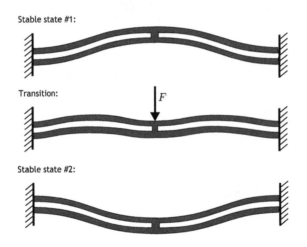

Figure 17.5: A bi-stable MEMS actuator with nonlinear springs. The actuator "snaps" from one stable position to another and it stays in a stable position even without power.

Key concepts

- Thermal actuator offers large forces but the displacements from the thermal expansion alone are small.

- The power consumption and response time for the thermal actuators are poor in comparison to capacitive and piezoelectric actuators.

- Asymmetric actuators and bi-morphs can be used to obtain larger displacements.

- Bi-stable actuators eliminate the dc power consumption needed to hold the actuator in the actuated state.

Exercises

Exercise 17.1
Assuming that the maximum temperature change for a thermal actuator is $\Delta T = 300$ K, calculate the displacement for a 300-μm long thermal actuator made of i. silicon and ii. nickel.

Exercise 17.2
Calculate the thermal expansion induced displacement for a nickel rod with the cross sectional area $A = 8$ μm^2 and the length $L = 200$ μm assuming temperature change $\Delta T = 100$ K. What is the force that the rod extends on boundaries if it is not allowed to expand? Compare your results to the silicon rod in Example 17.1.

Exercise 17.3
Design a thermal actuator that is based on the chevron shape to amplify displacement. Your actuator should achieve a 10-μm displacement with a $\Delta T = 150$ temperature change.

Exercise 17.4
In Example 17.2, the displacement magnification in a chevron shaped actuator is shown to be $\frac{L_0}{y_0}$. What limits the maximum magnification that can be obtained (=why y_0 cannot be arbitrarily small)?

Exercise 17.5
Modify the actuator in Example 17.3 to obtain a 5-μm amplified displacement without increasing the heating power. What are the heating and cooling time constants for the new actuator?

Exercise 17.6
An ideal thermal actuator material should maximize the displacement ΔL at given heating power while minimizing the response time constant $\tau = 1/\lambda$. What materials maximize the $\Delta L \cdot \lambda$-product?

18

Micro-optical devices

Many optical devices utilize mechanically tunable components. For example, cameras have adjustable lenses to change the image magnification and the focal length. Micromechanical components can be used to control light on a small scale. The most successful optical MEMS product has been the micromirror matrix from Texas Instruments that is used in projection displays. The device contains over one million tiny mirrors with a single mirror corresponding to a display pixel. Other optical MEMS applications include laser scanners, fiber optical networks, and interferometry.

In this chapter, the optical principles relevant to micro-optics are reviewed. We will first study the Huygens' principle used to qualitatively study the effect of small size on optical performance. Next, Gaussian beam optics is introduced to quantify how quickly light beams from small optical sources spread. The fundamental limitations of small optical elements are illustrated by analyzing several applications including displays, scanners, and optical switch matrices.

18.1 Huygens' principle

The Huygens' principle is a physically intuitive and mathematically accurate method to describe wave propagation. It states that each point of an advancing wave front acts as a point source for a new wave. Figure 18.1(a) illustrates the Huygens' principle for a plane wave that travels the distance $\Delta x = c\Delta t$ in time Δt. The new wave front is constructed by using the old wave front as a location for point sources. Each point source generates waves that travel spherically in all directions. By summing the spherical waves, a new wavefront is created. The Huygens' principle can also be applied to curved wavefronts as is illustrated on the right in Figure 18.1(a).

Planar wavefront:

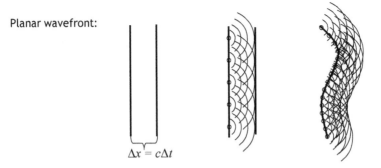

$\Delta x = c\Delta t$

(a) A new wavefront is obtained by summing the spherical waves from the previous wavefront.

Point source through
an aperture:

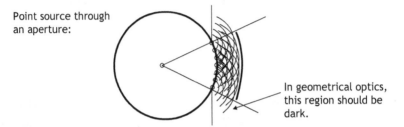

In geometrical optics,
this region should be
dark.

(b) Waves traveling through openings are analyzed by taking the opening as a source of new waveronts.

Figure 18.1: Huygens' principle: Every point on a wavefront may be regarded as a secondary source of wavelets.

Huygens' principle is useful for analyzing the wave propagation (usually light or radio waves) through an opening as illustrated in Figure 18.1(b). In this case, the opening is analyzed as a collection of point sources that emit waves in all directions. The Huygens' principle shows that some light will also enter into the "shadow region" which in classical ray optics is completely dark. This explains how radio waves can "bend" around the corner and demonstrates how the edge of a shadow is never sharp.

The implications of Huygens' principle in micro-optics are illustrated in Figure 18.2. When the light source is large, the wave propagates as a uniform wave and ray optics is a good approximation. When the light source is small, however, it looks more like a point source that spreads light in all directions. The significance to micro-optics are obvious; the smaller the (micro)mirror, the less it acts as a mirror and more as a light scatterer.

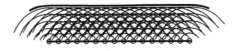

Wave from a large light source
such as mirror is close to plane
wave and ray optics apply.

A small light source acts like
a point source.

Figure 18.2: A small light source such as small mirror acts more like a point source than plane wave source.

18.2　Gaussian beam optics

The Huygens' principle covered in the previous section is an accurate but mathematically tedious method for analyzing the light propagation. Calculations based on the Huygens' principle requires integration of the spherical waves from the previous wave front. Even in simple geometries, the problems quickly become mathematically too complex to give physical insights.

Gaussian beam optics provide a mathematically simple way to analyze the wave propagation in microsystems. It is based on the assumption that the light beam intensity has a Gaussian distribution away from the beam center as shown in Figure 18.3. The beam spreads as it propagates away from the source; the shape of the beam, however, remains a Gaussian.

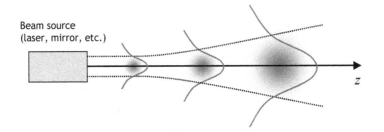

Beam source
(laser, mirror, etc.)

z

Figure 18.3: Many light sources such as laser diodes emit light that has Gaussian intensity distribution. The Gaussian beams have the property that while the beam spreads as the light travels, the shape of the beam remains Gaussian.

Figure 18.4(a) shows the distribution of the electric field $\mathcal{E}$ in a Gaussian beam. The electric field at distance r from the beam center is

$$\mathcal{E} = \mathcal{E}_0 \exp\left(-\frac{r^2}{w^2(z)}\right), \tag{18.1}$$

where $\mathcal{E}_0$ is the electric field at the beam center and w is the beam width defined as the distance where the electric field $\mathcal{E}$ drops to $1/e$ [136]. Recalling that the

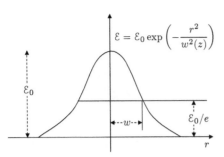

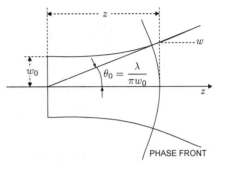

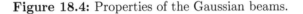

(a) Amplitude distribution of the Gaussian beam.

(b) The beam width w increases with the distance z from the beam waist.

Figure 18.4: Properties of the Gaussian beams.

intensity of the light is $I \sim \mathcal{E}^2$, the intensity of the Gaussian beam is

$$I = I_0 \exp\left(-\frac{2r^2}{w^2(z)}\right), \tag{18.2}$$

where I_0 is the intensity at the beam center.

As the beam travels, it expands and its width w grows. As is illustrated in Figure 18.4(b), the beam is the narrowest at the beam waist where $w = w_0$. At distance z from the waist, the beam width is

$$w^2(z) = w_0^2\left[1 + \left(\frac{\lambda z}{\pi w_0^2}\right)^2\right], \tag{18.3}$$

where λ is the wavelength.

Due to the spreading of the beam, the electric field at the beams center $\mathcal{E}_0$ decreases. At distance z from the waist, the electric field at the center of the beam is

$$\mathcal{E}_0(z) = \mathcal{E}_0(0)\frac{w_0}{w}, \tag{18.4}$$

where $\mathcal{E}_0(0)$ is the electric field at the beam center at the waist $(z = 0)$.

Referring to Figure 18.4(b), the angle at which the beam waist spreads is

$$\theta_0 = \tan^{-1}\frac{w}{z} \approx \frac{\lambda}{\pi w_0}, \tag{18.5}$$

where the approximation is valid for $z \gg \pi w_0^2/\lambda$. As Equation (18.5) shows, a small waist w_0 means that the angle θ_0 is large and the beam spreads quickly. This is in agreement with the with Huygens' principle that results in small sources behaving more like point sources that spread light in all directions.

By combining Equation (18.1) and (18.5), the electric field can be written in terms of the angle as

$$\mathcal{E} = \mathcal{E}_0 \exp\left(-\frac{\theta^2}{\theta_0^2}\right). \tag{18.6}$$

Equation (18.6) will be used in subsequent sections for analyzing the micromirror scanners.

Example 18.1: Gaussian beam waist size

Problem: Laser beam with waist $w_0 = 30\ \mu$m is steered with a micromechanical mirror. Show that a circular mirror with a radius of $R = 3w_0/2 = 45\ \mu$m is sufficient to steer 99% of the beam power if the waist is located at the mirror. Calculate the beam width at a distance of $z = 10$ cm from the mirror.

Solution: From Equation (18.2), the beam intensity at distance r from the beam center is

$$I(r) = I_0 \exp\left(-\frac{2r^2}{w^2(z)}\right).$$

Integrating the beam intensity over the area gives the total beam power

$$P_{\text{tot}} = \int_A I(r)\mathrm{d}A = \int_0^\infty I_0 \exp\left(-\frac{2r^2}{w_0^2}\right) 2\pi r \mathrm{d}r \approx I_0 1.414 \cdot 10^{-9}\mathrm{m}^2$$

and the power of the beam hitting the mirror is

$$P_{\text{m}} = \int_0^R I_0 \exp\left(-\frac{2r^2}{w_0^2}\right) 2\pi r \mathrm{d}r \approx I_0 1.398 \cdot 10^{-9}\mathrm{m}^2.$$

The ratio is $P_{\text{m}}/P_{\text{tot}} = 0.99$ showing that 99% of the total beam power hits the mirror.

From Equation (18.3), the beam width w at distance $z = 10$ cm from the mirror is

$$w(z) = \sqrt{w_0^2\left[1 + \left(\frac{\lambda z}{\pi w_0^2}\right)^2\right]} \approx \frac{\lambda z}{\pi w_0} \approx 680\ \mu\text{m}.$$

This example shows that mirror does not need to be large to accommodate the Gaussian beam: a mirror diameter of $D = 3w_0$ is enough to capture 99% of the light if the waist is located that mirror. The beam width from a small source, however, increases quickly. We will see that this has significant implications for many optical devices such as scanners and optical switches.

18.3   Micromirrors

Movable optical mirrors are used for example in laser scanners, projection displays, and adaptive telescopic mirrors. Micromachining is well suited for cost effective manufacturing of miniature movable mirrors at a large scale. This is exemplified by optical displays where more than 100,000 mirrors are needed in a single system. Without the batch manufacturing, the integration of thousands of mirrors would not be cost effective.

In this section, we will focus on two consumer applications for the micromirrors: laser scanners and optical displays. Both applications are mainly consumer market driven. More demanding but historically promising MEMS applications in fiber optical networks are reviewed in Section 18.4.

18.3.1   Optical scanners

Laser scanners are used to read the ubiquitous bar codes. Practically every product including this book has a bar code that is used for the price scan at the register and for controlling the store and supply chain inventory. The laser scanners typically employ either a reciprocating mirror or a rotating prism to move the laser beam back and forth across the bar code. The reflected light is measured with photo diodes to interpret the code.

Micromirrors seem like an ideal solution for low cost laser scanners: The scanning frequency and displacement does not need to be accurately controlled as long as the laser beam moves across the bar code. The hardest part is obtaining sufficient resolution, as the laser beam from the small MEMS mirrors spreads quickly.

Figure 18.5 illustrates the scanner geometry [137–141]. The bar code is located at distance $z = L$ from the mirror. To discern the bar code, the beam size at the bar code has to be smaller than the bar code line spacing. Gaussian beam optics mandates that in order to obtain the small spot size, the beam waist at the mirror needs to be large. Thus, the resolution requirement sets the minimum size for the mirror. The second challenge for a micromirror-based scanner is that the laser needs to scan over the entire bar code, which sets the requirement for the minimum mirror rotation. These challenges are illustrated in Example 18.2.

Currently at least two companies are commercializing MEMS based laser scanners [142, 143]. In addition to the conventional laser scanners, the competition to the MEMS scanners include CCD based solid-state optical imagers that do not require any moving parts. It will take a few years to see whether MEMS scanners are as cost effective as bar code readers.

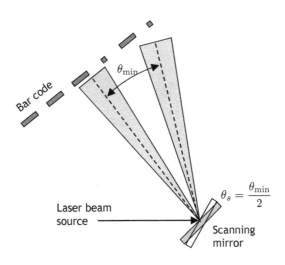

Figure 18.5: Principle of micromirror based bar code scanner. The laser beam is steered over the bar code to deduce the thin and thick lines. The laser beam width at the bar code has to be smaller than the line spacing.

Example 18.2: Bar code laser scanner

Problem: A typical bar code is 4 cm wide and has a 0.5-mm line spacing. Estimate the mirror size and scan displacement requirements if the bar code is to be read from distances of 10 cm to 15 cm with a laser beam that has the wavelength $\lambda = 640$ nm.

Solution: The mirror size requirement is set by the 0.5-mm resolution necessary to read the bar code lines. Requiring that the beam width at distance $z = 15$ cm is $w = 0.5$ mm, the minimum waist size from Equation (18.3) is

$$w_0 \approx \frac{\lambda z}{\pi w} \approx 60 \ \mu\text{m} \tag{18.7}$$

The minimum mirror diameter is roughly $D = 3w_0 = 180 \ \mu$m. If the mirror is smaller, the beam will diverge too quickly and the scanner will not have sufficient resolution.

The minimum scan angle is set by the requirement to scan across the 4 cm bar code placed at distance $z = 10$ cm. From the problem geometry illustrated in Figure 18.5, the beam has to scan an angle of

$$\theta_{\text{min}} = \pm \tan^{-1} \frac{2 \text{ cm}}{10 \text{ cm}} \approx \pm 11.3^\circ. \tag{18.8}$$

The minimum mirror angle change is $\theta_s = \theta_{\text{min}}/2 \approx \pm 5.7^\circ$ and the minimum mirror edge displacement corresponding this scanning angle is $\Delta h = \theta_s D/2 \approx 9 \ \mu$m.

This example shows that it is possible to make a micromirror based scanner, but the size of the mirror and the required scan displacement are better suited for bulk micromachining than surface micromachining.

18.3.2 Displays

The MEMS based displays are the biggest optical MEMS success story and possible the most lucrative MEMS product to date. The market is dominated by Texas Instruments (TI) who currently holds the key patents in the field. The TI's digital micromirror device (DMD) is shown in Figure 18.6. Each pixel in the display is formed by a small mirror that switches between two positions: In one position, the light is steered to the display screen and the pixel appears bright. In the other position, the light is steered away and the pixel is black. To produce varying shades, the mirror is toggled on and off up to 1000 times per second. The ratio of on-time to off-time determines the lightness of the pixel. Different colors are produced by filtering a white light source or by using colored light sources such as LEDs.

The TI's DMD has a long history. The MEMS development at TI started in 1977 and in early 80's TI was already developing micromirror arrays for printers [144]. The first devices were based in analog mirrors but this approach proved difficult. As we learned in Chapter 15, the parallel plate actuator displacement is limited to one third of the electrode gap before the pull-in voltage causes the gap the collapse. If large analog displacements are necessary, the electrode gap and consequently the actuation voltage need to be large. Actuation voltages as high as 30 V were used and deflection was still not sufficient [144].

In 1987 Larry Hornbeck, invented the DMD that was based on binary switching using voltages greater than the pull-in voltage [144]. As the whole electrode gap could now be utilized for mirror deflection, large displacement was possible at low voltages. The first product, the DMD2000 airline ticket printer, went to market in 1990. The devices were further developed, and in 1996 TI introduced their first display product. Currently, TI is the market-share leader in the professional digital movie projection arena, largely because of the high contrast ratio offered by the MEMS mirror matrix. In the consumer market, TI competes with LCD projector manufacturers and the market is more evenly divided.

The TI's mirrors are made of aluminum and are 16 μm across which is approximately one fifth human hair diameter [144–146]. Each mirror is mounted on a yoke, which in turn is connected to two support posts by torsion hinges. During the operation lifetime, the torsion hinges endure billions of rotations. Perhaps surprisingly, no metal fatigue has been observed indicating the fatigue

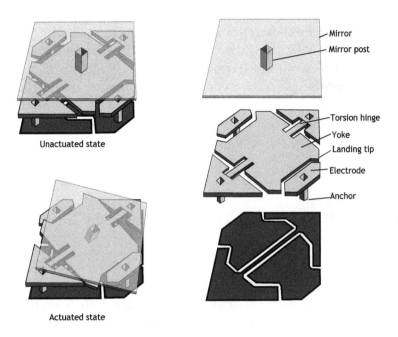

Figure 18.6: The digital micromirror device (DMD) from Texas Instruments is electrostatically actuated to modulate the light. More than a million mirror are laid out in a matrix to form an image. Note that in the actuated state, only the landing tip makes a contact with the substrate surface.

physics in microstructures are different from macro-scale.

The complex structure shown in Figure 18.6 is a result of the long development work. The mirror is mounted on the yoke so that the mirror surface is maximally uniform and the mirror remains flat during the rotation. Early prototypes had problems of mirrors getting stuck in the actuated positions. To eliminate this "stiction" problem where the mirror gets permanently stuck in the actuated state, the yoke is designed to contact the substrate only through the sharp landing tips. A similar approach of minimizing the contact area is routinely used in other MEMS components such as accelerometers.

Since the display matrix contains over one million pixels, on-chip integration of control electronics is required. To move the mirrors, the required state is first loaded into a memory cell beneath each pixel. Each mirror has two electrodes to rotate the mirror $\pm 10\text{-}12°$ and the change in the mirror angle between the "light" and "dark" state is therefore 20-24°. Example 18.3 further illustrates the physics of micromirror operation.

Example 18.3: TI micromirror operation

Problem: Estimate how much the light intensity for the display pixel changes when the TI's micromirror moves from "light" position to "dark" position.

Solution: The light propagation from the mirror is modeled using the Gaussian beam model. We use $\lambda = 700$ nm for the longest wave length of interest and set the waist size at the mirror to $w_0 = W/3 = 16\ \mu\mathrm{m}/3 = 5.3\ \mu\mathrm{m}$ after Example 18.1. The far field angle from Equation (18.5) is

$$\theta_0 = \frac{\lambda}{\pi w_0} \approx 2.4°. \tag{18.9}$$

Using $I \sim |E|^2$ and Equation (18.6), the light intensity in terms of angle is

$$I = I_0 \exp\left(-2\frac{\theta^2}{\theta_0^2}\right). \tag{18.10}$$

Since the mirror moves by total of $20°$ which translates to beam angle change of $\theta = 2 \cdot 20° = 40°$ (see also Figure 18.5), and the intensity in the dark position is $I \approx 3 \cdot 10^{-243} I_0$. Thus, in practical terms, the pixel is completely black.

As this example illustrates, even a small mirror is suitable for on/off light modulation. The scattered light from the mirror edges, from the underlying substrate, and from the mirror via will cause the "dark" state intensity to be higher than the estimated here. Even with these effects, the performance the micromirror displays is still excellent and a full on/off contrast ratio of 400 is obtained [147]. The high light/dark contrast ratio of micromirror display is a significant advantage in comparison to competing technologies such as LCD.

18.4   MEMS for fiber optical communications

Fiber optical communication requires means to physically route the optical signals and to control the signal intensity. The switch systems that connect optical fibers together are called optical cross connects [148, 149]. The cross connects are usually based on first converting the optical signal to electronic signal with a photodiode, routing the signal electrically, and finally converting the electrical signal back to optical signal with a laser diode connected to the fiber optical cable. The electro-optical conversion and routing requires substantial hardware complexity and consumes power. "All optical switching" uses micromirrors to route the signal to provide a potentially cheaper, more robust, and lower power alternative to the electrical switches.

Another MEMS application in fiber optical communication is the variable optical attenuator (VOA) [148]. In fiber optical systems, the signal intensity is adjusted to keep the signal at optimal level for optical amplifiers and detectors. Typical optical amplifiers have fixed gains and the laser intensity is adjusted with the VOAs. Mechanical attenuators have excellent performance that make them attractive alternatives to the other existing solutions.

In this section, the MEMS solutions for optical attenuation and switching are reviewed. First, the operation principles of optical attenuators are covered. Next, the beam optics for NxN all optical switch network is analyzed.

18.4.1 Attenuators

Variable optical attenuators (VOAs) are used to control the light intensity in fiber optical communication systems. Mechanical attenuators based on motor or solenoid driven mirrors offer great optical performance but leave room for improvement in terms of the device size, power consumption, mechanical reliability, and cost. Micromechanical attenuators have been demonstrated that offer a good overall performance [150–152]. A typical attenuator can control the light intensity with a range form -1 dB to -50 dB which on a linear scale is from 0.8 to 10^{-5}! This type of dynamic range is hard to achieve with non-mechanical attenuators.

Figure 18.7 shows three different MEMS attenuator implementations. The shutter type actuator shown in Figure 18.7(a) is based on blocking the light path between two closely based optical fibers. The design is straightforward to implement but it suffers from polarization dependent losses. Also, the required shutter displacement is fairly large, $\sim 10~\mu$m, which requires a large actuator area. The rotating mirror based actuator shown in Figure 18.7(b) is based on steering the light away from the fiber. The fibers can be inserted in grooves that are etched onto the silicon as shown in Figure 18.7(b). While the idea of integration is appealing, the etch grooves require a large silicon area and it may be more cost effective to implement the fiber holders with conventional mechanical clamps. Clamping the fibers outside the MEMS chip increases the spacing between the mirror and fibers necessitating the use of a collimator shown in Figure 18.7(c). At least one MEMS attenuator based on using micromirror together with a collimating lens is commercially available [153]. The commercialization of MEMS VOAs is still in the early stages, but it appears that in this application the MEMS technology offers real performance advantages over conventional technologies [154].

Example 18.4: Optical attenuator made of SOI
Problem: Figure 18.8 shows a shutter type optical attenuator that is electrostatically actuated [150]. To obtain a clean over damped mechanical response

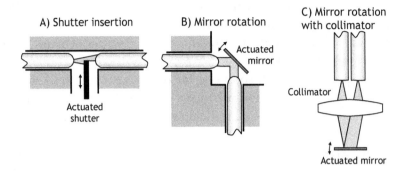

Figure 18.7: Three different implementations for MEMS VOAs. (a) A microactuated shutter is inserted between fibers placed in a groove. (b) Micromirror is rotated between fibers placed in grooves. (c) Coupling of fibers through a collimator lens is controlled by a tilt mirror (After [151]).

and to reduce the insertion loss, the attenuator is operated in an oil filled package. The oil also increases the actuator capacitance which lowers the needed actuation voltage. The actuator has a total of $N = 104$ finger electrode overlaps. Each electrode is $h = 75$ μm tall and the electrode gap is $d = 3$ μm. The device has a 1.5-dB insertion loss and maximum attenuation is 57 dB. Assuming that the total spring constant is $k = 3$ N/m and the oil permittivity is $\epsilon_R = 3$, how large voltage is needed to move the shutter by 20 μm to block the light path?

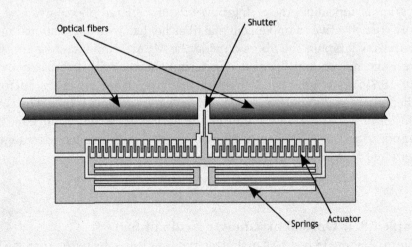

Figure 18.8: Variable optical attenuator (After [150]).

Solution: The actuation force from (15.24) is

$$F = N\epsilon_R\epsilon_0 \frac{h}{d} V^2$$

and from the Hooke's law $F = kx$, we obtain that $V = 30$ V is needed to obtain $x = 20$ μm displacement.

18.4.2 Optical switches

Optical switches are the answer to the perceived need to connect optical signals between optical fibers without intermediate conversion from the optical to an electrical signal. The first MEMS optical switches in the 90's were designed for switching between two fibers [155, 156], but at the turn of the decade, there was a big development effort to develop a switch that is capable of optically cross connecting a large number of fibers [148, 149].

A typical implementation for an "all optical switch" is shown in Figure 18.9. Optical fibers are arranged in a 2D NxN matrix and N^2 mirrors are used to couple the light between any input and output fiber. Although a large number of mirrors are needed, the architecture has the advantage that the mirrors need to switch between two positions (in the light path and away from the light path). At the turn of the decade, several 2D switch prototypes were demonstrated, but despite large investments in the development, the market adaption has been disappointing.

To appreciate the difficulties in making a good optical switch, we will study the scaling of the 2D switch architecture. The size occupied by one mirror is D_M. In practice, each mirror may require some additional space D_f for example for the actuator. For simplicity, we will study the optimistic scaling where the addition area is zero ($D_f = 0$). The minimum mirror size D_M is set by the beam size w at the mirror. We write $D_M = k_M w_{max}$ where k_M is a constant with value of $k_M \approx 3 - 4$ to capture most of the Gaussian beam and w_{max} is the largest beam size that the mirror needs to accommodate. If the mirror is smaller, the insertion loss will increase as a portion of the beam is lost.

Referring to the Figure 18.9, the longest possible distance from fiber to the mirror is the distance from fiber to the furthest mirror. We write this "array size" as

$$z_A = N(D_M + D_f) = Nk_M w_{max}, \qquad (18.11)$$

where we have used $D_f = 0$ and $D_M = k_M w_{max}$. The smallest spot size is when the mirror is located at the beam waist and $w = w_0$. Referring to the Figure 18.9, the waist located at the mirror furthest from the fibers.

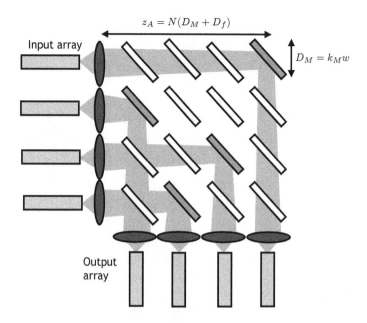

Figure 18.9: 2D optical switch matrix. The light from any input fiber can be switched to any output fiber by positioning a mirror in the light path. Here the dark mirrors are actuated and the light mirrors are placed away from the light path.

The spot size on the mirror is the biggest when the mirror is closest to the fiber. This point is located at the distance z_A from the waist location (at mirror furthest away from the fiber). Thus, the largest beam width is $w_{\max} = w(z_A)$. From Equation (18.3), the beam width w at distance z_A from the waist is

$$w_{\max} = w(z_A) = \sqrt{w_0^2\left[1 + \left(\frac{\lambda z_A}{\pi w_0^2}\right)^2\right]}. \qquad (18.12)$$

By combining Equations (18.11) and (18.12), the number of mirrors in the array is

$$N = \frac{z_A}{k_M w_0 \sqrt{1 + \left(\frac{\lambda z_A}{\pi w_0^2}\right)^2}}. \qquad (18.13)$$

Equation (18.13) gives the number of mirrors as a function of the mirror and array dimensions. To find the optimal design, we will look for the maximum N with the smallest possible total size z_A. Deriving Equation (18.13) with respect to z_A and solving for $\frac{\partial N}{\partial z_A} = 0$, the maximum number of mirrors N with the minimal size z_A is

$$N = \frac{1}{k_M}\sqrt{\frac{\pi z_A}{2\lambda}} \qquad \text{(optimal design).} \qquad (18.14)$$

Solving Equation (18.14) for z_A, the smallest possible array size as a function of N is

$$z_A = \frac{2N^2 k_M^2 \lambda}{\pi} \qquad \text{(optimal design)}. \qquad (18.15)$$

Equation (18.15) is the scaling law for 2D optical switch arrays: The array size increases as a square of number of inputs/outputs. As example 18.5 illustrates, this unfavorable scaling limits the maximum array sizes to less than 32×32 or the dimensions become uneconomical or even impossible to manufacture.

Example 18.5: Scaling of 2D optical switches
Problem: Calculate the smallest possible dimensions for 10×10, 30×30, 100×100 and 1000×1000 optical switch. Assume that the wave length is $\lambda = 1.55$ μm, which is typical for fiber optical communication. What is the practical size limit for a MEMS device and how large switch (N×N) would that enable?
Solution: To minimize the insertion loss, we wish to capture 99% of the beam with the mirrors and therefore set $k_M = 3$. From Equation (18.15), the array dimension is

$$z_A = \frac{2N^2 k_M^2 \lambda}{\pi} \approx N^2 \cdot 9 \quad \mu\text{m}$$

giving 0.9 mm, 8 mm, 90 mm, and 9 m for the 10×10, 30×30, 100×100 and 1000×1000 array, respectively. The practical size limit for a commercially viable device is measured in centimeters. Thus, the maximum switch array is less than 100×100 and probably closer to 30×30 or the cost would be prohibitively high.

As the 2D switch architecture is limited to relatively small number of fibers, 3D switch matrices have been investigated [149, 154, 157]. Figure 18.10 shows the architecture for a 3D optical switch. The light is steered between any input and output fibers using two sets of mirrors. Instead of requiring N² mirrors, the 3D architecture scales much better and requires only 2N mirrors for N input and output fibers. The switching, however, is not a simple binary operation between on and off states, but accurate analog control is needed to steer the light from one mirror to another and to the correct output fiber.

Currently, the interest in optical switches is low (see also the next section "Vanishing optical MEMS"). There are many reasons for this. In addition to financial difficulties after the stock market crash in 2000, the slow adoption is partially due to the technology itself. It turned out to be much more difficult than expected to make a good optical switch that operates reliably for several decades. The digital 2D architecture does not scale to a large number of fibers

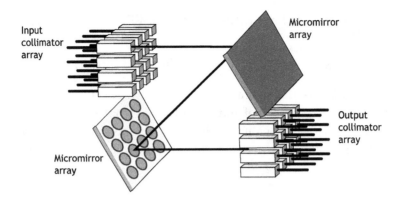

Figure 18.10: 3D optical switching architecture based on steering each light beam with two accurately controlled mirrors.

and the accurate analog steering of 2N mirrors is much more difficult to implement than digital control. Finally, there is a persistent question about the need for mechanical switches. Typically, the communication router needs to be able direct data packets from one fiber to another at rate much faster than possible with mechanical switching. Thus, the MEMS switches are only suitable for semi-stationary connections for which there may not be sufficient demand.

18.4.3 Vanishing optical MEMS

In 2000, the network business was skyrocketing and optical MEMS switches were poised to become the next big thing out of Silicon Valley. In the US alone, there were 30+ start-ups working on optical switches, and many big companies such as Lucent had internal development programs. Everyone was excited but that did not mean that the business and technology was sound. A few years later, much of the MEMS switch work was abandoned. The most famous example is the JDS Uniphase venture into MEMS by a purchase of a small MEMS manufacturer Cronos Microsystems. The saga is illustrated in Figure 18.11. In a little over two years, the value of the Cronos investment was reduced to mere 1% of the original acquisition price. Perhaps the moral to the story is that if MEMS technology can shrink optical devices, the MEMS business can excel with shrinking money!

Currently, there is very little talk about optical switching using MEMS components. It is tempting to conclude that the entire field is a failure but this is not entirely fair. The demise was partly a result of unrealistic expectations, immaturity of the technology, and the stock market crash and high-tech recession in 2000-2003. Recalling that it took more than ten years for Texas Instrument to develop the digital micromirror device (DMD), it would not be surprising to see MEMS eventually making a comeback in optical switching.

JDS Uniphase, a large optical network manufacturer, buys the small MEMS start-up, Cronos, for **$750 million**. JDS Uniphase chief executive Keven Kalkhoven says, "This is one of the most important single decisions we've made in the company's history." JDS Uniphase plans to use optical MEMS switches in fiber optic networks.

JDS Uniphase annual report states that the company is developing MEMS switches "that use MEMS technology to provide advanced switching speeds and improved optical performance bundled in a condensed, hermetically sealed package".

JDS Uniphase annual report states that "the MEMS 2x2 switch development has been terminated ... and transferred to another division."

Later in 2002, JDS Uniphase announces that it will sell Cronos in a deal valued at **$7.7 million**.

2000	2001	2002

Figure 18.11: The saga of how a network manufacturer JDS Uniphase lost massive amount of money in its MEMS venture.

18.5 Interferometry

The Fabry-Pérot interferometers are based on the interference of multiple reflections of light. Figure 18.12 shows the geometry for a typical Fabry-Pérot interferometer. The device is also known as the etalon and it consist of two partially reflective surfaces spaced a distance t apart. The material between the reflecting surfaces has the refractive index n. To illustrate the multiple reflections in the Figure 18.12, the light is shown to enter into etalon in a small angle θ. For the subsequent analysis, we assume this angle is sufficiently small so as not to affect the light path distance between the two surfaces.

Depending on the wavelength of light and the distance between the surfaces,

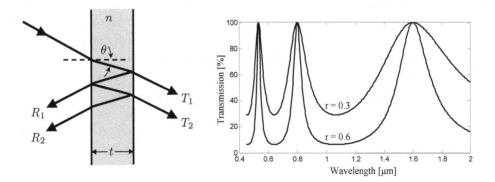

Figure 18.12: Schematic of Fabry-Pérot interferometer (etalon) and example light transmissions for $t = 0.8$ μm and $n = 1$.

the interference between multiple reflections can be constructive or destructive. The maximum transmission occurs when the etalon thickness t is an exact multiple of half wavelength

$$t = m \frac{\lambda}{2n}, \tag{18.16}$$

where m is an integer multiple, λ is the wavelength in air and λ/n is the wavelength in the etalon. In this case the reflections of light are in-phase and constructive interference occurs. The condition for minimum transmission occurs when the reflections of light are 180° out-of-phase and etalon thickness is

$$t = \left(m + \frac{1}{2} \right) \frac{\lambda}{2n}. \tag{18.17}$$

A more accurate analysis takes into account the reflection coefficient r of the two etalon surfaces. The resulting transmission function for light intensity through the Fabry-Pérot interferometer is

$$T = T_1 + T_2 + \ldots = \frac{(1-r)^2}{1 + r^2 - 2r \cos(4\pi n t/\lambda)} \tag{18.18}$$

where r is the reflection coefficient of the etalon surface. The etalon reflection function is

$$R = 1 - T, \tag{18.19}$$

which provides the relative intensity of the reflected light. Figure 18.12 shows the transfer function for an example etalon with the thickness $t = 0.8$ μm, the refractive index $n = 1$, and the reflection coefficients $r = 0.3$ or $r = 0.6$. We see that the etalon selectivity increases with the increasing surface reflectivity r and the transmission peaks become sharper.

MEMS technology allows tunable Fabry-Pérot filters where the distance between the filter surface is changed. A commercial application for a tunable Fabry-Pérot interferometer is the carbon dioxide detector shown in Figure 18.13 by Vaisala [158]. The device consists of an infrared source at the end of the measurement gas chamber. Any carbon dioxide gas present absorbs a part of the light at its characteristic wavelength. At the end of the chamber, there is an electronically tunable MEMS interferometer and an infrared detector. The filter is essentially a surface micromachined membrane that can be electrostatically actuated to change the optical path length between the two partially reflective surfaces. The MEMS filter is adjusted between two wavelengths, CO_2 absorption wavelength and reference wavelength where the absorption by CO_2 is not significant. The detected signal strengths are compared and the ratio of detected light intensities indicates the degree of light absorption in the gas. The use of reference signal compensates sensor aging and contamination making the sensor very stable over time.

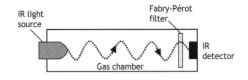

Figure 18.13: Carbon dioxide sensor based variable MEMS Fabry-Pérot interferometer (After [158]).

A second commercial application example is the interferometric modulator display (IMOD) developed by Iridigm Display Corporation and commercialized by Qualcomm [159]. Figure 18.14 shows the operating principle for the display. Instead of light transmission, the display uses the etalon to modulate the light reflection. Each pixel in the display consists of two mirrors, opaque back membrane and partially reflective front surface. When the pixel is in open state, the reflections from the pixels surface and the back membrane are in-phase and the constructive interference makes the pixel appear bright. In the collapsed state, the air gap between the two mirror surfaces is so small that the constructive interference occurs only at non-optical wavelengths and the pixel appears dark.

The interferometric modulator display is attractive for portable devices as it operates with ambient light. In comparison to other displays such as LCD, the IMOD does not require back light and works well with strong ambient light, for example in outdoors where LCD displays lose their contrast. The pixels are electrostatically actuated and the total power consumption is very small. The first commercial monochrome displays are already on market but the IMOD does not yet compete with the color LCDs.

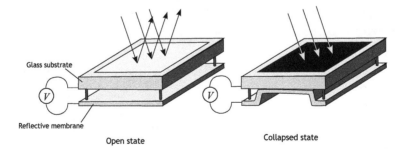

Figure 18.14: Operation principle for the interferometric modulator display pixel (After [159]).

Key concepts

- The divergence of optical beams depends on the source size. The smaller the beam waist, the faster the beams spreads.

- The beam spreading angle is $\theta_0 = \frac{\lambda}{\pi w_0}$, where λ is the wavelength, and w_0 is the waist width.

- The beam size limits optical scanner resolution. Even low resolution applications such as bar code scanners can require mirrors that are over 200 μm in diameter in order to obtain sufficiently small beam divergence.

- Micromirrors are well suited for display applications as the beam size and scanning resolution are not critical.

- Micromechanical variable optical attenuators (VAOs) find applications as in fiber optical communications. Optical switching holds promise but the application is very demanding.

- Tunable optical filters can be used as low performance interferometers for chemical analysis and reflective displays.

18.6   Exercises

Exercise 18.1
Torsional mirror has dimensions of 500 μm × 500 μm and it is to be used as a laser scanner. The laser wavelength is 670 nm. Assume that the propagation of beam can be modeled as a Gaussian beam. Calculate the spot size at distance of 10 m from the mirror.

Exercise 18.2
Laser pointer waist size is $w_0 = 200$ μm and the beam width at distance of 3 m is $w = 3$ mm. How much is the beam intensity I reduced at the beam center compared to the original intensity I_0?

Exercise 18.3
Show that for the geometry in Figure 18.5, the beam movement angle is twice the mirror rotation angle.

Exercise 18.4
Calculate the minimum dimensions for a 16×16 optical switch.

Exercise 18.5
Consider the interferometric modulator display (IMOD). Assume that the reflectivity of the etalon surfaces is $r = 0.2$ and the distance between the reflective surface is $t = 0.4$ μm in the bright state and $t = 0.01$ μm in the dark state. Plot the etalon reflection R as a function of wavelength for the two states. What

color is reflected most in the bright state? Integrate the total reflected light in the visible wavelength range and calculate the contrast ratio (the total reflected light intensity ratio for the light and dark stages). How can the contrast ratio be increased?

Exercise 18.6

Estimate the feasibility to make a micromirror based bar code scanner that reads the bar code from a distance of 1 m.

Exercise 18.7

Calculate the minimum rotation angle for the TI micromirror display that results in a theoretical contrast ratio of 10,000 between the light and dark states.

19

RF MEMS

Radio frequency microelectromechancial systems (RF MEMS) is the emerging technology platform of mechanical resonators, switches, and tunable capacitors (varactors) designed to operate in wireless communication frequencies (MHz to GHz). The field is often compared to the optical MEMS industry. The comparison works as both optical MEMS and RF MEMS expand microsystems beyond the classical sensing applications. Both optical MEMS and RF MEMS also promise multi-billion dollar market potentials, and both technologies have been slow to deliver on the promises. The comparison fails in that the RF MEMS was never over-hyped to the same extent that the optical MEMS was before the stock market crash of 2000 to 2002.

The main interest in RF MEMS stems from the high performance of the mechanical components in comparison to their solid-state counterparts. The mechanical switches are much more expensive than the solid-state relays, but applications such as radars and test equipment require the performance that only mechanical switching can provide. MEMS components are also considered for cell phones where they can simplify the design of multi-band radios. Modern cell phones operate on three or more frequency bands. Currently, the switching between the different bands is achieved with solid-state switches that are lossy, difficult to design, and consume power. With the introduction of even higher number of communication bands, there is a real need for high performance mechanical switches in cell phones. Cell phones will also benefit from the linearity of the mechanical varactors but the demand is not as urgent.

This chapter will focus on the RF MEMS switches and tunable capacitors. Both components move slowly in comparison to the high frequency electrical signal. The micromechanical RF resonators that vibrate at the signal frequency are discussed in Chapters 20 and 21. To make the case for RF MEMS, we will first review the limitations of existing solid-state solutions for switches and capacitors. Next, we will review the mechanical varactors and switches with the focus on cell phone applications. Finally, the benefits of manufacturing RF inductors using microfabrication techniques are discussed.

19.1   Solid-state switches and varactors

To appreciate the benefits and drawbacks of micromechanical RF components, we will first review properties of the solid-state counterparts. Both switches and varactors can be realized by biasing a diode with a dc voltage. Diodes are based on the solid-state pn-junction that passes current in forward direction but allows only limited current in the reverse direction. The large signal transfer function for a diode is

$$I = I_s(e^{V/nV_T} - 1), \tag{19.1}$$

where I_s is the saturation current, V is the voltage over the pn-junction, $V_T = k_B T/q$ is the thermal voltage, and n is the ideality factor. The ideality factor varies from about 1 to 2 depending on the fabrication process [160].

For small signals, the diode can be represented with a resistor and capacitor in parallel as is illustrated in Figure 19.1. The small signal resistance is obtained by deriving Equation (19.1) giving

$$r = \frac{\partial V}{\partial I} = \frac{nV_T}{I_s} e^{-\frac{V}{nV_T}} = \frac{nV_T}{I_s} \left(\frac{I}{Is} + 1 \right)^{-1}. \tag{19.2}$$

As Equation (19.2) shows, the diode resistance is high for negative bias voltages V (reverse bias) but decreases exponentially for positive bias voltages (forward bias). For forward bias where $I \gg I_s$, Equation (19.2) can be approximated as

$$r \approx \frac{nV_T}{I}. \tag{19.3}$$

Equation (19.3) shows that to obtain a low resistance r, the dc current I through the diode must be high. This is a serious drawback, as switch applications require a low impedance which translates to a high power consumption.

Besides modulating the small signal diode resistance, the bias voltage changes the small signal pn-junction capacitance given by

$$C_j = \frac{C_{j0}}{\sqrt{1 - \frac{V}{V_{bi}}}}, \tag{19.4}$$

where C_{j0} is the junction capacitance at zero bias $V = 0$ and V_{bi} is the build-in voltage for the diode. The typical build-in voltages range from 0.5 V to 0.7 V. As Equation (19.4) and Figure 19.2 shows, the capacitance is small for

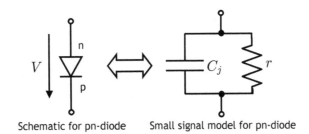

Schematic for pn-diode   Small signal model for pn-diode

Figure 19.1: A solid-state alternative to RF MEMS: pn-diode resistance and capacitanse can be adjusted with the bias.

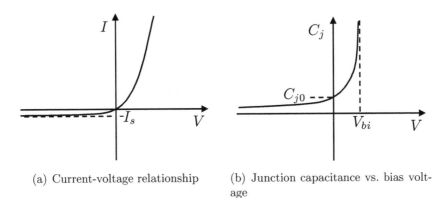

(a) Current-voltage relationship   (b) Junction capacitance vs. bias voltage

Figure 19.2: Current and junction capacitance for a pn-diode.

large reverse biases but increases quickly as the bias voltage approaches the build in voltage. Physically, the reverse bias increases the depletion width of the pn-junction, thus reducing the parallel plate capacitance between the p- and n-regions. Forward biasing reduces the depletion width and at $V = V_{bi}$ the depletion width becomes zero.

19.1.1   Solid-state switches

A solid-state diode can be used as a switch by alternating between forward bias (low small signal resistance, switch "on") and reverse bias (high small signal resistance, switch "off"). Figure 19.3 shows a practical implementation of a diode based RF switch. The switch is dc isolated from the rest of the circuit with coupling capacitors C_{ac} and biased with a control voltage $v_{control}$ to switch between "on" and "off" states. Inductors are placed in series with the control voltage and diode. The inductors allow dc current to flow through the diode but block the RF signal from coupling with the control voltage.

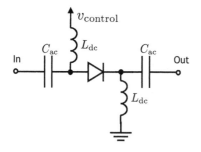

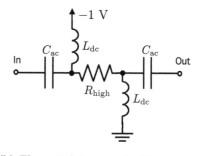

(a) Schematic for the switch showing the diode, biasing inductors, and bypass capacitors.

(b) The switch is turned off by reverse biasing the diode to make the small signal resistance high.

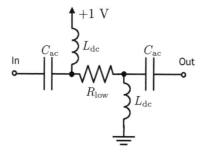

(c) The switch is turned on by forward biasing the diode to make the small signal resistance low.

Figure 19.3: Operation of pn-diode-based switch.

Example 19.1: Diode switch

Problem: A pn-diode with $I_s = 0.1$ nA, $n = 1.3$, $C_{j0} = 0.5$ pF, and $V_{bi} = 0.7$ V is used as a RF switch. What are the "on" and "off" resistances, if the diode is biased with $V = 0.7$ V and $V = -2$ V, respectively. (1) What is the impedance of the junction capacitance in the "off" position at $f = 1$ GHz? (2) What is the "off" state isolation in a 50 Ω-system? (3) How large signals can the switch block? (4) What is the current consumption in the "on"-state?

Solution: (1) From Equation (19.2)

$$r = \frac{nV_T}{I_s}e^{-\frac{V}{nV_T}}$$

we have $r_{on} = 0.3$ Ω and $r_{off} = 2 \cdot 10^{34}$ Ω.

The junction capacitance at $V = -2$ V from Equation (19.4) is

$$C_j = \frac{C_{j0}}{\sqrt{1 - \frac{V}{V_{bi}}}} \approx 0.25 \text{ pF},$$

which at $f = 1$ GHz is $|Z_j| = 1/\omega C_j \approx 690\ \Omega$.

(2) Since the "off" resistance is very high, the isolation is set by the junction capacitance. The fraction of transmitted power in "off"-state is (see also Example 19.7 on page 299)

$$|S_{21,\text{off}}|^2 = \left| \frac{v_{\text{out}}}{v_{\text{in}}} \right|^2 = \left| \frac{2Z_0}{2Z_0 + 1/j\omega_0 C_j} \right|^2 = -17\text{ dB}.$$

This isolation alone is not sufficient for mobile communication systems that require ~ 30 dB isolation. To obtain better isolation performance, two diodes are typically used for a single switch.

(3) When the diode is biased at $V = -2$ V, signals with amplitudes > 2 V will start to turn the diode on. The maximum signal power that can be blocked is therefore $P = V^2/2/Z_0 \approx 40$ mW or 16 dBm. For mobile handsets that have maximum transmit power of 33 dBm, higher reverse bias voltage is needed to make sure the diode is always "off".

(4) The "on" state current from Equation (19.1) is

$$I = I_s(e^{V/nV_T} - 1) \approx 110\text{ mA}. \tag{19.5}$$

This is significant for a portable device and a major shortcoming for a pn-diode switch. As shown by Equation (19.3), the resistance is inversely proportional to the bias current. Thus, reducing the current by $10\times$ will increase the "on" state resistance by $10\times$. Transistor based switches have been used to reduce the power consumption but they also suffer from trade-offs between linearity and insertion loss.

The diode based switch is simple to implement, does not require any special packaging, and can have low "on" state resistance which translates into a low insertion loss. The "off" state resistance is high but the RF signal can also pass through the junction capacitance which decreases the isolation at high frequencies. To obtain high "off" state isolation, two switches are often combined to direct the RF signal to the ground in the "off" state as shown in Figure 19.4. One diode is in series with the signal path and one shunts the signal to ground. Similar series/shunt switch approach can also used with the RF MEMS switches to increase the isolation.

The drawbacks of the diode based switch are the high power consumption in the "on" state, the large number of external bias components necessary, and the poor linearity. The last shortcoming is easy to understand, as the small signal model is valid only for signals that are small in comparison to bias voltage. Large signals can self-modulate the diode operation point, which leads to signal distortion.

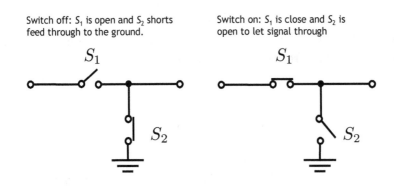

Figure 19.4: Two switch implementation to increase the total isolation.

19.1.2 Solid-state varactors

A diode can also be used as a variable capacitor or varactor by reverse biasing it as shown in Figure 19.5. Depending the bias voltage v_{bias}, the junction
capacitance can be adjusted as shown by Equation (19.4) and illustrated in Example 19.2. Since the reverse bias currents are small, it is also possible to bias
the varactor with a large resistor instead of the inductor shown in Figure 19.5.
As with the switch, the diode varactors are limited to signals that are much
smaller than the bias voltage. Large signals will self-modulate the capacitance,
which causes distortion and aliasing.

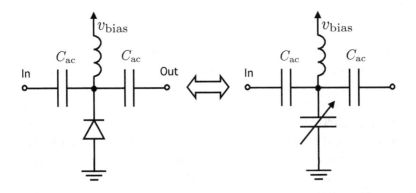

Figure 19.5: Diode based variable capacitor is based on change in junction capacitance
with varying reverse bias voltage v_{bias}. The bias is ac isolated with an inductor and the
dc bias voltage is blocked with ac capacitors C_{ac}.

Example 19.2: Diode varactor
Problem: A pn-diode with $C_{j0} = 2$ pF and $V_{bi} = 0.7$ V is used as an RF varactor. How much the junction capacitance changes when the bias is adjusted between -10 V and -5 V?
Solution: From Equation (19.4) we have the junction capacitance

$$C_j = \frac{C_{j0}}{\sqrt{1 - \frac{V}{V_{bi}}}} \approx 0.51 \text{ pF and } 0.70 \text{ pF}$$

for the -10 V and -5 V bias, respectively. The capacitance chance is 0.19 pF and the capacitance tuning ratio is 1:1.4.

19.2   Micromechanical varactors

MEMS based variable capacitors are an emerging, high performance alternative to the low cost, solid-state varactors. The main advantage of the mechanical capacitor is the linearity: The MEMS capacitors have mechanical response bandwidths of $1 - 100$ kHz limited by the mechanical resonance frequency, hence the capacitors do not move with the RF signal at $f > 100$ MHz. In comparison, the solid-state varactors are based on controlling the charge carriers that can move with the RF signal resulting in intermodulation as the RF signal itself changes the junction capacitance. The linearity of MEMS varactors opens new applications such as tuning of the RF transmitters where large signals are present and the linearity requirements prevent the use of solid-state varactors [161].

The second advantage of the mechanical capacitor is that the losses can be minimized by fabricating the devices of well conducting metals. The ohmic losses for metal-based capacitors can be more than ten times smaller than for solid-state varactors that are based on semiconducting materials. In addition, the MEMS varactors can be integrated directly on top of signal lines whereas the connection inductances limit the operation frequency of discrete solid-state varactors. The wire bonding inductances become significant at frequencies above 1 GHz and MEMS varactors have a significant advantage at higher frequencies.

Figure 19.6 illustrates a typical structure for a MEMS varactor. The design is based on a parallel plate capacitor supported by four beams. The varactor is biased with a resistor that blocks the RF signal and the device is ac coupled with a capacitor. The dc bias generates electrostatic force that moves the plates closer together thus increasing the parallel plate capacitance. The linear tuning range is limited to 1:1.3 by the pull-in effect that will snap the plates together if the bias voltage exceeds the pull-in voltage as analyzed in Examples 15.2 and 15.7 on pages 229 and 236, respectively. Larger tuning range is possible with

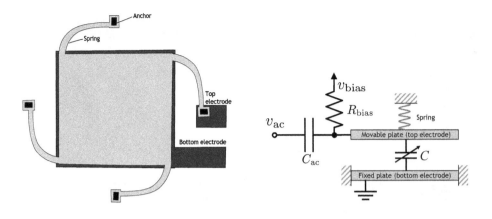

Figure 19.6: Top view and biasing schematic for RF varactor

more complex design, for example by using separate capacitors for actuation as illustrated in Example 19.3, or changing the geometry from parallel plate to comb-drive that does not have pull-in effect [162].

Although the MEMS varactor is a relatively simple MEMS component, it is significantly more expensive than the solid-state varactors. This is because the MEMS varactor requires complex packaging to protect it from the environment, and it cannot be integrated with CMOS without changes in the fabrication and/or packaging process. Clearly, MEMS varactors will not be used for low-end applications such as Bluetooth transreceivers where the varactor is implemented within the standard CMOS IC for a fraction of a cent. The MEMS varactors are therefore only competitive in applications that require the high performance of the mechanical structure.

Example 19.3: Extended tuning range MEMS varactor
Problem: Figure 19.7 shows a schematic for a wide tuning range MEMS varactor that is suitable for applications up to 60 GHz [163]. The tuning range is extended beyond the capacitor pull-in displacement by using separate actuation electrodes with a wide gap. Estimate the actuation voltage and the capacitance range assuming that the spring constant is $k = 15$ N/m, the capacitor area is $A_{\mathrm{cap}} = 140~\mu\mathrm{m} \times 140~\mu\mathrm{m}$, the capacitor gap is $d_{\mathrm{cap}} = 1.5~\mu\mathrm{m}$, the total tuning electrode area is $A_{\mathrm{el}} = 2 \times 140~\mu\mathrm{m} \times 50~\mu\mathrm{m}$, and that the tuning electrode gap is $d_{\mathrm{el}} = 2.5~\mu\mathrm{m}$.

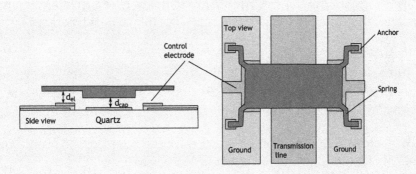

Figure 19.7: A micromechanical varactor with wide tuning range has separate capacitor and actuator electrodes (After [163]).

Solution: The pull-in voltage for the tuning electrodes from Equation (15.17) is

$$V_P = \sqrt{\frac{8}{27}\frac{kd^3}{\epsilon_0 A}} \approx 24 \text{ V}$$

and the tuning range is approximately 0-24 V. The capacitor bridge displacement at the pull-in point is $\Delta x = d_{\text{el}}/3 = 0.83$ μm. The RF capacitor gap varies from $d_{\text{cap}} - \Delta x$ to d_{cap} and the capacitance range is then from

$$C = \epsilon_0 \frac{A}{d_{\text{cap}}} = 116 \text{ fF}$$

at zero voltage to

$$C = \epsilon_0 \frac{A}{d_{\text{cap}} - \Delta x} = 260 \text{ fF}.$$

at $V = 24$ V. The theoretical tuning range is more than 1:2, which compares well with the solid-state varactor in Example 19.2. The actual tuning rate from the prototype device was significantly less due to fabrication stress induced curvature of the capacitor bridge [163].

19.3 Micromechanical switches

The interest for RF MEMS switches has been much bigger than for any other RF MEMS component. The mechanical switches are currently used in measurement and instrumentation systems, but the macroscopic mechanical relays are too bulky and expensive for wider use. Solid-state switches are used extensively in portable electronic devices such as cell phones but the diode based switches suffer from poor linearity, high power consumption, and resistive and inductive losses. In comparison, RF MEMS switches offer much better electrical performance than the solid-state solutions [164–166]. The advantages of MEMS switches are:

High linearity: Ideally, the mechanical relays are perfectly linear and the demonstrated MEMS switches have been 1000x more linear than solid-state switches.

High isolation: The isolation is based on an air gap which results in very small off state parasitic coupling capacitance.

Low insertion loss: The switches are typically made of metal and the resistive losses are small. Integration on top of an IC or transmission line eliminates the parasitic contact inductances that limit the operation frequency of discrete switches.

Low cost: While the cost is currently much higher than for solid-state switches, it is comparable or smaller than that of macroscopic mechanical relays.

Easy of design: The mechanical relay behaves much like an ideal on/off switch and does not require complex biasing schemes. As dc current is not needed to bias the switch, it can be biased with resistors. As large inductors are not needed, the area consumption is reduced and the design floor planning is simplified.

Low power consumption: Ideally, the capacitive actuation does not consume any current. Even considering the circuitry to generate high voltage drive bias needed for actuation, the total power consumption is smaller than for diode switches.

The disadvantages for the MEMS switches are:

Low switching speed: Since the switches are based on mechanical operation, the switching speed is lower than for solid-state relays.

Poor reliability: The switches are based on mechanical components that can fail more easily than the solid-state switches. The reliability is constantly improving but lifetime data of several years of operation is just recently becoming available.

The MEMS switches are either based on ohmic contact between two metal conductors (ohmic switch) or by changing capacitance by a large factor (capacitive switch). These approaches are reviewed in the next section.

19.3.1 Capacitive switches

The capacitive switches are based on a large change in capacitance that modulates the signal transmission. A typical shunt switch is illustrated in Figure 19.8 where a capacitive bridge is placed above a transmission line [164, 165, 167–169]. When the bridge is up, the capacitance to the ground is small and the signal passes through. When the bridge is down, the large capacitance effectively shorts the signal line to the ground and the signal does not pass the switch. The structure can also be used as a varactor integrated directly on top of a transmission line [170].

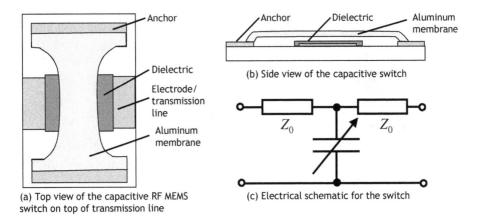

(a) Top view of the capacitive RF MEMS switch on top of transmission line

(b) Side view of the capacitive switch

(c) Electrical schematic for the switch

Figure 19.8: Capacitive RF MEMS switch. The transmission line is denoted with Z_0 (After [164, 165, 169]).

The required "off/on" capacitance ratio depends on the maximum allowed signal loss in the "on"-position (insertion loss) and the minimum signal isolation in the off state. Using the equivalent circuit in Figure 19.8(c), the magnitude of the transmitted signal power relative to the incoming signal power is

$$|S_{21}|^2 = \left|\frac{v_{\text{out}}}{v_{\text{in}}}\right|^2 = \frac{1}{1 + (\omega C Z_0/2)^2} \tag{19.6}$$

where ω is the signal frequency, C is the bridge capacitance, and Z_0 is the characteristic impedance to the transmission line [171].

As Example 19.4 illustrates, the small insertion loss and high isolation requirements lead to large "off/on"-capacitance ratios. For mobile phones where isolation greater than 30 dB is necessary, multiple switches may be required to obtain acceptable "off" state isolation.

Example 19.4: Capacitive switch
Problem: The capacitive switch in Figure 19.8 has the capacitor area $A = (110 \ \mu m)^2$. Calculate the maximum dielectric thickness ($\epsilon_R = 7.5$ for Si_3N_4), the bridge height in open position, and the "off/on" capacitance ratio for a switch with 15 dB "off" state isolation and 0.05 dB insertion loss at $f = 5$ GHz. What is the actuation voltage if the bridge spring constant is $k = 20$ N/m?
Solution: In the "off" state, the bridge is down and the "off" state capacitance is

$$C_{off} = \epsilon_R \epsilon_0 \frac{A}{d_1}$$

where d_1 is the dielectric thickness and $A = (110 \ \mu m)^2$ is the electrode area. Using Equation (19.6), we can estimate the isolation with different dielectric thicknesses. For example, $d_1 = 100$ nm gives

$$|S_{21}|^2 = \left| \frac{v_{out}}{v_{in}} \right|^2 = \frac{1}{1 + (\omega C_{off} Z_0 / 2)^2} \approx -16.1 \text{ dB} \qquad (19.7)$$

which satisfies the isolation specifications at $f = 5$ GHz.

In the "on" state, the bridge is in the up position. The "on" state capacitance is a series combination of the air capacitance and the dielectric capacitance given by

$$C_{on} = \left(\left(\epsilon_R \epsilon_0 \frac{A}{d_1} \right)^{-1} + \left(\epsilon_0 \frac{A}{d_2} \right)^{-1} \right)^{-1} \approx \epsilon_0 \frac{A}{d_2}$$

where d_2 is the air gap when the bridge is up. Using Equation (19.6) and $d_2 = 1 \ \mu m$ gives

$$|S_{21}|^2 = \left| \frac{v_{out}}{v_{in}} \right|^2 = \frac{1}{1 + (\omega C_{on} Z_0 / 2)^2} \approx -0.03 \text{ dB} \qquad (19.8)$$

which satisfies the insertion loss specifications. The capacitances are $C_{off} = 8.0$ pF and $C_{on} = 106$ fF, which translates to "off/on"-capacitance ratio $C_{off}/C_{on} = 76$.

The pull-in voltage from Equation (15.17) is

$$V_P = \sqrt{\frac{8}{27} \frac{k d_2^3}{\epsilon_0 A}} \approx 7.4 \text{ V}$$

and the actuation voltage should be higher than this.

Example 19.5: Hold down voltage for the capacitive switch

Problem: Since the electrode dielectric in capacitive switch is thin, only a small voltage is necessary to hold down the capacitive switch. Estimate the dc and ac voltages that will hold the bridge down for the switch geometry in Example 19.4. What signal power would prevent the switch from opening if the transmission line impedance is 50 Ω?

Solution: The spring displacement $\Delta x = d_2$ results in restoring force $|F_k| = kd_2$. This force needs to be countered with the electrostatic force that holds the plate down. From Equation (15.9), the dc voltage on the electrode needs to be at least

$$V_{dc} = \sqrt{\frac{2|F_k|d_1^2}{\epsilon_R \epsilon_0 A}} \approx 0.7 \text{ V.}$$

The hold-down voltage is significantly less than the pull-in voltage $V_P = 7.4$ V. Thus, a smaller voltage can be used for holding down the switch, which can reduce the dielectric charging effects and increase the device lifetime.

From Equation (15.32), the ac voltage that will hold the bridge down in absence of any dc voltage is

$$V_{ac} = \sqrt{\frac{4|F_k|d_1^2}{\epsilon_R \epsilon_0 A}} \approx 1 \text{ V,}$$

which for $Z_0 = 50$ Ω transmission line corresponds to signal power of $P = 0.5 V_{ac}^2/Z_0 \approx 10$ mW. The power in dBm units is $P_{dBm} = 10 \log_{10} \frac{P}{1 \text{ mW}} = 10$ dBm (see Appendix H). For systems such as cell phones where larger signal powers are used, the switch cannot change the state unless the ac signal is turned off, as the ac signal will hold the bridge in the down position.

A big problem with the capacitive switches is that the large capacitive contact area can result in the bridge becoming permanently stuck in the down position. The "stiction" is due to the thin dielectric becoming charged over time and the charges can hold the bridge electrostatically in the down position. As was shown by Example 19.5, the hold-down voltage necessary to keep the bridge in the actuated state is much lower than the pull-in voltage needed to actuate the switch. This allows the reduction of the actuation voltage in the down position to reduce the charging effect. Lowering the hold-down voltage has been shown to reduce but not completely eliminate the charging related stiction problems [172]. Increasing the bridge spring constant increases the restoring forces, which improves the switch reliability, but also increases the required actuation voltage. Pull-in voltages greater than 100 V have been used in part to increase the switch reliability.

Example 19.6: Dielectric charging of capacitive switches
Problem: Estimate the charge density required to permanently hold down the capacitive switch bridge of Example 19.4.
Solution: The hold-down voltage from Example 19.5 is $V \approx 0.7$ V and the dielectric capacitance is $C \approx 8.0$ pF. The capacitor charge needed to hold down the bridge is

$$Q = CV \approx 5.7 \text{ pC}.$$

The equivalent electron charge density is

$$\frac{Q}{Aq_e} \approx 2.9 \cdot 10^{11} \text{ electrons/cm}^2.$$

This is in the same order-of-magnitude as the typical charge trap densities for amorphous films such as silicon nitride and silicon dioxide used as the switch dielectric. The trapping of this charge in the dielectric can permanently hold down the bridge.

19.3.2   Ohmic switches

The capacitive switch performance is limited by the achievable "off/on" capacitance ratios. Especially at relatively low frequencies ~ 1 GHz, very large capacitance values are needed to obtain low enough impedance. A solution to this is to use a resistive switch with metal/metal contact that allows low closed state resistance from dc to microwave frequencies. Typical contact resistances are less than 0.5 Ω which compare very well to a 10-pF capacitor which has impedance of 16 Ω at 1 GHz. The low on resistance of the ohmic switches makes them especially suitable for cell phones operating in the frequency range of $0.9 - 1.8$ GHz.

Figure 19.9 shows a schematic for an electrostatically actuated ohmic switch. The switch manufacturing process is slightly more complex than for capacitive switches due to the need to make a good metal to metal contact. Separate well-defined contact points are formed to control the contact area and the contact force. Too large contact force will decrease the contact lifetime, and insufficient force will result in a large contact resistance. Typical contact forces are ~ 200 μN which requires large actuation voltages. Multiple contacts may be used in parallel to further decrease the contact resistance [173].

The first MEMS ohmic switches were demonstrated in the 1970s [174]. One reason for the slow adoption has been the relatively poor lifetime and reliability. The ohmic switch lifetime is limited by the deterioration of the switch contact resistance, which is a fundamentally different failure mechanism than

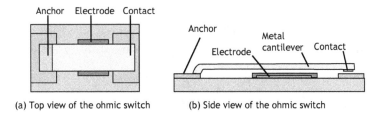

(a) Top view of the ohmic switch   (b) Side view of the ohmic switch

Figure 19.9: Electrostatically actuated ohmic switch (After [164]).

the charging related stiction observed with the capacitive switches. The impacts due to the switch closings will deform the contact, which will damage and harden the contact metals. Over time, this will increase the contact resistance. The demonstrated contact lifetimes have typically been ~100 million cycles with few switches demonstrating over 10 billion cycles when switching with no signal flowing through the contact (cold switching) [164, 173]. Lifetime of the metal to metal contact remains a big issue for ohmic switches and no switch has been able to demonstrate "hot switching" with current flowing through the contact while switching.

Example 19.7: Insertion loss and isolation in ohmic switch
Problem: Calculate the insertion and isolation for an ohmic switch with $R_{on} = 0.5\ \Omega$, $C_{off} = 50$ fF, $Z_0 = 50\ \Omega$ at 1 GHz. The equivalent circuit for the system is shown in Figure 19.10 with source and load impedances of $R_s = R_L = 50\ \Omega$. Compare to the capacitive switch in Example 19.4.

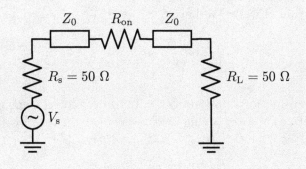

Figure 19.10: Equivalent circuit for ohmic switch.

Solution: The power delivered to the load with no switch losses is $P_0 = |V_s|^2/8R_L = |V_s|^2/8Z_0$ where $|V_s|$ is the peak magnitude of the voltage source. With the switch resistance R_{on} included, the power delivered to the load is $P = R_L I^2/2$, where the current through the load resistance is $|I| = |V_s|/(R_s + R_{on} + R_L) = |V_s|/(2Z_0 + R_{on})$. The "on" state insertion loss is

$$|S_{21,on}|^2 = \frac{P}{P_0} = \left|\frac{2Z_0}{2Z_0 + R_{on}}\right|^2 = -0.04 \text{ dB.}$$

The "off" state isolation is obtained by replacing the switch resistance in Figure 19.10 with a capacitor C_{off}. We have

$$|S_{21,off}|^2 = \left|\frac{2Z_0}{2Z_0 + 1/j\omega_0 C_{off}}\right|^2 \approx -30 \text{ dB.}$$

In comparison, the "off" state isolation for the capacitive switch in Example 19.4 is only -4.1 dB at 1 GHz .

Example 19.8: Switch contact force

Problem: Calculate the minimum actuation voltage for an ohmic switch with the actuation electrode area $A = (50 \ \mu m)^2$, the initial electrode gap $d_1 = 1 \ \mu m$, the electrode gap $d_2 = 0.5 \ \mu m$ in the closed position, and the spring constant $k = 40$ Nm. What is the contact force in the closed position if the actuation voltage is 1.5 higher than minimum voltage required?

Solution: The minimum actuation voltage is set by the pull-in voltage and is given by Equation (15.17). We have

$$V_P = \sqrt{\frac{8}{27} \frac{k d_1^3}{\epsilon_0 A}} \approx 23 \text{ V.}$$

The restoring spring force for the $\Delta x = 0.5 \ \mu m$ displacement is $F_k = k\Delta x = 20 \ \mu N$. The actuation force acting on the electrode in the closed position from Equation (15.9) is

$$F_{el} = \frac{1}{2}\epsilon_0 \frac{A}{d_2^2} V^2 \approx 53 \ \mu N,$$

where we have used $V = 1.5 V_P$. The net contact force is $F = F_{el} - F_k = 33 \ \mu N$. Typically larger contact forces are desired, which require a larger actuation voltage or a smaller actuation gap.

19.3.3 Switching speed

Since the MEMS switches are mechanical components, the switching speed is limited by the mechanical resonance frequency. The typical switching speeds are $1 - 100$ μs which are much slower than for solid-state switches. The step response of a harmonic resonator is reviewed in Appendix B. Figure B.2(c) on page 412 shows the step response for over and under damped structures. The response of the under damped structure is faster but the under damped resonator vibrates before settling. Similar behavior is observed with under damped switches that quickly make the initial contact but bounce before settling. To reduce the contact bounce, the quality factors for the RF MEMS switches are usually $0.2 < Q < 5$ [164].

Calculation of the exact switching time is difficult for electrostatically actuated switches as the actuation force changes with the gap and the air damping for the large displacement is nonlinear. A simple expression for the switching time that is accurate to within 10% for $Q > 1$ and $V_S > 1.3 V_P$ is

$$t = \sqrt{\frac{27}{2}} \frac{V_P}{V_s \omega_0}, \tag{19.9}$$

where V_S is the applied voltage and V_P is the pull-in voltage [164]. As Equation (19.9) shows, the fast response require high resonant frequency and high actuation voltage in comparison to pull-in voltage.

A slightly more complex expression for the switch closing time is

$$t_{\text{close}} = t_m + t_\gamma \tag{19.10}$$

where

$$t_m = \sqrt{\frac{27}{32}} \frac{\pi}{\omega_0} \frac{V_P}{V}. \tag{19.11}$$

and

$$t_\gamma = \frac{V_P^2}{Q \omega_0 V^2} \left(2.25 + 1.52 \frac{V_P^2}{V^2} + 1.22 \frac{V_P^4}{V^4} \right). \tag{19.12}$$

Equation (19.10) is accurate to 20% for $V > 1.15 V_P$ and $Q > 0.1$ [175]. If the actuation voltage is close to the pull-in voltage, the net force acting on the switch is small and the switching time is dramatically larger.

The switch opening time is

$$t_{\text{open}} = \begin{cases} \frac{\pi}{2\omega_0} & \text{for } Q > 1 \\ 1.6 \frac{1}{\omega_0 Q} & \text{for } Q < 1 \end{cases} \tag{19.13}$$

Equation (19.13) is accurate to 20%.

Example 19.9: MEMS relay switching speed
Problem: Calculate the switching speed for a switch with $f_0 = 20$ kHz, $V_S = 1.4V_P$, and $Q = 2$.
Solution: From Equation (19.9), the switch closing speed is

$$t = \sqrt{\frac{27}{2} \frac{V_P}{V_s \omega_0}} \approx 19 \ \mu s.$$

19.3.4 Cost and reliability

The capacitive switches are remarkably simple components and attractive for many systems. Although Example 19.4 showed that multiple switches may be needed to increase the isolation, the isolation for a single switch is comparable to the solid-state switches and ohmic switches offer even better performance. Given the simplicity of the switch, the good electrical properties, and more than $100 million spend on the RF switch development [164], it is perhaps a surprise that the switches have not been more widely adopted. The bottlenecks for wider acceptance are the cost and the reliability.

The MEMS fabrication is simple but the reliable switch operation requires hermetic packaging. In small-scale production, the price has been close to $10, which may be acceptable for aerospace and test equipment applications but is too expensive for consumer applications such as cell phones. Large-scale adoption by the cell phone module manufacturers is needed to get the MEMS switches to the consumers.

The reliability has been the biggest obstacle in adoption of MEMS switches. As we saw in the previous sections, the capacitive switches suffer from the stiction failure and ohmic switches suffer from contact degradation. The switch performance is steadily improving and operation lifetimes of over 100 billion cycles have been demonstrated in cold switching [176]. However, the long term reliability is still a question mark and no one can guarantee operation after ten years as this type of test data does not exist. This is a problem especially in high-end applications where the MEMS devices compete with existing solid-state and macroscopic relay solutions that have a track record from several decades of field use.

The challenges are illustrated by the MEMS start-up and RF MEMS developer Teravicta that closed operations and laid of 55 employees in 2008 after raising at least $22 million of venture funding over the course of seven years. Another well-known switch start-up Wispry refocused its efforts to varactors after several years of switch development. RF switches are still under active in development in many companies, but as is often the case for MEMS, the path to commercial success is longer than the initial rosy expectations.

19.4   RF inductors

Inductors at GHz frequencies suffer from two shortcomings: the parasitic capacitances limit the operation frequencies to a few GHz and the ohmic losses limit the inductor quality factor. To address these shortcomings, micromachining techniques have been used to make better inductors on top of ICs.

The basic idea in improving the inductors is to use better metals such as gold to reduce ohmic losses and to lift the inductors above the lossy IC substrate to reduce the parasitic capacitance to the substrate. Figure 19.11 illustrates the geometries for improved RF inductors. Typically, the inductor quality factors can be increased by 2× compared to the regular RF inductors.

Despite the performance increase and the relative simplicity of the fabrication process, the RF MEMS inductors have not been adopted in RF IC manufacturing [170]. This suggests that benefits do not outweigh the additional cost of processing and the yield reduction associated with the MEMS processes.

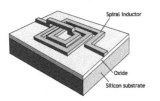 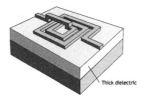 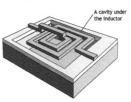

(a) A regular spiral inductor on top of silicon IC.

(b) The inductor/substrate coupling can be decreased by increasing the dielectric thickness.

(c) The inductor/substrate coupling can be decreased by removing the substrate under the inductor.

Figure 19.11: Methods to increase the performance of spiral inductors.

Key concepts

- Miniaturized mechanical RF switches and varactors have very good electrical performance but the cost and the reliability problems have slowed the adoption.

- The mechanical switches and varactors can be biased with resistors as opposed to inductors used for solid-state diode based components.

- The RF MEMS varactors have a wide tuning range and excellent linearity in comparison to the solid-state varactors.

- Capacitive MEMS switches are based on changing the capacitance by a large factor to block or pass the signal. The obtainable capacitance ratio limits the electrical performance of the capacitive switches.

- Ohmic switches are based on metal to metal contact that has a low resistance in the closed state. Consequently, the ohmic switches have a low insertion loss.

- The lifetime of MEMS switches remains an issue. The capacitive switches typically fail due to dielectric charging that causes the bridge to get stuck in the low position. The ohmic switches fail due to wear or hardening of the metal contacts, which increases the contact resistance.

- Inductors fabricated using MEMS techniques can have smaller parasitic capacitance to the substrate and higher quality factors. The overall performance gain has not been sufficient to justify the increased cost.

- Packaging of the RF MEMS components is a major cost issue. Solid-state devices that do not require complex protective packaging are cheap and replacing them is difficult in consumer applications. MEMS devices have a more level playing field against macroscopic mechanical relays in high-end applications where the linearity of the mechanical components is a major advantage.

Exercises

Exercise 19.1
Use Google Scholar (`scholar.google.com`, see also Exercise 1.7) to rank RF MEMS varactors, capacitive switches, ohmic switches, and inductors in the order of academic interest (the most interesting RF MEMS application first).

Exercise 19.2
Which of the following terms better describe diode based varactors and which RF MEMS varactors: fast, linear, low-loss, cheap, robust, CMOS compatible, and high bandwidth.

Exercise 19.3
Design a parallel plate MEMS varactor with a capacitance range of 1 pF to 1.2 pF and power handling capacity of 33 dBm in 50 Ω-system. Optimize your design for the lowest possible tuning voltage. Hint: use Appendix H to convert the dBm level to voltage and make sure that the tuning range can be achieved even with this voltage.

Exercise 19.4
A pn-diode with $I_s = 0.4$ nA, $n = 1.3$, $C_{j0} = 2$ pF, and $V_{bi} = 0.7$ V is used as a RF switch. (1) What is the "on" and "off" resistance if the diode is biased with on-state current $I = 1$ mA and off-state reverse bias $V = -10$ V. (2) What is the junction capacitance impedance in the off-state at $f = 1$ GHz? (3) What is the off-state isolation in a 50 Ω-system? (4) What is the on-state insertion loss?

Exercise 19.5

RF MEMS switches has experienced several "false starts" with several start-ups exiting the business without making profit. On November 21, 2007, however, a global cell phone chip manufacturer RF Micro Devices announced that they will introduce ohmic RF MEMS switches for 3G cell phones. When this happens, RF MEMS has become a mainstream consumer application. See what you can find out about the current status of the project.

Exercise 19.6

The Raytheon capacitive MEMS switch has the following properties: the spring constant is $k = 6 - 20$ N/m, the gap height in the up state is $d = 3 - 5$ μm, the dielectric thickness is $d_1 = 100$ nm, the relative permittivity for the dielectric is $\epsilon_R = 6 - 9$, and the electrode area is $A = 8,000$ μm^2 [170]. Calculate the range for the pull-in voltage and the off/on-capacitance ratio. How large is the isolation at 10 GHz in the off-state in a 50-Ω system?

Exercise 19.7

Derive the insertion loss (Equation (19.6)) for the capacitive switch in Figure 19.8 assuming that the load and source impedances are both $R_S = R_L = Z_0$.

Exercise 19.8

In the capacitive switch example (Example 19.4), we assumed that the pull-in voltage equation is not affected by the dielectric ($d_1 = 100$ nm, $\epsilon_R = 7.5$) under the air gap. This is justified if most of the applied voltage is over the air gap and the voltage drop over the thin dielectric can be ignored. Investigate this assumption.

Exercise 19.9

What signal level would cause self-actuation of the capacitive switch in Example 19.4. Assume 50-Ω system and express your answers in dBm-units.

20

Modeling microresonators

Resonators are used as sensor elements, filters, and frequency references. In this chapter, we will cover the mechanical and electrical modeling of MEMS resonators. The two commercially used actuation methods, electrostatic and piezoelectric, are investigated and compared. Using the theory and models developed in this chapter, the different microresonator applications are covered in Chapter 21 and gyroscopes are covered in Chapter 22.

Unlike the accelerometers that have clearly identifiable mass and spring, the resonating structures in this chapter have distributed mass and elasticity. For example, in thin film resonators, the film acts both as the "mass" and the "spring". This complicates the resonator analysis as vibration mode shape has to be included into the resonator model.

In this chapter, we will develop lumped electrical models for the distributed mechanical vibrations. First, the 3D vibrations will be modeled with a lumped 1D model that accounts the vibration mode shape. Next, we develop piezoelectric and capacitive electromechanical transduction models that are suitable for analyzing resonators. Finally, the lumped mechanical model and the transducer models are combined to develop an electrical equivalent circuit for microresonators.

20.1 Lumped model for mechanical vibrations

Until now, we have dealt with devices that have clearly identifiable mass and spring structures. This has made the analysis simple, as the system is characterized with just two parameters: the mass velocity and displacement. In typical resonating structures, however, the mass and the spring are distributed over the entire resonator. A good example is a guitar string where the string is both the mass and the spring.

Fortunately, the distributed model is not required for using the resonator as an electrical component and significant simplifications are obtained by developing a lumped model for the distributed vibrations. This can be compared to the circuit design where it is not required to master the physical details of transistor operation in order to use lumped circuit elements.

In this section, we will develop the lumped model from a continuum vibration mode shape for a distributed resonator. Figure 20.1 shows our example resonator that is used to illustrate the analysis steps. The resonator is a half wavelength beam resonator developed for reference oscillator applications [90]. As the figure shows, the vibration amplitude depends on the location X. The beam ends move in the opposite direction and the beam is anchored from the middle, which is the vibration nodal point. Anchoring at the nodal point is desirable as it minimizes the energy losses to the substrate and therefore maximizes the quality factor. We wish to replace the continuum model in Figure 20.1 with an equivalent lumped mass-spring-dash pot resonator to hide the location dependency.

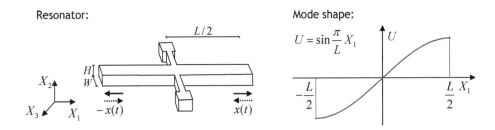

Figure 20.1: Schematic of longitudinal mode beam resonator and the vibrational mode shape [90].

20.1.1 Vibration mode

To analyze the beam resonator in Figure 20.1, we start with the Hooke's equation for stress and strain

$$T_{XX} = ES_{XX}, \tag{20.1}$$

where T_{XX} is the normal stress in X-direction, E is Young's modulus, and S_{XX} is the longitudinal strain in X-direction due to the beam displacement. Writing the beam displacement in X-direction at X as $u_X(X)$, the strain is $S_{XX} = \frac{\partial u_X(X)}{\partial X}$. Since the resonator is essentially one dimensional, we will drop the subscripts 'X' and 'XX' in the subsequent analysis steps to simplify the notation.

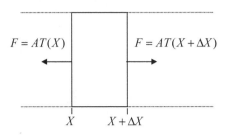

$F = AT(X)$ $F = AT(X + \Delta X)$

X $X + \Delta X$

Figure 20.2: Stress forces acting on an infinitesimal beam segment.

Figure 20.2 shows a small segment ΔX for the beam. The equation of motion for the beam is obtained by summing the forces acting on the beam segment. The force is the product of the normal stress T and the beam cross sectional area A. The net force is the difference of the forces acting on the left $(F = AT(X))$ and right boundary $(F = AT(X + \Delta X))$. We have

$$F = A(T(X + \Delta X) - T(X)) = A\frac{T(X + \Delta X) - T(X)}{\Delta X}\Delta X \approx A\frac{\partial T}{\partial X}\Delta X. \quad (20.2)$$

If the net force is not zero, the beam segment will accelerate. The Newtons equation for the displacement u is

$$F = m\frac{\partial^2 u}{\partial t^2} = \rho A \Delta X \frac{\partial^2 u}{\partial t^2}, \quad (20.3)$$

where we have used the beam mass $m = \rho A \Delta X$. Combining Equations (20.1), (20.2), and (20.3) gives

$$\rho A\frac{\partial^2 u}{\partial t^2} = EA\frac{\partial^2 u}{\partial X^2}. \quad (20.4)$$

This is the wave equation for a longitudinal motion in the beam.

Equation (20.4) is a partial differential equation for which the solution can be searched by the method of separation of variables by substituting $u(X, t) = U(X)x(t)$ into Equation (20.4). The separation of variables allows splitting the system into the mode shape $U(X)$ that depends only on the location and the vibration amplitude $x(t)$ depending only on time. Since the solutions to the wave equation are harmonic sinusoidal waves, we will use this information and use

$$u(X, t) = U(X)x(t) = (a\sin\beta X + b\cos\beta X)e^{j\omega_0 t} \quad (20.5)$$

as our trial solution. Substituting Equation (20.5) into Equation 20.4 gives

$$-\omega_0^2\rho A(a\sin\beta X + b\cos\beta X)e^{j\omega_0 t} = -\beta^2 EA(a\sin\beta X + b\cos\beta X)e^{j\omega_0 t}, \quad (20.6)$$

which simplifies to

$$\rho\omega_0^2 = E\beta^2. \quad (20.7)$$

The boundary conditions are given by the requirement that there is no stress and no stress gradient on the free beam ends. We have

$$T = E\frac{\partial u}{\partial X} = 0 \text{ and}$$
$$\frac{\partial T}{\partial X} = 0 \tag{20.8}$$

at the beam ends $(X = \pm L/2)$. Applying the boundary conditions given by Equation (20.8) to Equation (20.5) gives

$$\beta_n = (2n - 1)\pi/L, \tag{20.9}$$

$$U(X) = \sin(\beta_n X), \tag{20.10}$$

and

$$u(X, t) = \sin(\beta_n X)e^{j\omega_0 t}, \tag{20.11}$$

where $n = 1, 2 \ldots$ is the mode number. We are mostly interested in the first solution $\beta_1 = \pi/L$, which has the lowest resonance frequency

$$\omega_0 = \sqrt{\frac{E}{\rho}\beta_1^2} = \frac{\pi}{L}\sqrt{\frac{E}{\rho}} \tag{20.12}$$

or

$$f_0 = \frac{1}{2L}\sqrt{\frac{E}{\rho}}. \tag{20.13}$$

In the next section, we will use Equation (20.10) as the basis for a lumped model for the beam resonator.

20.1.2   Effective mass and spring for the lumped resonator

Before proceeding to obtain the effective mass and spring for the lumped resonator model, we will add damping to the beam resonator model. Equation (20.4) becomes

$$\rho A\frac{\partial^2 u}{\partial t^2} + bA\frac{\partial u}{\partial t} - EA\frac{\partial^2 u}{\partial X^2} = 0, \tag{20.14}$$

where b models the viscous damping losses. The work done in the previous section will help us in obtaining a solution to Equation (20.14). Equation (20.11) does not work right away but we'll take it as a starting point and assume that the *mode shape* given by Equation (20.10) remains the same and only the time behavior changes. We thus write the solution to Equation (20.4) as

$$u(X, t) = x(t)U(X) = x(t)\sin \beta X, \tag{20.15}$$

where $x(t)$ is the beam tip displacement at time t and $U(X) = \sin \beta X$ is the mode shape with a maximum amplitude of one.

Substituting Equation (20.15) into (20.14) leads to

$$\rho A \frac{\partial^2 x}{\partial t^2} \sin \beta X + bA \frac{\partial x}{\partial t} \sin \beta X + EA\beta^2 x \sin \beta X = 0. \tag{20.16}$$

Next, we'll multiply Equation (20.16) with the mode shape $\sin \beta X$ and integrate over the beam length:

$$\int_{-L/2}^{L/2} \left(\rho A \frac{\partial^2 x}{\partial t^2} \sin^2 \beta X + bA \frac{\partial x}{\partial t} \sin^2 \beta X + EA\beta^2 x \sin^2 \beta X \right) dX = 0. \tag{20.17}$$

Using $\int_{-L/2}^{L/2} \sin^2 \beta X \, dX = L/2$, Equation (20.17) leads to

$$\frac{\rho A L}{2} \frac{\partial^2 x}{\partial t^2} + \frac{bAL}{2} \frac{\partial x}{\partial t} + \frac{EA\beta^2 L}{2} x = 0. \tag{20.18}$$

By recognizing the effective mass, damping coefficient, and the spring constant in Equation (20.18) as

$$\begin{aligned} m &= \rho AL/2, \\ \gamma &= bAL/2, \\ &\text{and} \\ k &= EA\beta^2 L/2 = \pi^2 EA/2L, \end{aligned} \tag{20.19}$$

respectively, we obtain the familiar equation of motion for a lumped resonator given by

$$m \frac{\partial^2 x}{\partial t^2} + \gamma \frac{\partial x}{\partial t} + kx = 0. \tag{20.20}$$

The actual value for the effective mass and spring is arbitrary as the mode shape $U(X)$ can be scaled by any factor, which will scale Equation (20.16) as well. Our normalization choice of $U_{\max}(X) = 1$ is convenient as the lumped resonator displacement corresponds to the physical displacement of the beam end $u(L/2, t) = x(t)$.

It is interesting to compare the effective mass and spring to the physical mass $m = \rho AL$ and the spring constant $k = EA/L$ for a uniform beam elongation (see Section 4.3.1 on page 57). The effective mass from Equation (20.19) is half of the physical mass and the spring constant is about five times higher than for the uniform rod extension. This makes sense, as only portion of the beam is moving – the mass is reduced and the spring is effectively shorter.

Example 20.1: Effective mass and spring constant for the longitudinal mode beam resonator

Problem: Calculate the effective mass, the effective spring constant, and the resonant frequency for the silicon beam resonator in Figure 20.1, if the beam width and height are $H = 10$ μm and $W = 10$ μm, respectively, and the beam length is $L = 400$ μm.

Solution: From Equation (20.19), the effective mass and the effective spring constant are

$$m = \rho A L/2 \approx 46.6 \text{ pkg}$$

and

$$k = \pi^2 E A/2L \approx 210 \text{ kN/m},$$

respectively, where we have used $A = WH$, $E = 170$ GPa, and $\rho = 2329$ kg/m^3. The resonance frequency is

$$f_0 = \frac{1}{2\pi}\sqrt{\frac{k}{m}} \approx 11 \text{ MHz}.$$

20.1.3   General calculation of the lumped model parameters

In the previous section, we obtained the effective mass and spring by directly solving the partial differential equation that governs the vibrations. While this approach was illustrative, it is cumbersome for more complex geometries when vibrations are two- or three-dimensional. The analysis can be simplified by noting that all we need to know for the effective mass and the effective spring constant are the mode shape $\vec{U}(\vec{X})$ and the resonance frequency ω_0. Extending the 1D analysis to 3D, we write the effective mass as

$$m = \int \rho |\vec{U}(X, Y, Z)|^2 \mathrm{d}V. \tag{20.21}$$

With the effective mass and the resonance frequency known, the effective spring constant is obtained from

$$k = \omega_0^2 m. \tag{20.22}$$

Equations (20.21) and (20.22) allow easy calculation of the effective mass and spring constant based on the known mode shape and resonance frequency. For example, Appendix H tabulates several often encountered resonators and further tables are found in reference [34]. Example 20.2 illustrates how the effective mass and spring constants are obtained for a clamped-clamped beam based on the known mode shape. For less common geometries, the mode shape and resonant frequency can be estimated using Rayleigh-Ritz method [6] or finite element modeling (FEM).

Example 20.2: Clamped-clamped beam

Problem: Figure 20.3 shows a clamped-clamped or fixed-fixed beam. The beam is solidly supported from both ends. The width of the silicon beam is $w = 4$ μm, the height is $h = 10$ μm, and the length is $L = 240$ μm. Calculate the resonant frequency, the effective mass, and the effective spring constant for the first resonator vibration mode.

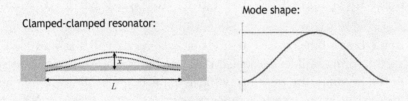

Figure 20.3: Clamped-clamped beam resonator. Both ends of the beam are fixed and only the middle portion vibrates.

Solution: From Appendix H, the resonant frequency for the clamped-clamped beam is

$$\omega_0 = \frac{\beta_1}{\sqrt{12}}\sqrt{\frac{E}{\rho}}\frac{w}{L^2}.$$

Using $\beta_1 = 4.730041$ from Table H.3 for the first mode shape, we obtain $\omega_0 \approx$ 9.6 MHz and $f_0 \approx 1.5$ MHz. The mode shape for the first resonance is plotted in the Figure 20.3 and it is given by

$$U_1(X) = C_1[\sinh \beta_1 X/L - \sin \beta_1 X/L + \alpha_1(\cosh \beta_1 X/L - \cos \beta_1 X/L)],$$

where $\alpha_1 = \frac{\sinh \beta_1 - \sin \beta_1}{\cos \beta_1 - \cosh \beta_1}$. Requiring that the maximum of the mode shape is one ($U_1(L/2) = 1$), we obtain the normalizing constant as $C_1 = 1.62$. The effective mass is obtained using Equation (20.21):

$$m = \int \rho U^2 \mathrm{d}V = \rho A \int_0^L U_1^2(X)\mathrm{d}X = 0.40\rho AL \approx 8.8 \text{ pkg},$$

where numerical integration has been used. The effective spring constant from Equation (20.22) is

$$k = \omega_0^2 m \approx 819 \text{ N/m}.$$

20.2   Electromechanical transduction

The two main methods to actuate and sense resonators are the piezoelectric and capacitive transduction. While other transduction methods have been demonstrated, they have not had commercial traction.

The piezoelectric transduction is the dominant actuation method for high frequency resonators used in communication systems. The piezoelectric transduction provides strong electromechanical coupling which is especially important for filter applications. Piezoelectric coupling also competes with capacitive actuation in gyroscopes. A disadvantage of the piezoelectric coupling is that the additional processing steps for the piezoelectric material increases processing cost and may limit the resonator geometries.

The capacitive coupling is easy to implement and is well suited for resonators made of any conductive material. Silicon based gyroscopes are especially popular as the modern fabrication methods enable very small size and low cost in comparison to often bulkier piezoelectric gyroscopes. Capacitive actuation also has potential in miniature resonators suitable for frequency and timing applications. In this section, we cover the linear electromechanical transduction models for the capacitive and the piezoelectric actuation.

20.2.1   Transduction factor

Before analyzing a specific transduction mechanism, we will define a generic model for the transducer. In resonators, we are interested in how the motional current i_{mot} relates to the vibration velocity and how the actuation force relates to the ac drive voltage. For piezoelectric and capacitive actuation, both relationships can be expressed in terms of an electromechanical transduction factor η that relates the vibration velocity to the motional current

$$i_{\mathrm{mot}} = \eta \dot{x} \qquad (20.23)$$

and the ac voltage to the actuation force

$$F = \eta v. \qquad (20.24)$$

In the following section, we will first investigate how the transduction factor depends on the transducer location. Next, we will develop expressions for the transduction factor for the electrostatic and capacitive actuation.

20.2.2   Transduction in distributed resonators

The efficiency of the electromechanical transduction in distributed resonators depends on the location of the transducer. For example, the beam resonator in Figure 20.1 is most efficiently actuated from the beam ends that have the highest vibration amplitude. Conversely, a time harmonic force at the location of the anchor (the nodal point) will not excite the resonant mode.

To model the location dependency of the transducer, we will define an effective transduction factor that is normalized by the mode shape amplitude at the location of the transducer. Recall that in our model, the resonator displacement at location $\vec{X} = (X, Y, Z)$ is $\vec{u}(\vec{X}, t) = x(t)\vec{U}(X, Y, Z)$, where $x(t)$ is the vibration amplitude at the location of maximum displacement and $\vec{U}(X, Y, Z)$ is the resonator mode shape normalized to have the maximum amplitude of one. The effective transduction factor for a transducer placed at the location (X, Y, Z) is

$$\eta = \eta_0 U(X, Y, Z), \tag{20.25}$$

where the η_0 is the transduction factor and $U(X, Y, Z)$ amplitude of the mode shape in the direction of the transducer force [177]. As the transduction factor η_0 is normalized by the mode shape, placing the transducer at the location other than the maximum displacement reduces the both force and the motional current.

The use of Equation (20.25) is best illustrated with an example. Figure 20.4 shows three ways for actuating the beam resonator. In Figure 20.4(a), the resonator is actuated with a single capacitive transducer that generates force $F_0 = \eta_0 v$. As the transducer is located at the right beam end where $X = L/2$, the effective transduction factor is

$$\eta = \eta_0 U(X) = \eta_0 U(L/2) = \eta_0, \tag{20.26}$$

where we have used $U(L/2) = \sin \beta L/2 = 1$. The effective force that excites the vibration mode is $F = \eta v = \eta_0 v$.

In Figure 20.4(b), an additional transducer is placed at the left beam end where $X = -L/2$. This generates a force $F = -\eta_0 v$ which has equal magnitude but the opposite direction. The effective transduction factor is

$$\eta = \eta_0 U(L/2) - \eta_0 U(-L/2) = \eta_0 \sin(\beta L/2) - \eta_0 \sin(-\beta L/2) = 2\eta_0. \quad (20.27)$$

The effective force that excites the vibration mode is $F = \eta v = 2\eta_0 v$.

In our last example in Figure 20.4(c), the beam is actuated with a piezoelectric thin film on the top of the beam. The electrode on top of the film does not cover the entire surface and only the excited portion of the film actuates beam. As discussed in Chapter 16, the piezoelectric actuator can be modeled as a lumped force placed at the actuator edge. Writing the force at the right electrode edge location as $F_0 = \eta_0 v$ and left edge location as $F_0 = -\eta_0 v$, the effective force for the two transducers edges placed at $X = L_1$ and $X = -L_1$ is

$$\eta = \eta_0 U(L_1) - \eta_0 U(-L_1) = 2\eta_0 \sin \beta L_1. \tag{20.28}$$

The effective coupling for the thin film transducer is further explored in Example 20.3.

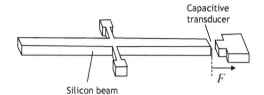

(a) Beam resonator actuated with one transducer at $X = L/2$.

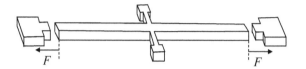

(b) Beam resonator actuated with two transducers.

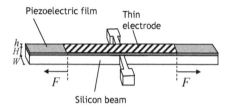

(c) Beam resonator actuated with a piezoelectric film.

Figure 20.4: The location of the transducer affects the effective electromechanical transduction factor.

Example 20.3: Effective transduction factor on the beam transducer
Problem: Calculate the effective transduction factor for resonator in Figure 20.4(c) if the transducer locations are $-L/3$ and $L/3$.
Solution: From Equation (20.28), the effective transduction factor is

$$\eta = \eta_0 U\left(\frac{L}{3}\right) - \eta_0 U\left(\frac{-L}{3}\right) = \eta_0 \sin\frac{\pi}{L}\frac{L}{3} - \eta_0 \sin\frac{\pi}{L}\frac{-L}{3} = \sqrt{3}\eta_0 \approx 1.73\eta_0.$$

If the metal electrode had covered the entire beam surface, the effective transduction factor would have been $2\eta_0$. However, increasing the electrode length from $L/3$ to $L/2$ would increase the capacitance between the electrodes by 50%. For example in oscillators, the smaller electrode area may be preferred as the benefits of the lowered capacitance may offset the undesirable loss in electromechanical coupling.

20.2.3　Capacitive transduction

The capacitive sensing was covered in Chapter 6 and actuation was covered in Chapter 15. Before proceeding further, we will summarize the key results relevant to the resonator transduction. The current through the capacitor is

$$i = i_{ac} + i_{mot}, \tag{20.29}$$

where $i_{ac} = C\dot{v} = sCv$ is the normal ac current through the capacitance due to the ac voltage v. The motional current i_{mot} is given by Equation (6.5) rewritten here for convenience:

$$i_{mot} = V_{dc}\frac{\partial C}{\partial t} = V_{dc}\frac{\partial C}{\partial x}\frac{\partial x}{\partial t} = \eta\dot{x}, \tag{20.30}$$

where V_{dc} is the dc bias voltage, C is the capacitance, x is the capacitance plate displacement, and η is the electromechanical transduction factor that is defined as

$$\eta = V_{dc}\frac{\partial C}{\partial x}. \tag{20.31}$$

The force generated by the capacitive actuator depends on the dc bias voltage V_{dc} and ac voltage v. For resonators, we are interested in the time harmonic force given by Equation (15.29) rewritten here as

$$F_{ac} = V_{dc}\frac{\partial C}{\partial x}v = \eta v. \tag{20.32}$$

Equations (20.29), (20.30) and (20.32) can be written more efficiently as

$$i = sCv + \eta\dot{x} \tag{20.33}$$

and

$$F_{ac} = \eta v. \tag{20.34}$$

For parallel plate capacitors, the electromechanical transduction factor from Equation (15.30) is

$$\eta_P = V_{dc}\frac{\epsilon A}{(d - x_0)^2} = V_{dc}\frac{C_0}{(d - x_0)}, \tag{20.35}$$

where $C_0 = \epsilon A/(d - x_0)$ and ϵ is the permittivity, A is the electrode area, d is the nominal electrode gap, and x_0 is the electrode displacement at the operating point. For a comb finger capacitor, we have

$$\eta_C = V_{dc}\frac{\epsilon h}{d} \tag{20.36}$$

per electrode overlap. Here h is the electrode height and d is the electrode gap.

Example 20.4: Electrostatic actuation of longitudinal mode beam resonator
Problem: Consider the beam resonator in Example 20.1 and Figure 20.1. The beam is actuated capacitively with electrodes on both ends. Assume the electrode areas are $A = WH$, the electrode gaps are $d = 1$ μm, the resonator quality factor is $Q = 100,000$, the bias voltage is $V_{dc} = 50$ V, and the ac excitation voltage is $v_0 = 100$ mV at the resonance frequency. Calculate the electromechanical transduction factor, the vibration amplitude, the motional current, and the ac current through the electrodes.
Solution: From Equations (20.31) and (15.30), we have

$$\eta_0 = V_{dc}\frac{\epsilon A}{(d - x_0)^2} \approx V_{dc}\frac{\epsilon A}{d^2} \approx 44.25 \text{ nN/V}.$$

Since we have two transducers located at the beam ends, the effective transduction factor for both transducers is $\eta = 2\eta_0 \approx 89$ nN/V. We see that the capacitive force generated $F_{ac} = \eta v$ is fairly small, only 8.8 nN for $v = 100$ mV.
From Appendix B, the vibration amplitude at $v_0 = 100$ mV is

$$|x_0| = \frac{QF}{k} = \frac{Q\eta}{k}v = 4.2 \text{ nm},$$

where we have used $k = 210$ kN/m from Example 20.1.
From Equation (20.30), the motional current is

$$|i_{mot}| = \eta\omega x_0 \approx 25 \text{ nA}.$$

The ac current through the electrode capacitance $C_0 = 2\epsilon_0 WH/d \approx 1.77$ fF is $i_{ac} = 12$ nA.

20.2.4   Piezoelectric transduction

The piezoelectric sensing was covered in Chapter 7 and the actuation was covered in Chapter 16. In this section, we will collect the results from these chapters to develop a generic model for the piezoelectric transducer. We will first consider the longitudinal configuration where the strain and electric field are aligned in the same direction between the two parallel electrodes located $X = h/2$ and $X = -h/2$ as shown in Figure 7.1(a) on page 111.

The electrodes form a capacitor and the current through the capacitor will have a piezoelectric component in addition to the regular capacitance current.

From Equation (7.8), the current through the capacitor is

$$i = sCv + s\frac{e_{33}A}{h}\left(u_3(h/2) - u_3(-h/2)\right) \equiv i_{ac} + i_{mot}, \qquad (20.37)$$

where u_3 is the electrode displacement. The first term in Equation (20.37) is the regular current due to the capacitance between the electrodes and the second term is the motional current. By defining the electromechanical transduction factor as

$$\eta_0 = \frac{e_{33}A}{h} \qquad (20.38)$$

and using $u_3 = xU(X)$ where x is the lumped displacement and U is the mode shape, we can rewrite Equation (20.37) as

$$i = C\dot{v} + \eta\dot{x} \qquad (20.39)$$

where the effective electromechanical transduction factor is $\eta = \eta_0(U(h/2) - U(-h/2))$.

The model for the piezoelectric transducer force is shown in Figure 16.1(a) on page 242. The piezoelectric transducer force is given by Equation (16.5) which we rewrite for resonators as

$$F_0 = \frac{e_{33}A_3}{h}v = \eta_0 v. \qquad (20.40)$$

Accounting both electrodes and the effect of the mode shape, the effective force acting on the resonance mode is

$$F = \eta_0(U(h/2) - U(-h/2))v = \eta v. \qquad (20.41)$$

Equations (20.39) and (20.41) are identical to the respective equations for the capacitive transduction.

Example 20.5: Transduction factor for a piezoelectric film bulk acoustic resonator (FBAR)
Problem: An AlN FBAR resonator in its simplest approximation is a thin film that vibrates in the thickness mode along X_3-axis. Referring to Figure 16.1(a) on page 242, the device has the area $A = (100\ \mu\text{m})^2$ and the thickness $h = 2\ \mu\text{m}$. Assuming that the mode shape is $u(X_3) = \sin(kX_3)$ where $k = \pi/h$, calculate the piezoelectric electromechanical transduction factor. Compare to the capacitive transduction factor if there is a 1,000 V dc bias between the top and bottom electrodes.

Solution: Table 7.1 list the properties for the AlN. From Equation (20.38), we have

$$\eta_0 = \frac{e_{33}A}{h} = 0.052 \text{ N/V}$$

for the electromechanical transduction factor and the effective electromechanical transduction factor is $\eta = \eta_0(U(h/2) - U(-h/2))v = 2\eta_0 = 0.10$ N/V. With the 1,000-V dc bias between the electrodes, the capacitive transduction factor from Equation (20.31) is

$$\eta_{C,0} = \epsilon_R\epsilon_0\frac{A}{h^2}V_{\text{dc}} = 0.65 \cdot 10^{-4} \text{ N/V}.$$

The effective electromechanical transduction factor for the two electrodes is $\eta_C = 2\eta_{C,0} = 1.7 \cdot 10^{-4}$ N/V. Even with the extremely high bias voltage, the capacitive transduction factor is three orders-of-magnitude smaller than the piezoelectric transduction factor. This is the fundamental reason why piezoelectric actuation is dominant in the commercial high frequency filters where strong electromechanical coupling is needed.

In the transverse transducer configuration shown in Figure 7.1(b) on page 111, the strain and electric field are orthogonal. The electrode edges are located at $X_1 = -L/2$ and $X_1 = L/2$ and the total current through the electrodes is

$$i = sCv + s\frac{e_{31}A_3}{h}\left(u_1(L/2) - u_1(-L/2)\right), \tag{20.42}$$

where $A_3 = wL$ is the electrode area and h is the transducer thickness. Using $u_1 = xU(X)$ where x is the lumped displacement and U is the mode shape, we can rewrite Equation (20.42) as

$$i = C\dot{v} + \eta\dot{x} \equiv i_{\text{ac}} + i_{\text{mot}}, \tag{20.43}$$

where the effective electromechanical transduction factor is $\eta = \eta_0(U(h/2) - U(-h/2))$.

The transverse transducer force is shown in Figure 16.1(b) on page 242. The generated transverse force at the edge of an electrode is

$$F_0 = \frac{e_{31}A_1}{h}v = e_{31}wv \equiv \eta_0 v \tag{20.44}$$

where

$$\eta_0 = e_{31}w. \tag{20.45}$$

Accounting for the force on both the electrode edges and the mode shape, the effective force acting on the resonator is

$$F = \eta_0(U(L/2) - U(-L/2))v = \eta v. \tag{20.46}$$

In conclusion, the currents and forces in capacitively and piezoelectrically actuated resonators can be analyzed using the transduction factor. The formalism for the analysis is identical and only the transduction factor is different for different transducers.

Example 20.6: Piezoelectric actuation of beam resonator
Problem: Figure 20.5 shows the beam resonator actuated with a piezoelectric AlN film on top of the resonator. Assume that the film covers the whole beam and that the film does not affect the mode shape. The silicon beam width and height are $H = 10$ μm and $W = 10$ μm, respectively and the beam length is $L = 400$ μm. The metal electrode has length $L_e = 2L/3$ μm and it is thin enough to be ignored for this problem. Use the film thickness $h = 1$ μm, and the resonator quality factor $Q = 100,000$. Calculate (1) the effective mass, the spring constant, and the transduction factor for the resonator, (2) the vibration amplitude for $v_0 = 100$ mV excitation voltage at resonance frequency, and (3) the motional and the ac current through the electrodes. (4) Compare the coupling to the electrostatic actuation analyzed in Example 20.4.

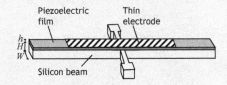

Figure 20.5: Beam resonator actuated with piezoelectric film.

Solution: (1) From Equation (20.19) and Example 20.1, the effective mass and spring constants for the silicon beam are

$$m = \rho AL/2 \approx 46.6 \text{ pkg}$$

and

$$k = EAL/2 = \pi^2 EA/2L \approx 210 \text{ kN/m},$$

respectively, where we have used $A = WH$, $E = 160$ GPa, and $\rho = 2329$ kg/m^3. The values for the film are

$$m_f = \rho WhL/2 \approx 6.4 \text{ pkg}$$
$$k_f = EWhL/2 = \pi^2 EA/2L \approx 49.3 \text{ kN/m}.$$

The resonant frequency is

$$f_0 = \frac{1}{2\pi}\sqrt{\frac{k + k_f}{m + m_f}} \approx 11 \text{ MHz}.$$

The transduction factor for one electrode edge from Equation (20.45) is

$$\eta_0 = e_{31}w \approx -4.8 \ \mu\text{N/V}.$$

From Example 20.3, the effective transduction factor that accounts for the two electrode edges and their location is

$$\eta = 1.73\eta_0 \approx -8.3 \ \mu\text{N/V}.$$

(2) From Appendix B, the vibration amplitude at $v = 100$ mV is

$$|x_0| = \frac{QF}{k} = \frac{Q\eta}{k + k_f}v_0 \approx 321 \text{ nm}.$$

(3) From Equation (20.30)

$$|i_{\text{mot}}| = |\eta\omega x_0| \approx 19 \ \mu\text{A}.$$

The electrode capacitance $C = \epsilon_R\epsilon_0 L_{el}W/h \approx 0.15$ pF and the ac current through the capacitance $|i_{ac}| = \omega C v_0 \approx 1.1 \ \mu\text{A}$.

(4) Compared to electrostatic transduction in Example 20.4 with one micron gap and 50 V bias, the piezoelectric coupling is about 100× larger. This large coupling translates into a significantly higher vibration amplitude and motional current.

The conclusion that the electrostatic coupling is weaker that the piezo-electric coupling seems to hold in general and may prevent the electrostatic actuation from being successful in filters. In applications such as oscillators and gyroscopes where the weaker coupling is acceptable, the simpler fabrication process can provide an edge for the electrostatic actuators.

20.3   Electrical equivalent circuit

Until now, we have described the resonators in terms of the mechanical parameters (mass, spring, damping) and electrical transduction factor. The lumped model was a significant simplification in comparison to solving differential equations for the mechanical vibrations, but our model is still not ideal for electrical design. To make the resonator transparent to electrical engineers who are used describing the resonators in terms of resistance, inductance, and capacitance, we will describe the mechanical resonator in terms of circuit parameters.

To develop an equivalent circuit for the mechanical resonator, we start by substituting $\dot{x} = \frac{\partial x}{\partial t} = i_{\text{mot}}/\eta$ from Equation (20.23) into Equation (20.20) to obtain

$$\frac{m}{\eta}\frac{\partial i_{\mathrm{mot}}}{\partial t} + \frac{\gamma}{\eta}i_{\mathrm{mot}} + \frac{k}{\eta}\int i_{\mathrm{mot}}\mathrm{d}t = F, \qquad (20.47)$$

where η is the effective electromechanical transduction factor for the vibration mode. Next, using $F = \eta v$ from Equation (20.23), we write Equation (20.47) as

$$\frac{m}{\eta^2}\frac{\partial i_{\mathrm{mot}}}{\partial t} + \frac{\gamma}{\eta^2}i_{\mathrm{mot}} + \frac{k}{\eta^2}\int i_{\mathrm{mot}}\mathrm{d}t = v. \qquad (20.48)$$

Equation (20.48) is identical to the current-voltage relationship in a series RLC-resonator

$$L_m\frac{\partial i_{\mathrm{mot}}}{\partial t} + R_m i_{\mathrm{mot}} + \frac{1}{C_m}\int i_{\mathrm{mot}}\mathrm{d}t = v, \qquad (20.49)$$

if we define the motional resistance, the motional capacitance, and the motional inductance as

$$\begin{aligned} R_m &= \gamma/\eta^2 = \sqrt{km}/Q\eta^2, \\ C_m &= \eta^2/k, \text{ and} \\ L_m &= m/\eta^2. \end{aligned} \qquad (20.50)$$

Figure 20.6 shows the complete resonator model. The motional arm is represented by the series RLC-network and the capacitance C_0 represents the regular current path so that the total current through the circuit is $i = i_{\mathrm{mot}} + i_{\mathrm{ac}}$. The electrical representation is convenient for a designer who wants to use the resonator in an electrical system such as a radio receiver. The usage of the RLC-model is illustrated in Example 20.7.

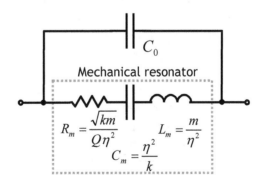

Figure 20.6: Electrical equivalent circuit for microresonator.

Example 20.7: Equivalent model for the beam resonator
Problem: Calculate the electromechanical electrical equivalents for the beam

resonator in Examples 20.1 and 20.4 when the resonator is excited with both electrodes. How large is the motional current for $v = 100$ mV excitation signal? How does it compare to capacitance current through C_0?

Solution: From Example 20.1 the effective mass and spring constants are

$$m = \rho AL/2 \approx 46.6 \text{ pkg}$$

and

$$k = EAL/2 = \pi^2 EA/2L \approx 210 \text{ kN/m},$$

respectively. From Example 20.4, the electromechanical transduction factor at 50 V bias voltage is

$$\eta = 2V_{\text{dc}}\frac{\epsilon A}{d^2} \approx 89 \text{ nN/V}.$$

From Equation (20.50), the electrical equivalents are

$$R_m = \sqrt{km}/Q\eta^2 \approx 4.0 \text{ M}\Omega,$$
$$C_m = \eta^2/k \approx 3.7 \cdot 10^{-20} \text{ F, and}$$
$$L_m = m/\eta^2 \approx 5.9 \cdot 10^3 \text{ H}.$$

At resonance, the C_m and L_m cancel. Thus, the motional current due to the $v = 100$ mV excitation signal at resonance frequency is $i_{\text{mot}} = v/R_m \approx 25$ nA in agreement with Example 20.4. The electrode capacitance is $C_0 = \epsilon_0 WH/d \approx 1.77$ fF and the current through the two capacitances is $|i_{\text{ac}}| = 2\omega_0 C_0 v \approx 24$ nA.

This example illustrates the challenge inherent with capacitive actuation of high frequency resonators. Even with the high bias voltage of 50 V, the motional resistance is very high and the current through the capacitance is larger than the motional current. This device has little practical use as an electrical resonator or filter.

A straightforward way to reduce the motional resistance and bias voltage is to reduce the electrode gap. For example, a gap of $d = 100$ nm and bias voltage of $V_{\text{dc}} = 5$ V would lead to $R_m = 39$ kΩ. This is a 100× improvement but resistance is still a fairly high value. To lower the motional resistance even further, resonator geometries with large electrode areas have been developed.

20.4 Nonlinear effects in microresonators

The maximum vibration amplitude for resonators is set by nonlinear effects. As the signal current is proportional to the vibration amplitude, nonlinear effects set the maximum signal amplitude in oscillators and filters. Typically, the signal is reduced when the resonator is scaled to a smaller size, and it therefore becomes important to operate the microresonators at the nonlinear performance limit to obtain performance comparable to macroscopic resonators.

The nonlinearities in resonators can be modeled by including nonlinear springs k_1 and k_2 into the harmonic resonator:

$$m\frac{\partial^2 x}{\partial t^2} + \gamma\frac{\partial x}{\partial t} + kx + k_1 x^2 + k_2 x^3 = F_\omega \cos\omega t. \tag{20.51}$$

Here F_ω is amplitude of the excitation force and k_1 and k_2 are the first- and second-order nonlinear springs that cause the response to deviate from the linear spring force $F = -kx$.

The nonlinear vibrations are reviewed in Appendix C. Due to nonlinear springs, the resonant frequency depends on vibration amplitude and is

$$\omega_0' = \omega_0 + \kappa x_0^2, \tag{20.52}$$

where

$$\kappa = \frac{3k_2}{8k}\omega_0 - \frac{5k_1^2}{12k^2}\omega_0. \tag{20.53}$$

If the vibration amplitude is increased sufficiently, the frequency response will show hysteresis as is illustrated in Figure C.2 on page 423. The maximum vibration amplitude before hysteresis is

$$x_c = \sqrt{\frac{4\omega_0}{3\sqrt{3}Q|\kappa|}}. \tag{20.54}$$

The maximum hysteresis-free vibration amplitude x_c is a good measure of the nonlinearity and can be used to compare microresonator designs [178].

The mechanical signal-to-noise ratio for resonators is $W/k_B T$ where W is the energy stored in the resonator and $k_B T$ is the thermal rms-noise. The maximum energy stored in the resonator at the critical vibration amplitude x_c is

$$W_c = \frac{1}{2}k_0 x_c^2. \tag{20.55}$$

As a general trend, the maximum vibration amplitude x_c and energy W_c decrease with decreasing device size. As the thermal noise energy $k_B T$ remains unchanged, this will limit the signal-to-noise ratio in microresonators. Of practical interest is the maximum power dissipated in the resonator or the "drive level"

$$P_c = \frac{\omega_0 W_c}{Q}. \tag{20.56}$$

Equation (20.56) also provides the maximum output power that can be obtained from the resonator when the resonator is matched to an optimal load; for the optimal power matching, the source and load resistances are equal.

20.4.1 Case study: Nonlinearity in a clamped-clamped beam

To illustrate the nonlinear effects in resonators, we'll consider the clamped-clamped beam resonator shown in Figure 20.7. As is typical in MEMS, the spring forces have both mechanical and electrical origin. We therefore write the spring constant as $k = k_m + k_e$, $k_1 = k_{m1} + k_{e1}$, and $k_1 = k_{m2} + k_{e2}$ where m and e refer to mechanical and electrical origin. The nonlinear mechanical spring constants are obtained from the large deformation analysis. Typically, finite element simulations are used, but for our purposes, we use the first-order approximation for the nonlinear spring constant

$$k_{m1} = 0$$
$$k_{m2} = \frac{k_m}{\sqrt{2}w^2}, \tag{20.57}$$

where w is the beam width and k_m is the linear mechanical spring constant [178]. As Equation (20.57) indicates, the nonlinear spring constant k_{m2} is positive, meaning that the restoring force increases with the displacement. This additional force is due to the stretching of the beam and is in addition to the linear spring due to the beam stiffness.

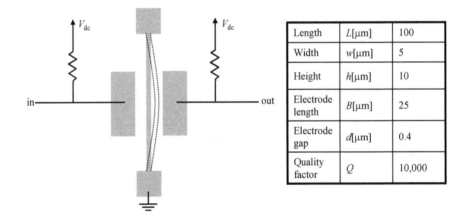

Length	$L[\mu m]$	100
Width	$w[\mu m]$	5
Height	$h[\mu m]$	10
Electrode length	$B[\mu m]$	25
Electrode gap	$d[\mu m]$	0.4
Quality factor	Q	10,000

Figure 20.7: Clamped-clamped beam example.

The electrical nonlinearity arises from the inverse relationship between the displacement and the parallel plate capacitance. From Equation (15.21), the electric spring is

$$k_e(x) = k_{0e} + k_{1e}x + k_{2e}x^2, \tag{20.58}$$

where the electrical spring constants are

$$k_{0e} = -\frac{V^2 C_0}{d^2}, \quad k_{1e} = \frac{3}{2d}k_{0e}, \quad \text{and } k_{2e} = \frac{2}{d^2}k_{0e}. \tag{20.59}$$

Here C_0 is the sum of the both electrode capacitances and d is the electrode gap. Notice that the linear electrostatic spring is negative thus lowering the resonator resonance frequency. This can be used to electrically tune the resonance frequency. The electrical adjustment of the spring constant is sometimes used in gyroscopes to fine tune the spring constants.

Figure 20.8 shows the analytical and simulated resonator responses at different bias voltages. At low bias voltages, the mechanical spring constant dominates and at the high drive levels the resonant peak shifts to a higher frequency. At higher bias voltages, the capacitive nonlinearity dominates, and the peak frequency tilts toward lower frequencies. To the first-order, we can even compensate for the mechanical nonlinearity with the capacitive nonlinearity [179]. Figure 20.8 also shows that with the increasing bias voltage, the resonance frequency shifts to a lower frequency as the electrical spring constant k_{e0} grows. This effect can be used to tune the resonator resonance frequency.

As shown in Figure 20.8, the vibration amplitude of the resonator is limited to about 0.1 nm before the nonlinear effects take effect. From Equations (20.55) and (20.56) we can estimate the maximum energy stored in the resonator $E = 1/2k_0x_c^2 \approx 0.17$ pJ and the maximum power output $P = \omega_0 E/Q \approx 0.46$ nW. We see that the nonlinearity seriously limits the resonator power handling capacity.

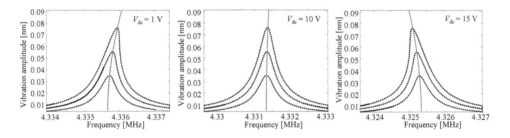

Figure 20.8: Simulated (dotted lines) and analytical (solid lines) responses for the beam example.

Key concepts

- Distributed resonators can be modeled with a lumped mass and spring.

- The effective electromechanical transduction factor depends on the location of the transducer. The maximum coupling is obtained by placing the transducers at the locations of the maximum resonator displacements.

- Electrostatic and capacitive actuation can be modeled with the electromechanical transduction factor that relates current to resonator velocity and force to excitation voltage.

- Electrical equivalents can be used analyze the resonator's electrical properties.

- For oscillator and filters, low motional resistance is needed. This requires a high Q-value and a high effective electromechanical transduction coefficient.

- Nonlinear effects limit the microresonator power handling and energy storage capacity.

Exercises

Exercise 20.1
Calculate the ratio of effective mass and geometrical mass for a cantilever resonator. Use the mode shape in Table H.3 on page 445 as your starting point.

Exercise 20.2
The clamped-clamped beam resonator in Example 20.2 is excited with a parallel plate transducer with the area $A = 500$ μm^2 and the gap $d = 1$ μm. Calculate the transduction factor, the vibration amplitude, the motional current, and the motional resistance for $V = 15$ V bias voltage and $v_0 = 1$ mV excitation signal. Assume that the quality factor is $Q = 1,000$.

Exercise 20.3
For the clamped-clamped beam resonator in Exercise 20.2, calculate the maximum vibration amplitude and the motional current before hysteresis due to the nonlinear capacitive springs. Repeat analysis if the electrode gap is $d = 200$ nm.

Exercise 20.4
Consider a longitudinal mode beam resonator similar to Figure 20.1 but made entirely of PZT5. The beam is excited with metal electrodes covering the entire bottom and top surfaces. Assume that the quality factor is $Q = 10,000$ and calculate the motional resistance, inductance, and capacitance. Compare your results to the silicon resonator in Example 20.7.

Exercise 20.5
A "square-extensional" plate resonator is made of a square plate which has the thickness h, and the X and Y dimensions are L. The resonant frequency is $f_0 = \frac{1}{2L}\sqrt{\frac{Y_{2D}}{\rho}}$ and the mode shape is $\vec{u} = \sin(\pi X/L)\vec{u}_X + \sin(\pi Y/L)\vec{u}_Y$. Derive expression for the effective mass and the effective spring constant.

Exercise 20.6
The microresonator shown schematically in Figure 20.9 has the resonant frequency $f_0 = 32$ kHz, the mass $m = 22.4$ pkg, the quality factor $Q = 1,500$, the height $h = 3$ μm, the electrode finger length $L = 40$ μm, the finger overlap $l = 20$ μm, the finger width $w = 3$ μm and the gap between fingers $d = 2\mu m$.

One side of the resonator is dc biased with 5 V and 5 mV ac signal is applied to it. The other comb drive is also biased to 5 V to measure the vibrations. Calculate (1) the dc force and displacement, (2) the electromechanical transduction factor, (3) the ac force, (4) the ac displacement and motional current for actuation at 10 kHz, (5) the ac displacement and motional current for actuation at the resonance, and (6) the motional resistance of the comb drive.

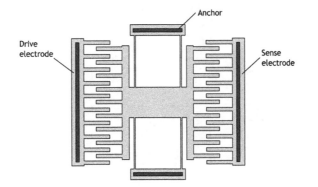

Figure 20.9: Comb-drive resonator.

21

Microresonator applications

Resonators are ubiquitous in modern electronics. For example, a typical cell phone contains more than a dozen piezoelectric resonators used as frequency references and as filters. Macroscopic resonators are typically made of piezoelectric crystals. Quartz is an unusual material as the resonant frequency is insensitive to temperature, which makes quartz ideal for making stable frequency references. For filters at GHz-frequencies, materials with stronger electromechanical transduction such as lithium niobate crystals are used. More recently, microfabrication techniques have been used to make resonators. Film bulk acoustic resonators (FBARs) are made of a thin film of piezoelectric material. FBARs are used mainly in communication systems but they also have potential as sensitive mass sensors. Silicon resonators are currently used as gyroscopes, but they also have potential to replace quartz references at frequencies less than 100 MHz.

In this chapter, the resonator applications are reviewed with the focus on using MEMS components in portable devices, where the size is important. Another important microresonator application, gyroscopes, is covered in Chapter 22. We will first look at clock oscillators operating in the kHz-range. Next, we will investigate the feasibility of making silicon based reference oscillators operating in the MHz-range. Finally, we will cover the filtering applications at GHz-frequencies.

21.1 Clock oscillator

Most electronic devices have an integrated clock that keeps track of time and maintains rudimentary processor functions when the devices is in "sleep mode".

Currently, the clock is realized with a 32,768-kHz quartz crystal tuning fork resonator. This frequency is convenient for clocks as it is obtained by multiplying twos (2^{15} = 32,768). Conversely, a simple digital division by twos will yield one second pulses. The quartz tuning forks can be considered the highest volume MEMS component in the world as over one billion lithographically defined quartz resonators are sold every year. Traditionally, quartz resonators have not been considered "MEMS" as quartz crystal market was established before the word MEMS was even invented; recently, however, a major quartz resonator manufacturer labeled their devices "quartz MEMS" [180]. This may be confusing, but it is quite natural since MEMS is a loosely defined acronym that has been used to lump together a number of different industries.

The quartz technology, whether it is called MEMS or not, has one serious drawback: It is very difficult to further reduce the size of quartz tuning forks. The smaller the resonators are, the higher the absolute manufacturing tolerances are to obtain the desired final frequency. Currently, the smallest packaged tuning fork resonators are 2.05 mm × 1.25 mm × 0.6 mm [181] and a smaller size will reduce the yield thus increasing the cost. This is the opposite of the IC and MEMS technologies where smaller die size translates to smaller costs.

Silicon resonators are a potential alternative to the quartz technology. The first MEMS resonator (resonant gate transistor) was demonstrated in the 1960s [182, 183]. The effort showed that it is possible to fabricate resonators in micro-scale but the development was abandoned due to stability problems and a low resonator quality factor. In the 70's and 80's, there was no significant research effort in microresonators.

In the nineties, the interest in microresonators was rekindled and several micromachined resonators were demonstrated [166]. Figure 21.1 shows an example of a 16-kHz polycrystalline silicon resonator that is integrated on the same die with the CMOS circuitry to drive the resonator [184]. The resonator has two comb drive electrodes. One electrode is used for excitation and the other side is used for sensing. The quality factor of the resonator is $Q = 23{,}400$ at 20 mTorr (2.7 Pa), the effective mass is $m = 5.73 \cdot 10^{-11}$ kg, the effective spring constant is $k = 0.65$ N/m, the structural layer thickness is $h = 2$ μm, the electrode spacing is $d = 2$ μm, and the number of finger overlaps on one comb-drive is $N = 60$. Given these values, the electromechanical transduction factor from Equation (20.36) is

$$\eta = \frac{\partial C}{\partial x} V_{\text{dc}} = \epsilon \frac{Nh}{d} V_{\text{dc}} \approx 1.86 \cdot 10^{-8} \text{ N/V},$$

when the resonator is biased at $V_{\text{dc}} = 35$ V. From Equation (20.50), the motional resistance is

$$R_m = \sqrt{km}/Q\eta^2 \approx 755 \text{ k}\Omega.$$

The motional resistance can be compared to a "typical" parasitic capacitance $C_P \sim 0.1$ pF. We want the motional resistance to be smaller than the parasitic impedances. At 16 kHz, the impedance for a 0.1-pF capacitor is $|Z| = 1/\omega C \approx 100$ MΩ, which is 100× larger than R_m indicating that making a good oscillator is relatively easy. The bias voltage, however, is impractically large. Modern CMOS processes have a typical maximum voltage of $1.8 - 5$ V. Higher voltages will break the circuit. At $V_{dc} = 3$ V, the motional resistance becomes $R_m = 100$ MΩ which is too high for a good oscillator. In summary, the work demonstrated that it is possible to integrate a micromechanical resonator together with electronics to make an oscillator but more engineering is needed to reduce the bias voltage.

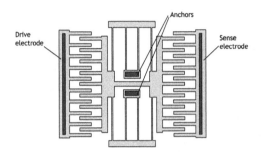

Figure 21.1: A 16-kHz comb drive resonator with sense and drive electrodes (After [184]).

Recently, a US based start-up Discera demonstrated a 32-kHz oscillator that has overcome the problem of a high bias voltage by increasing the resonator and the electrode size [185]. The resonator motional impedance was $R_m = 400$ kΩ at 2.5 V dc bias showing that it is possible to make a low frequency silicon resonator that operates with low voltage electrostatic actuation. However, two technical challenges still remain with the resonator:

1. The initial accuracy for quartz tuning fork crystals is typically specified to be ±20 ppm (part per million). To achieve this accuracy, the quartz tuning forks are physically large and are individually laser trimmed. In comparison, the MEMS fabrication variations are ~ 100 nm which for a 10-μm beam translates to 30,000 ppm variations in resonance frequency. Recall that one of the limitations for miniaturizing the quartz tuning fork crystals was the finite manufacturing tolerances. Innovations and development are needed to obtain micro-scale parts with ppm-level accuracy.

2. The temperature coefficient of frequency for silicon resonators is more than 100× larger than for quartz crystals. Silicon, like most materials, becomes softer when heated. As a result, the resonant frequency decreases about 25-30 ppm/K. Quartz is a special material that can show positive temperature coefficients (resonant frequency increases with temperature). By orienting the resonator in certain crystal directions, the quartz res-

onator resonance frequency is almost constant over a wide temperature range. Figure 21.2 shows the temperature dependencies for two common quartz "cuts" or resonator orientations. The Z-cut is the basis for the 32-kHz tuning fork resonators and offers good performance around room temperature. The AT-cut is popular for MHz resonators and is stable over even wider temperature span. In comparison, the silicon resonators exhibit a huge temperature coefficient that needs to be compensated in clock and reference applications. This issue is further noted in the next section where the MHz-frequency reference oscillators are reviewed.

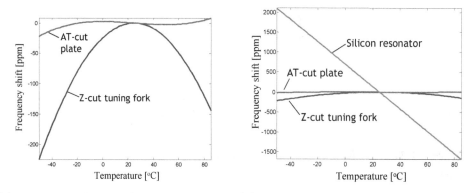

(a) Z-cut tuning fork and AT-cut shear mode resonators.

(b) Silicon resonator in comparison to quartz crystals.

Figure 21.2: The temperature dependencies for two common quartz crystal orientations and single crystal silicon resonators.

21.2 Reference oscillators

Electronic communication requires accurate frequency references that are stable and have low noise. For example, high speed data transfer within a computer can happen at a clock rate close to 1 GHz. To eliminate the bit errors due to clock fluctuations, the clock cycle to cycle variation or jitter has to be in the picosecond range. Wireless communication systems have even more stringent requirements for the frequency reference. In GSM, the cell phones have a reference oscillator that is accurate to ± 10 ppm to enable fast locking to the base station signal. This reference has to have signal-to-noise ratio 150 dB/Hz, which means that the noise level is 10^{15} times smaller than the oscillator signal level! Currently, millimeter-sized quartz crystal resonators are used to meet these requirements.

Achieving the GSM noise specifications with micro-scale silicon resonators is not trivial but it can be done [95]. As we saw in Chapter 8, CMOS amplifier noise depends on the transistor bias current. Achieving noise densities lower than 1-2 nV/$\sqrt{\text{Hz}}$ would require milliamperes of current, which is not an option for portable electronics. Thus, to achieve a 150-dB/Hz signal-to-noise ratio, the resonator needs to provide a 100-mV signal amplitude or more. The maximum vibration amplitude is limited by the nonlinear effects in the resonator and we can use Equations (20.55) and (20.56) as guidance for designing high power output resonators. Basically, the resonator should have high spring constant k_0 and high maximum vibration amplitude x_c.

One candidate for the high power output resonator is the 1D beam resonator in Figure 20.1. Compared to the flexural resonators, the bulk resonator has a large effective mass and a large spring constant. Moreover, a large oscillation amplitude x_c is possible as mechanical nonlinearities are minimized. Fundamentally, the oscillation amplitude is limited by material nonlinearities. This is a potential advantage for silicon resonators as silicon is a more linear material than quartz. The demonstrated attainable *energy densities* for silicon resonators are 1000× higher than for AT-cut quartz crystal resonators [178].

Extending the 1D resonator concept to 2D results in the "square-extensional" resonator shown in Figure 21.3. The resonator mode shape can be characterized as the square plate extending and contracting [95]. The resonator resonance frequency is $f = 13$ MHz, the quality factor is $Q = 130,000$, and the effective mass and spring constant are $m = 2.39$ nkg and $k = 16.2$ MN/m, respectively. Four electrodes are placed around the resonator and the total electrode area is $A = 4Wh = 4 \cdot 10 \ \mu\text{m} \cdot 280 \ \mu\text{m}$, which is more than 100× larger than for the 1D beam resonator. This will enable lower motional resistance which significantly helps in oscillator design.

At 13 MHz, the typical parasitic capacitance of 0.1 pF has the impedance

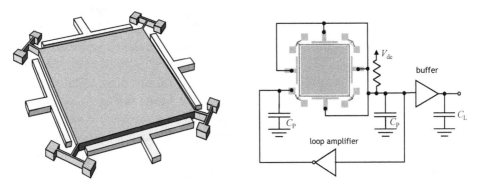

(a) The four electrodes around the plate resonator are used to excite and measure the extensional vibrations.

(b) The oscillator consists of the resonator, loop amplifier that provides positive feedback to sustain vibrations, and buffer amplifier.

Figure 21.3: A 13 MHz square-extensional oscillator [95, 186].

$Z = 1/\omega C \approx 120$ kΩ. To make a good oscillator, the motional resistance R_m should be at least ten times smaller. This is achieved by using narrow gaps ($d = 200$ nm). Using bias voltage of $V_{dc} = 20$ V, we obtain the effective transduction factor as

$$\eta = \frac{\partial C}{\partial x} V_{dc} = \epsilon \frac{A}{d^2} V_{dc} \approx 5.0 \cdot 10^{-5} \text{ N/V}.$$

From Equation (20.50), the motional resistance is

$$R_m = \sqrt{km}/Q\eta^2 \approx 616 \ \Omega.$$

For the narrow gap resonators, the maximum current through the resonator is limited by the capacitive nonlinearity. As shown by Equation (C.16) on page 422, the frequency shift due to nonlinearities is proportional to the square of the vibration amplitude $\Delta\omega = \kappa x_0^2$, where the nonlinearity factor κ given by Equation (C.17) is

$$\kappa = \frac{3k_2}{8k} \omega_0 - \frac{5k_1^2}{12k^2} \omega_0.$$

From Equation (15.21), the nonlinear spring constants due to the capacitive spring forces are

$$k_{1e} = -\frac{V^2 C_0}{d^2} \frac{3}{2d} \approx -1.5 \cdot 10^{11} \text{ N/m}^2 \text{ and } k_{2e} = -\frac{V^2 C_0}{d^2} \frac{2}{d^2} \approx -1.0 \cdot 10^{18} \text{ N/m}^3$$

giving $\kappa \approx 1.9 \cdot 10^{18}$ Hz/m^2.

For a reference oscillator, we want to frequency shift be small to obtain good stability as a large shift would mean that the vibration frequency is amplitude dependent. Requiring that the frequency shift is less than $|\Delta\omega/\omega_0| = 1$ ppm, the maximum vibration amplitude is $|x_{max} = |\sqrt{10^{-6}/|\kappa|}\omega_0 = 6.6$ nm and the maximum signal current is $|i_{max}| = \omega_0\eta x \approx 27$ μA. This current is converted by the capacitance $C_P = 2$ pF into voltage $|v| = i/\omega C \approx 160$ mV. Assuming that the buffer noise $\bar{v}_n = 4$ nV/$\sqrt{\text{Hz}}$ sets the noise floor, the signal-to-noise ratio is $20 \log v/\bar{v}_n \approx 150$ dB/Hz, which meets the wireless specifications [95].

The square-extensional resonator demonstrates very good raw noise performance but the device requires narrow gaps that are not easy to fabricate, and the bias voltage was still quite high. One way to lower the motional resistance is to use a flexural resonator to reduce the effective spring and mass. In addition, the low-frequency resonators are larger providing a larger electrode area, which helps in efficient coupling. The first commercially available silicon reference oscillator shown in Figure 21.4 from a start-up company SiTime uses both effects. The resonator can be thought of as made up of four clamped-clamped beams connected in a square "ring". The resonance frequency is 5.1 MHz allowing for a relative large size. Moreover, the innovative design allows a total of eight electrodes to be placed around the resonator. This enables an adequate electromechanical coupling coefficient η without requiring narrow gaps.

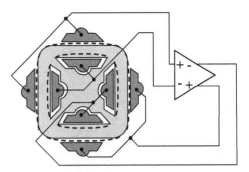

(a) Eight electrodes are placed around the flexural resonator for differential excitation and measurement.

(b) The oscillator is sustained with a differential amplifier with differential outputs.

Figure 21.4: A 5.1 MHz flexural resonator from SiTime [187].

The full technical data for the resonator is proprietary but the important parameters can be estimated based on the publicly available information and the resonator data sheet [187]. The published resonator quality factor is $Q = 80,000$, the resonance frequency is $f_0 = 5.1$ MHz, and the bias voltage is $V_{dc} = 4.6$ V. From the published die photos, the resonator side length is about $L = 225$ μm and the resonator height is $h = 10$ μm [188]. To estimate the lumped mass and spring, we model each resonator side as a clamped-clamped beam. Based on the formulas in Appendix H and the given length and resonance frequency, the estimated beam width is $w = 30$ μm. Following Example 20.2, the effective mass for the four beams combined is $m = 4 \cdot 0.4\rho hwL \approx 0.25$ nkg and the effective spring constant is $k = \omega_0^2 m \approx 253$ kN/m.

The electrode length is estimated to be $L_{el} = 140$ μm and the electrode gap is $d = 0.4$ μm [188]. Half of the electrodes are used for driving the resonator and half for sensing. The electromechanical transduction factor for the drive and sense is

$$\eta = \frac{\partial C}{\partial x} V_{dc} = \epsilon \frac{4L_{el}h}{d^2} V_{dc} \approx 7.1 \cdot 10^{-7} \text{ N/V}$$

for four electrodes biased at 4.6 V. The motional resistance is

$$R_m = \sqrt{km}/Q\eta^2 \approx 200 \text{ k}\Omega.$$

In comparison, at this frequency 0.1 pF capacitance has resistance of only 315 kΩ. The high motional resistance is probably one reason why the differential drive is used for the resonator.

The recommended drive level is specified as $v = 1$ V which translates to current of $i = v/R_m \approx 5$ μA. With a 2-pF load, the signal voltage is $|v_{sig}| = i/\omega C \approx 80$ mV. If the buffer amplifier noise is $\overline{v}_n = 4$ nV/$\sqrt{\text{Hz}}$, the oscillator has signal-to-noise ratio of $20 \log v_{sig}/\overline{v}_n \approx 146$ dB/Hz. However, the temperature and fabrication offsets need to be compensated which will increase the oscillator noise. The signal-to-noise ratio for a compensated oscillator is specified to be 115 dB/Hz.

The frequency compensation is achieved with a synthesizer circuit that multiplies the oscillator frequency with a temperature dependent constant $N(T)$. Figure 21.5(a) shows the block diagram for the compensation circuit. A voltage controlled oscillator (VCO) is locked to the MEMS reference with a phase locked loop. The chip also contains a temperature sensor for measuring the temperature and memory to calibrate initial offsets due to process variations. The lock ratio $N(T)$ is adjusted so the VCO frequency $f_{VCO} = N(T) \cdot f_0(T)$ stays constant even though the MEMS reference frequency $f_0(T)$ drifts with temperature. The total power consumption for the silicon based oscillator is 19 mA. In comparison, a quartz based oscillator shown schematically in Figure 21.5(b) is much simpler and consumes less than 4 mA of power [189]. Due to the compensation circuitry, the MEMS implementation is also noisier than the quartz oscillator and the complex circuit requires large die size, increasing the cost. We conclude that as a standalone, this MEMS oscillator does not appear to offer compelling advantages over quartz but the small size makes the MEMS resonator attractive for system integration. SiTime also sells bare resonator dies and reference oscillator design for customers interested in integrating the oscillator into their systems.

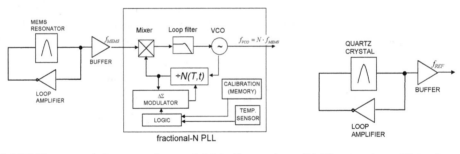

(a) MEMS resonator frequency has an initial offset and a temperature dependency that require complex compensation circuit.

(b) The quartz oscillator is accurate over a wide temperature range and does not require compensation.

Figure 21.5: Comparison of silicon MEMS and quartz oscillators.

Example 21.1: MEMS oscillator temperature compensation
Problem: To compensate the temperature induced change, accurate temperature measurement is necessary. If the frequency change for a silicon resonator

is $\frac{\partial f / f_0}{\partial T} = -28$ ppm/K, how accurate temperature measurement is needed to obtain accuracy of ± 10 ppm over the commercial temperature range of -45 °C to +85 °C?

Solution: The 10 ppm frequency shift corresponds to temperature change of

$$\Delta T = \frac{\Delta f}{\frac{\partial f / f_0}{\partial T}} = \frac{10 \text{ ppm}}{-28 \text{ ppm/K}} \approx -0.4 \text{ K}.$$

Thus, to obtain ± 10 ppm accuracy, the absolute temperature accuracy has to be ± 0.4 K over the temperature range of -45 °C to +85 °C. The actual accuracy will have to be even better to allow for other error sources. This measurement accuracy is difficult to obtain with solid-state temperature sensors without consuming significant current and/or doing expensive calibration at multiple temperatures. Current commercially available silicon oscillators have stability of ± 50 ppm over the temperature range [190] which is not sufficient for high end communication applications.

21.3   RF filters

Wireless communication systems operate in a very crowded and noisy radio environment. For example, a cell phone signal can be more than 100 dB weaker than interfering signals from, for example, a microwave oven. Band-pass filters are used to reduce the amplitude of the interferers that are located outside the frequency band of interest. A typical RF filter will reduce the amplitude of unwanted signals by 30 dB or more. In addition, it is important that that the filter does not significantly attenuate the wanted signal. The typical "insertion loss" is less than 1 dB.

A filter element should have a high electromechanical transduction factor coefficient and high quality factor to enable low motional resistance. Typically, the resonator is connected between the 50-Ω source and load impedances, which means that the filter resistance should be significantly smaller than 50 Ω or the insertion loss will be unacceptable high. In addition, a high Q-value is needed to make the filter selective. Commercial filters have resistance close to 1 Ω and quality factor greater than 1,000 at 1 GHz.

Example 21.2: Filter insertion loss
Problem: A filter is specified to have an insertion loss of 0.3 dB when connected

to a 50-Ω system. Calculate the maximum series resistance for the filter. If the filter is made of three resonators in series, what is the maximum motional resistance for the resonators?

Solution: Following Example 19.7 on page 299, the insertion loss is

$$|S_{21}|^2 = \left| \frac{2Z_0}{2Z_0 + R_s} \right|^2,$$

where $Z_0 = 50\ \Omega$ and R_s is the filter resistance in the pass band [191]. Requiring that the insertion loss is less than 0.3 dB, gives $R_s < 3.5\ \Omega$. If three resonators are connected in series, each should have motional resistance $R_m = R_s/3 \approx 1.2\ \Omega$.

21.3.1   FBAR filters

Until the 1990s, the GHz-frequency band-pass filters were based on surface acoustic wave (SAW) devices. Recently, thin film based FBARs (film bulk acoustic resonators) have captured a large market share. Sometimes this technology is referred as bulk acoustic wave (BAW) filters to differentiate them from SAW devices. The thin film resonators are especially attractive for frequencies higher than 1 GHz where SAW filters are difficult to manufacture. FBARs can be considered a MEMS success story as over one hundred million FBAR filters are sold every year and the market is growing. The FBAR filters, however, were developed outside the MEMS community and some of the FBAR developers do not regard it as MEMS but a distinct technology of its own. In the end, the name is irrelevant to the end-user such as a cell phone customer. FBARs are included in this book as they are currently the best resonators at GHz-frequencies and relevant when analyzing the performance of "silicon MEMS" filters.

The FBAR is based on a thin film approximately 1-2 μm thick vibrating primarily in the thickness mode. Currently, all commercial FBAR filters are manufactured of aluminum nitride (AlN) as it has been the only piezoelectric thin film material that can be reliably processed, has low losses, and a high electromechanical coupling. To eliminate the acoustic losses, the thin film resonator has to be mechanically isolated from the substrate. Ideally, the intrinsic material losses should dominate the quality factor, which translates to a quality factor greater than 2,000 at 2 GHz. Figure 21.6 shows the two commercially available FBAR types. In the membrane FBAR shown in Figure 21.6(a), the substrate losses are eliminated in a straightforward way by removing the substrate under the resonator. The substrate losses are almost completely eliminated apart from small losses at the resonator edges. Agilent Technologies first commercialized this type of resonator.

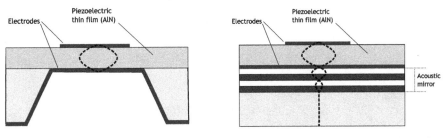

(a) A membrane FBAR resonator.   (b) Solidly mounted resonator (SMR).

Figure 21.6: Cross section for FBARs. Also shown is the stress fields inside the resonators (dotted lines).

An alternative approach is to mount the resonator on an acoustic mirror as shown in Figure 21.6(b). The acoustic mirror has 4-10 layers that are approximately a quarter wavelength thick to reflect the acoustic energy back to the resonator. Although this approach does not completely eliminate the substrate losses, the quality factor for a well designed mirror can be comparable to membrane devices [94]. Compared to the membrane resonator, the solidly mounted resonators have the advantages of a higher power handling capacity as the heat is transferred efficiently to the substrate and good mechanical robustness as no thin membranes are needed. The disadvantages of the solid mounting are that the quality is lower and electromechanical coupling is reduced because the mirror also contributes to the mechanical vibrations and the vibration is not completely confined to the piezoelectric thin film. Currently, both the membrane and mirror type device are commercially available, and although there are small differences in characteristics, both can meet the tight communication system specifications.

Example 21.3 illustrates the typical FBAR characteristics as derived from an ideal 1D model. Importantly, the motional resistance can be less than 1 Ω, which enables very low insertion loss. The 1D model is useful for understanding the device operation but the real thin film resonators vibrate in all directions. This results in multiple resonances around the desired resonance frequency. These so called "spurious modes" are detrimental to filter response and significant effort has been put into reducing or eliminating the spurious modes [94].

Example 21.3: Piezoelectric FBAR resonator
Problem: Calculate the electrical equivalents for the FBAR resonator of Example 20.5 on page 319. Recall that the electrode area is $A = (100~\mu\mathrm{m})^2$, the thickness is $h = 2~\mu\mathrm{m}$, and assume that quality factor is $Q = 1,000$.

Solution: For this example, we will assume that the loading due to electrodes will be negligible. This will simplify the analysis and will give quick order of magnitude estimates for the impedance. From Equation (20.19), the effective mass and spring constants are

$$m = \rho A L / 2 \approx 48 \text{ pkg}$$

and

$$k = \pi^2 E A / 2h \approx 6.6 \text{ GN/m},$$

respectively. The resonant frequency is

$$f_0 = \frac{1}{2\pi} \sqrt{\frac{k}{m}} \approx 1.86 \text{ GHz.}$$

From Example 20.5, we have $\eta = 2\frac{e_{33}A}{h} = 0.052$ N/V. The electrical equivalents are

$$R = \sqrt{km}/Q\eta^2 = 0.05 \ \Omega,$$

$$C = \eta^2/k = 1.6 \text{ pF},$$

and

$$L = m/\eta^2 = 4.5 \text{ nH}$$

for the motional resistance, the capacitance, and the inductance, respectively. The motional resistance is very small. In practice, ohmic losses, electrode mass, and spurious resonances will increase the resonator resistance.

The capacitance between the electrodes is

$$C_0 = \epsilon_R \epsilon_0 A / h = 0.25 \text{ pF.}$$

The capacitive impedance at resonant frequency is $|Z| = 1/\omega_0 C = 340 \ \Omega$ which is much larger than the motional resistance. This is a figure-of-merit for the coupling strength and important for good filter performance.

21.3.2   Silicon MEMS filters

Silicon based resonators have been investigated as a potential alternative to the current filter technologies. Micromechanical filters operating at kHz-frequencies were demonstrated in the 90's [192], but the real demand is in GHz-frequenies where SAW and FBAR resonator technologies are currently used. Modern cell phones can operate between two to five radio bands and each band requires its own filter. Since the MEMS filters are lithographically defined, multiple resonators with different resonance frequencies can be fabricated on the same chip, which could reduce the cost of multi-band radios [193]. In comparison, the FBAR resonance frequency is thickness dependent and it is not cost effective to process multiple film thicknesses on the same chip.

The state-of-the art high-frequency silicon resonator is shown in Figure 21.7. The disk resonator is made of polycrystalline silicon. In the fundamental mode, the disk expands and contracts uniformly. To reduce the direct capacitive coupling between the input and output, the resonator has separate drive and sense electrodes. The fundamental vibration mode is at 150 MHz and an overtone resonance is at 1.2 GHz [105]. The disk diameter is 20 μm and the thickness is 2.1 μm. The geometrical mass is $m = \rho\pi r^2 h = 1.53$ pkg. From the FEM mode shape, the effective masses are found to be $m = 1.2$ pkg and $m = 1.4$ pkg for the fundamental and overtone mode, respectively. The effective spring constants ($k = m\omega_0^2$) are 3.2 MN/m and 75 MN/m, respectively.

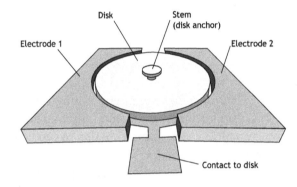

Figure 21.7: A polycrystalline silicon disk resonator with resonances at 150 MHz and 1.2 GHz [105].

To reduce the motional impedance, a narrow electrode gap of $d = 68$ nm is used. Using the electrode area of $A = 66 \cdot 10^{-12}$ m^2 and the bias voltage of $V_{\text{bias}} = 10.5$ V, we have motional resistance of $R_m = \sqrt{km}/\eta^2 Q \approx 132$ kΩ and $R_m \approx 2.3$ MΩ for the fundamental and the overtone, respectively. Clearly, these values are high for a filter. Approaches to lower the motional resistance include putting several resonators in parallel [194] but this will increase the resonator area and cost.

Given that there does not appear to be a straightforward way to increase the electrostatic transduction factor to obtain a single-digit motional resistance, the combination of piezoelectric actuation and silicon micromachining has received significant interest [195]. This approach could enable strong electromechanical coupling while maintaining the flexibility of lithographically defining the resonator resonant frequency. Currently, it is too early to tell whether this approach will be commercially successful.

Key concepts

- For oscillator and filters, low motional resistance is needed. This requires high Q-value and effective electromechanical transduction.

- Silicon based oscillators offer smaller size than quartz crystal based resonators but the silicon resonators require temperature compensation that increases total cost and oscillator noise.

- High-frequency piezoelectric thin film resonators are called FBARs (film bulk acoustic resonators) or BAWs (bulk acoustic wave). FBARs vibrate mostly in thickness mode and they are used extensively in communication systems. Silicon based resonators have equally high Q-values but the electrostatic transduction is not sufficiently strong to enable single digit motional resistances.

Exercises

Exercise 21.1
Micromechanical resonators have potential applications in chemical and biomedical sensing applications. In a common configuration a mass (for example an atom or a molecule) lands on a micromechanical beam. The added mass is detected by measuring the change in the resonance frequency. For a clamped-clamped flexural beam (14 μm $\times$ 200 nm $\times$125 nm), the resonance frequency and mass are $f_0 = 8$ MHz and $m = 1.4 \cdot 10^{-15}$ kg. Assuming that the smallest measurable change in resonance frequency is $\Delta f/f_0 = 10^{-5}$, what is the smallest measurable mass change? How many gold atoms this would be?

Exercise 21.2
How good fabrication tolerance is needed to make a 32-kHz clock resonator that is accurate to ± 20 ppm without using trimming. Assume that the spring beam width is $w = 10$ μm. If the typical manufacturing variations are ± 150 nm, what is the resonance frequency variations?

Exercise 21.3
Consider the RF resonator in Section 21.3.2 (Figure 21.7). If the bias voltage is kept constant, how small gap is needed to achieve the motional resistance $R_m = 50$ Ω?

Exercise 21.4
The first two MEMS resonator start-ups (Discera and SiTime) were alive and shipping products at the time of publishing this book. Do a web-search to see how they are doing now.

Exercise 21.5
What are the main challenges in commercializing silicon based resonators?

22

Gyroscopes

Angular velocity sensors, also known as gyroscopes or rotation rate sensors, measure how quickly an object turns. Gyroscopes are used for example for inertial navigation, automotive chassis control and rollover detection, and camcorder image stabilization. Historically, the angular rate has been measured with rotating wheel gyroscopes. The spinning wheel conserves the angular momentum resisting the change in the rotation axis orientation. The angular velocity can therefore be sensed by measuring the force on the spinning wheel due to the rotation.

Optical gyroscopes encompass more recent technology. They are based on measuring the phase difference of two laser beams traveling in opposing directions. The optical gyroscopes are not subject to a mechanical wear and are the most precise rotation rate sensors available. Typical optical gyroscope applications are the inertial navigation systems used for aircraft navigation and missile guidance. While the optical gyroscopes provide excellent performance, they are large and expensive. The automotive and consumer applications require cheaper and smaller solutions.

MEMS gyroscopes are light weight, small, and relatively low cost [28]. They therefore enable new applications that are not possible with the classic gyroscopes that are based on optical or macromechanical principles. Existing and emerging applications for MEMS sensors include image stabilization in cameras, input device for human-computer interaction for example in game consoles and mobile phones, and rudimentary inertial navigation ("dead reckoning") for aiding the GPS navigation systems while the contact to satellite is momentarily lost.

The micromachined angular velocity sensors are based on energy transfer between two orthogonal vibration modes, the drive-mode and the sense-mode.

In rest, the sensor vibration modes are not coupled and only the drive-mode is exited; however, when the sensor is rotated, the Coriolis force couples the modes and the drive-mode excites the sense-mode. The sense-mode vibration amplitude is used to measure the angular velocity.

The gyroscopes are the most complex inertial MEMS sensors. The fundamental challenge is that the Coriolis force is very small. A careful design is needed to measure the small Coriolis signal in the presence of much larger electrical signals. In this chapter, we will introduce the operation principles and design of MEMS gyroscopes. Electrostatic silicon MEMS gyroscopes are analyzed and the silicon gyroscopes are compared to piezoelectric rotation rate sensors.

22.1   Coriolis force

The Coriolis force affects objects moving in a rotating frame. Figure 22.1 illustrates how rotation affects the travel path of a freely moving object: A mass is thrown from the center of a spinning wheel toward the wheel edge. If no forces act on the mass, it will travel straight as expected. By the time the mass reaches the wheel edge, however, the wheel has rotated so that the mass will not hit the location 'B' that the mass was directed toward, but a point 'A' that has moved in front of the mass. In the eyes of an observer on the spinning disk, the path of the mass appears curved (dotted line). A practical example of the Coriolis effect is the long range artillery fire. If the gun is pointed directly to the target, by the time ammunition reaches the intended target, the target has moved due to Earth's rotation.

The vector formula for the magnitude and direction of the Coriolis acceler-

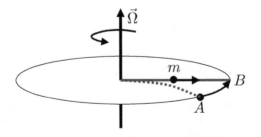

Figure 22.1: The origin of the Coriolis force. A mass is directed toward point 'B' on the rotating frame. Although the mass travels straight, it will not hit the point 'B' but a point 'A' that will move in front of the mass. In the eyes of an observer on the spinning disk, the path of the mass appears curved (dotted line) as if force was acting on it.

ation is

$$\ddot{\vec{x}} = -2\vec{\Omega} \times \dot{\vec{x}} \qquad (22.1)$$

where $\dot{\vec{x}}$ is the velocity of the object in the rotating system, and $\vec{\Omega}$ is the angular velocity vector [196]. The angular velocity vector $\vec{\Omega}$ is directed along the axis of rotation perpendicular to the rotating plane and it has the magnitude of the rotation velocity. According to the Newton's law, the Coriolis force is obtained by multiplying the Coriolis acceleration by the object mass. We have

$$\vec{F}_c = m\ddot{\vec{x}} = -2m\vec{\Omega} \times \dot{\vec{x}}. \qquad (22.2)$$

Going back to Figure 22.1, Equation (22.2) gives the apparent force that is acting on the object with the mass m. Since the Coriolis force is only seen in the rotating frame, it is sometimes called a "fictional force". Within the rotating system, however, the Coriolis force is measurable and there is nothing fictional about it. The use of Equation (22.2) in microresonators is illustrated in the Example 22.1.

Example 22.1: Coriolis force on vibrating mass

Problem: A microresonator is vibrating at the resonant frequency $f_0 = 10$ kHz with a vibration amplitude $x_0 = 1$ μm along X_1-axis. The resonator mass is $m = 0.6$ nkg. Calculate the Coriolis force acting on the resonator mass that is on a rotating platform with a rotation rate $\Omega = 2\pi$ rad/s around the X_3-axis.

Solution: The mass position can be written as

$$\vec{x} = x_0 e^{j\omega_0 t}\vec{a}_1,$$

where $\vec{a}_1$ is the unit vector in X_1-direction. The mass velocity is

$$\dot{\vec{x}} = \frac{\partial x}{\partial t} = j\omega x_0 e^{j\omega_0 t}\vec{a}_1 \approx j0.0628 e^{j\omega_0 t}\vec{a}_1 \text{ m/s}.$$

The resonator velocity and displacements are 90° out-of-phase meaning that the velocity is zero at maximum displacement and largest when the mass passes the rest position ($x = 0$).

The Coriolis force from Equation (22.2) and $\vec{\Omega} = 2\pi\vec{a}_3$ rad/s is

$$\vec{F}_c = -2m\vec{\Omega} \times \dot{\vec{x}} \approx -j0.24 e^{j\omega_0 t}\vec{a}_2 \text{ nN}.$$

The force is in X_2-direction and oscillates at the resonance frequency but is $-90°$ out-of-phase with the mass displacement. Although the rotation rate of one rotation per second is fairly high, the magnitude of the force $|F| \approx 0.24$ nN is very small.

22.2   Vibrating two-mode gyroscope

Nearly all MEMS gyroscopes are based on two orthogonal vibration modes. The operation principle is illustrated in Figure 22.2. The *drive-mode* is orthogonal to the *sense-mode* meaning that the two modes do not normally interact and the drive-mode movement does not result in movement to the sense-mode direction. The resonator is excited to vibrate in the drive-mode in X_1-direction, and with no angular rotation, the movement in the sense-mode direction is zero. The Coriolis force due to a rotation around X_3-axis, however, excites the resonator sense-mode in X_2-direction. Thus, the sense-mode vibration amplitude is proportional to the angular rotation rate.

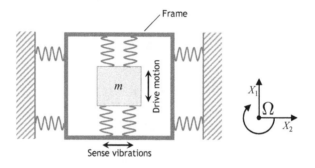

Figure 22.2: The operation principle of vibrating two-mode gyroscope. Due to the Coriolis force, the drive-mode vibrations will excite the sense-mode.

The first MEMS gyroscopes were introduced in the early 90's [197]. Since the mid-90's, the comb drive actuation and parallel plate sensing has been the prevalent design [198, 199]. Although nearly all MEMS gyroscopes are based on the two-mode vibration principle, the design details vary significantly. In some gyroscopes, there is just one proof mass that act as both the sense and drive mass as shown in Figure 22.2. Other designs have two or more masses to further decouple the drive- and sense-modes. Also, the choice for the drive- and sense-mode resonance frequencies varies between matching the frequencies to completely separating the drive- and sense-mode resonance frequencies.

To understand the design trade-offs, we will write the equations of motion for the coupled two-mode oscillator. The drive- and sense-mode displacements, x_d and x_s, respectively, are given by

$$x_d = H_d F_d \tag{22.3}$$
$$x_s = \epsilon_Q x_d + H_s F_c, \tag{22.4}$$

where H_d and H_s are the drive- and sense-mode response functions, respectively, F_d is the drive force that excites the drive motion, ϵ_Q is the fraction of the drive-mode that couples directly to sense-mode due to "quadrature error" covered in Section 22.4, and F_c is the Coriolis force exciting the sense-mode. From Equation (22.2), the Coriolis fore is $F_c = -2m_d\Omega\dot{x}_d$, where m_d is the drive-mode mass, Ω is the angular rotation rate, and $\dot{x}_d$ is the drive mode velocity.

From Appendix B, the drive- and sense-mode response functions are

$$H_d(\omega) = \frac{1}{m_d}\frac{1}{\omega_{0d}^2 - \omega^2 + j\omega\omega_{0d}/Q_d} \equiv H_{Rd} + jH_{Id} \tag{22.5}$$

and

$$H_s(\omega) = \frac{1}{m_s}\frac{1}{\omega_{0s}^2 - \omega^2 + j\omega\omega_{0s}/Q_s} \equiv H_{Rd} + jH_{Id}, \tag{22.6}$$

respectively. The subscripts d and s in the mass m, the quality factor Q, and resonant frequency ω_0 refer to drive- and sense-mode, respectively. The real and imaginary parts of the response function are given by Equation (B.8) and Equation (B.9), respectively. Both the drive and sense modes vibrate at excitation frequency ω but their amplitude and phase will differ.

In what follows, we will further explore the drive- and sense-mode displacements x_d and x_s. In most gyroscope circuits, the sense signal is measured with a phase locked-loop. Hence, both the phase and amplitude of the sense signal are important and we will write the phase relationship explicitly in our analysis.

22.2.1 Drive-mode vibrations

In MEMS gyroscopes, the drive-mode is usually excited via electrostatic actuation although magnetic actuation has also been demonstrated [200]. As the Coriolis force is proportional to the drive-mode vibration amplitude, capacitive comb-drives are preferred over parallel plate actuators to obtain vibrations amplitudes in the micrometer range. For macroscopic vibratory gyros, piezoelectric actuation is used.

The drive-mode is usually driven at the resonance to obtain the maximum displacement with a minimum excitation signal. Small drive-signal will save both power and reduce the parasitic coupling to the sense electrodes. Using Equations (22.3) and (22.5), the displacement at resonance ($\omega = \omega_{0d}$) is

$$x_d(\omega) = H_d(\omega)F_d = H_{Rd}(\omega)F_d + jH_{Id}(\omega)F_d = -j\frac{Q_d}{k_d}F_d, \tag{22.7}$$

where $H_{Rd} = 0$ and $H_{Id} = -Q_d/k_d$ are the real and imaginary parts of the response function H_d at the resonance frequency $\omega = \omega_{0d}$. The negative imaginary number $-j$ indicates that the displacement at the resonant frequency is 90° behind the drive force.

22.2.2 Sense-mode vibrations

The sense-mode vibration amplitude is obtained from Equation (22.4). We will first assume that the drive force does not couple directly to the sense-mode and the ϵ_Q-term in Equation (22.4) is zero. The effect of a non-zero ϵ_Q-term is analyzed in Section 22.4 where the "quadrature error" is discussed.

To explore the sense-mode design trade-offs, we will first obtain a general expression for the sense-mode vibration amplitude. From Equations (22.4) and (22.6), the sense-mode displacement due to the Coriolis force is

$$x_s(\omega) = H_s(\omega)F_c = -2m_d\Omega\dot{x}_dH_s(\omega) = -j2\omega m_d\Omega x_dH_s(\omega)$$
$$= 2\omega m_d\Omega x_d(H_{Is}(\omega) - jH_{Rs}(\omega)), \tag{22.8}$$

where we have used $\dot{x}_d = j\omega x_d$ and $H_s(\omega) = H_{Rs}(\omega) + jH_{Is}(\omega)$.

As shown by Equation (22.8), the sense-mode vibrations depend on the sense-mode response function H_s at drive-mode excitation frequency ω. The two interesting cases are the matched modes where the drive-mode and the sense-mode resonant frequencies are equal $\omega_{0s} = \omega_{0d}$ and the separated resonance frequencies where $\omega_{0d} \ll \omega_{0s}$. In both cases, the sense-mode vibrations are at the same frequency as the drive-mode but the amplitude and phase of the sense-mode relative to the drive-mode are different.

Matched modes ($\omega_{0s} = \omega_{0d}$)

Matching the drive- and sense-mode resonance frequencies amplifies the sense-mode vibration by the resonator quality factor Q_s. At resonance, we have $H_{Rs} = 0$ and $H_{Is} = -Q_s/k_s$. The displacement from Equation (22.8) is

$$x_s = -2\omega m_d\frac{Q_s}{k_s}\Omega x_d$$
$$= -2\frac{m_d}{m_s}\frac{Q_s}{\omega}\Omega x_d \tag{22.9}$$

where we used $\omega_0 = \omega_{0s} = \omega_{0d} = \sqrt{k_s/m_s}$. The sense-mode displacement x_s given by Equation (22.9) is in-phase with the drive-mode displacement.

As seen from Equation (22.9), for the matched modes, the signal displacement is amplified by the sense-mode quality factor Q_s. The sense-mode quality factor can be greater than 1,000, which gives a significant signal boost. This "extra gain" has made the matched mode operation the preferred design for piezoelectric gyroscopes.

The disadvantages for the matched mode operation are the requirement for the mode matching, which may require either electrical or physical trimming, and the reduced bandwidth due to the high sense-mode quality factor. For a resonator driven resonance, the time constant for building up the oscillation amplitude is $\tau = 2Q/\omega_0$, which increases linearly with Q (see page 416 in Appendix B). The effect of the quality factor Q on gyroscope bandwidth is illustrated in Example 22.2.

Example 22.2: Mode-matched resonator bandwidth
Problem: A piezoelectric gyroscope has a resonant frequency $f_s = f_d = 25$ kHz and quality factor $Q_s = 2,000$. Calculate the sensor bandwidth if it is limited by the mechanical element response time.
Solution: From page 416 in Appendix B, the time constant for the sense-mode amplitude change is

$$\tau = \frac{2Q_s}{\omega_0} \approx 0.0255 \text{ s}$$

and the bandwidth is $BW \approx 1/\tau = 40$ Hz.

Separated modes ($\omega_{0d} \ll \omega_{0s}$)

When the sense-mode resonant frequency is much higher than the drive-mode frequency, the sense-mode response function simplifies to $H_{sR} \approx 1/k_s$ and $H_{sI} \approx 0$. From Equation (22.8), we have

$$x_s \approx \frac{F_c}{k_s} = -j\frac{2\omega m_d x_d}{k_s}\Omega. \qquad (22.10)$$

The sense-mode displacement x_s is -90° out-of-phase with the drive-mode. This phase difference can be used, for example, to separate the Coriolis signal from the direct drive-mode coupling due to quadrature error. Also, by separating the drive- and sense-modes, the mode matching is not needed and the mechanical bandwidth of the sensor is high.

The approximate Equation (22.10) is only valid if the drive- and sense-mode frequencies are sufficiently separated. In general, the phase difference between the two modes is smaller than the perfect 90° phase difference. Figure 22.3 shows an example of the gyroscope sense displacement as a function of drive frequency. The gyroscope parameters are typical for a surface micromachined gyroscope. Due to relatively low value for the sense-mode quality factor, the mode separation has to be large to maintain a phase difference close to 90°. As the drive-frequency approaches the sense-mode resonant frequency, the sense-mode amplitude increases. At the sense-mode resonance, the sense-mode is exactly 180° behind the drive-mode as expected from Equation (22.9).

22.3   Capacitive gyroscopes

All commercial silicon gyroscopes are based on capacitive sensing and actuation. Example 22.3 illustrates several critical aspects of silicon micromechanical gyroscopes. The device is actuated with a comb drive to obtain large vibration

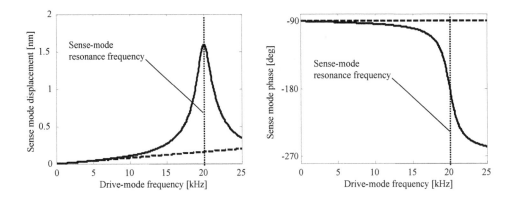

Figure 22.3: Example of the gyroscope response. The sense-mode displacement and phase relative to drive-mode displacement is plotted as a function of drive-mode frequency for a gyroscope with $f_{s0} = 20$ kHz, $Q_s = 10$, $m_s = m_d = 10$ nkg, $|x_d| = 10$ μm, and $\Omega = 1$ rad/s. The solid line is the exact solution from Equation (22.8) and dashed line corresponds to approximate Equation (22.10).

amplitudes. As the Coriolis force is directly proportional to the vibration velocity, the large amplitude helps in measuring the small rotation rates. However, the comb drives typically require large bias voltages. Even when exciting the device at the resonance, bias voltages $V_{dc} > 10$ V are usually needed, which is not trivial for a standard CMOS circuit. High voltage transistors need to be included with the circuit which increases the cost.

The second important observation is that even with the "large" $|x_d| \approx 4$ μm oscillation amplitude and rotation rate of $\Omega = 10$ rad/min, the Coriolis force is very small, only $|F_C| = 79$ pN. Given the $m_s = 6$ nkg proof mass, the Coriolis force equals to force from a 1-mG acceleration. We see that detecting the Coriolis force is a challenge and the small signal is easily lost under other effects such as noise or proof mass movement due to acceleration.

Finally, the proof mass displacement is measured with parallel plated electrodes. As the proof mass movement is small in the sense-mode direction, the parallel plate electrodes are a natural choice since they give better sensitivity than the comb transducers.

Example 22.3: Surface micromachined gyroscope

Problem: Figure 22.4 shows a schematic of a surface micromachined gyroscope that is based on two orthogonal vibration modes. The drive-mode is excited with the drive electrodes as shown in Figure 22.4(b). The electrodes are dc biased and the ac signal is added to the dc bias to excite the drive-mode motion. Different signal polarity is used for the left and right electrode so one electrode pushes while the other pulls the proof mass. This excites vibrations in the X_1-direction at the drive-mode resonant frequency $\omega_{0d} = \sqrt{k_d/m_d}$, where the total drive-mode spring constant for the four vertical beams is $k_d = 0.8$ Nm. The four horizontal beams in the figure are stiff in X_1-direction and do not contribute to the drive-mode but allow the sense-mode movement in X_2-direction (the total spring constant in X_2-direction is $k_s = 3.8$ Nm).

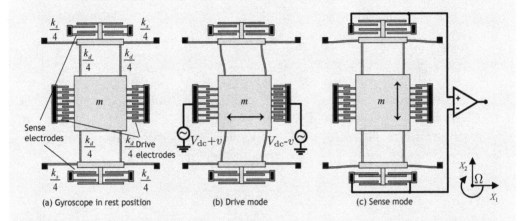

Figure 22.4: Two mode vibration gyroscope that measures rotations around X_3-axis.

The resonator is biased at $V_{dc} = 15$ V, the drive and sense electrodes are biased at ground potential, the drive voltage amplitude is $v_{ac} = 1.5$ V, the drive-mode and sense-mode quality factors are $Q_d = Q_s = 50$, and the drive- and sense-mode masses are $m_d = 5$ nkg and $m_s = 6$ nkg, respectively. The two drive electrodes are formed by comb drives that both have $N_d = 100$ electrode overlaps. The electrode gap and height are $d = 2$ μm and $h = 3$ μm, respectively. The two sense electrodes are made of twenty parallel plate capacitors connected in parallel and the electrode gap, width, and height are $d = 2$ μm, $w = 200$ μm, and $h = 3$ μm, respectively. Calculate (1) the drive-mode amplitude when the gyroscope is excited at the drive-mode resonance frequency, (2) the sense-mode vibration amplitude and phase for an angular rotation rate of $\Omega = 10$ rad/min, and (3) the current measured with the sense electrodes for the angular rotation rate of $\Omega = 10$ rad/min.

Solution: (1) The drive- and sense-mode resonance frequencies are $f_{0d} = 1/2\pi\sqrt{k_d/m_d} = 2$ kHz and $f_{0s} = 1/2\pi\sqrt{k_s/m_s} = 4$ kHz, respectively. The electromechanical transduction factor for one comb drive from Equation (15.31) is

$$\eta_d = N_d\epsilon\frac{h}{d}V_{dc} = 40 \text{ nN/V}.$$

The excitation force is $F = 2\eta_d v_{ac} = 60$ nN where the factor of two accounts for the two actuating comb drives. From Equation (22.7), the drive-mode displacement at resonance is

$$x_d = -j\frac{Q_d}{k_d}F_d \approx -j3.8 \ \mu\text{m},$$

where the negative imaginary number indicates that the displacement is $90°$ behind the excitation voltage.

(2) The Coriolis force is $F_C = j2\omega m_d \Omega x_d \approx -79$ pN. The sense-mode vibration amplitude from Equation (22.8) is

$$x_s(\omega_d) = H_s F_c = -j\omega_d 2 m_d H_s(\omega_d)\Omega x_d \approx (-28 + j0.37) \ \text{pm} = 28\angle179.2° \ \text{pm}.$$

The phase difference between the sense displacement and the Coriolis force is $-90.76°$ which is close to the ideal $-90°$ phase shift.

(3) From Equation (15.30), the electromechanical transduction factor for the sense electrode with $N_s = 20$ sense capacitors is

$$\eta_s = N_s \epsilon \frac{hw}{d^2} V_{\text{dc}} = 0.40 \ \mu\text{N/V}.$$

The differential motional current from the two sense electrodes is $i_{\text{mot}} = 2\eta_s \dot{x}_s \approx (-3.7 - j279)$ fA $\approx 279\angle - 90.8°$ fA.

22.4   Quadrature error

Ideally, the drive- and sense-mode are perfectly orthogonal, and without the Coriolis force, the sense-mode is not excited. In real devices, however, some coupling between the two modes occurs. For example, if one of the beams in Figure 22.4 is slightly stiffer than the others due to variations in the manufacturing, the drive-mode is not perfectly aligned to X_1-direction but moves also in the sense-mode X_2-direction. This movement is referred to as the quadrature error. Another common quadrature error source is illustrated in Figure 22.5. The etched beam sidewalls are not perfectly vertical but are angled. Typical plasma etch is not uniform and the etch angle can be several degrees at the edge of the wafer. This causes the resonator movement to be off the indented direction.

To illustrate the magnitude of the problem, we assume that the fraction of the movement in the sense direction is

$$x_Q = \epsilon_Q x_d. \tag{22.11}$$

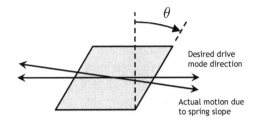

Figure 22.5: Example origin for the quadrature error: the sloping sidewall for the spring cross section causes coupling between the two ideally orthogonal drive- and sense-mode (After [201]).

Assuming that $|x_d| = 1$ μm and $\epsilon_Q = 0.01$ or 1%, which is realistic for batch fabrication process with no trimming, the displacement is $x_Q = 10$ nm. Compared to the typical Coriolis induced displacements that are in the picometer range or smaller (see Example 22.3), the quadrature error is orders of magnitude larger.

The name quadrature error derives from the phase difference between the Coriolis signal and quadrature error signal. By comparing Equation (22.11) to the ideal sense-mode movement due to the Coriolis force when the drive sense and drive-modes are separated (Equation (22.10))

$$x_s = -j\frac{2\omega_d m_d x_d}{k_s}\Omega, \qquad (22.12)$$

we see that the quadrature error induced displacement is ideally 90° out-of-phase with the Coriolis displacement. The Coriolis and the quadrature error signals can therefore be separated with synchronous demodulation that locks to the Coriolis signal but rejects the quadrature signal. However, the large quadrature signal is also problematic, as it can set unreasonable dynamic range requirements for amplifying the signals. It is therefore important to reduce the quadrature error at the sensor element level before the quadrature error signal is amplified by the electronics.

The quadrature error is often compensated by trimming the resonator. Physical trimming is commonly used in piezoelectric gyroscopes but it is not cost effective for silicon gyroscopes where electrical tuning is used. This active tuning makes use of the electrical spring forces in parallel plate capacitors. As covered in Section 15.2.2, the linear capacitive spring from Equation (15.21) is

$$k_{0e} = -\frac{V_{dc}^2 C_0}{d^2}, \qquad (22.13)$$

where C_0 is the dc capacitance and d is the electrode gap. By changing the bias voltage V_{dc}, the electrical spring constant can be adjusted to cancel the mechanical spring variations.

Figure 22.6 illustrates how the electrical tuning cancels the quadrature error. In a perfect device shown in Figure 22.6(a), the proof mass-motion is elliptical as the drive- and sense-mode displacements are 90° out-of-phase. Due to the quadrature error, the proof mass-will also move in the sense-mode direction in phase with the drive-mode as is illustrated in Figure 22.6(b). By adjusting the electrical bias as shown in Figure 22.6(c), the electrical springs can compensate the asymmetry in mechanical springs and the drive-mode can be perfectly aligned to the desired direction [202].

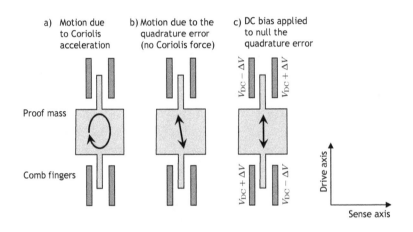

Figure 22.6: The quadrature error can be compensated by using electrical spring forces to trim the mechanical springs (After [202]).

Good design can also be used to alleviate the problems due to the quadrature error. Figure 22.7 shows the spring structure investigated by Analog Devices [203]. The selectivity of the suspension flexures is improved using levers. A simple folded spring allow movements in multiple degrees of freedom as shown in Figures 22.7(a) and 22.7(b). By bridging the beams with a pivoting beam as shown in Figure 22.7(c), the motion can be constrained to one direction. The beneficial effect of the pivoting beam can be understood from the following analysis: When the mass moves up and down as desired, the linkage in the pivoting beam simply bends to allow this motion. Rotation of the mass would result in an extension of the linking beam. As we recall from Chapter 4, the spring constant for extension is much higher than for bending and hence the linkage inhibits the rotation. The advantage of this technique is that it removes the problem at the source. A small quadrature error permits the use of simpler measurement circuit without the electrical trimming, which reduces the total cost.

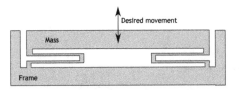

(a) The mass is anchored to a frame with folded beams.

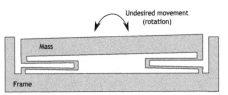

(b) In addition to the desired motion, the folded beams allow mass rotation.

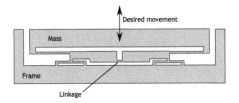

(c) A pivoting linkage between the springs inhibits rotating movement but allows the desired motion.

Figure 22.7: The springs can be designed to reduce quadrature error [203].

22.5   Measurement circuitry

The circuit design for micromechanical gyroscopes is a significant effort. Even a minimalistic system requires circuit blocks for maintaining oscillations, measuring the small Coriolis signal, generating the high voltage to bias the resonator, measuring the temperature to compensate for temperature offsets, and logic for controlling and calibrating the system.

The immediate challenge in measuring the Coriolis signal is that the signal current can be very small, typically in the femto-ampere range. A carefully designed low-noise transimpedance amplifier is used to convert the small current into voltage. In addition to the large gain to boost the small current, the amplifier needs a high dynamic range to tolerate often large quadrature error signals and the signal phase needs to be maintained in order to keep the Coriolis and quadrature signals separated. To prevent the signal phase shift, the amplifier must be biased with large valued resistors. For example, Analog Devices uses MOSFET transistors with equivalent resistance of > 2 GΩ to dc bias the amplifier.

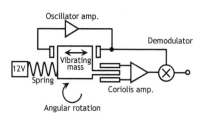

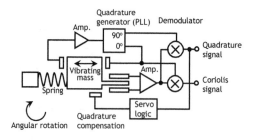

(a) A relatively simple gyroscope circuit from Analog Devices [204].

(b) More complex gyroscope that adaptively compensates the guadrature signal [205].

Figure 22.8: Gyroscope Coriolis for detection circuits.

The second challenge in the gyroscope circuit design is the separation of the Coriolis and quadrature signals. To understand how this is achieved, we will study two gyroscope circuits shown in Figure 22.8. Figure 22.8(a) shows the system level diagram for a commercial gyroscope by Analog Devices [204]. An oscillator circuit is used to excite the resonator at the drive-mode resonance. The drive and the quadrature error motion are both 90° out-of-phase with the drive signal. The Coriolis effect gives another 90° phase-shift and the Coriolis induced vibrations are therefore 180° out-of-phase with the drive signal and 90° out-of-phase with the quadrature signal. The sense amplifier picks up the Coriolis signal and the quadrature error signal. Due to relatively small quadrature error, the amplifier dynamic range is sufficient and the quadrature signal does not saturate the amplifier.

To separate the Coriolis and quadrature signals, the signal from the amplifier output is multiplied with the drive signal and low pass filtered to demodulate the Coriolis signal. Recalling that the time average of $\cos \omega t \cos \omega t$ is $1/2$ and average of $\cos \omega t \sin \omega t$ is zero, we see that the quadrature error that is 90° out-of-phase with the drive signal is removed by the synchronous demodulation. The demodulator output is therefore proportional to the Coriolis signal.

In the Analog Devices gyroscope, the relatively small quadrature error is the key to successfully rejecting the quadrature error signal within the electronics. If the quadrature error was too large, it would saturate the amplifier circuit and synchronous demodulation would not be possible.

The quadrature error can also be removed via active compensation as shown in Figure 22.8(b). The signal is demodulated with two signals that have a 90° phase-shift [205]. In communication systems this is known as the I/Q demodulation where the I/Q stands for in-phase and quadrature. The I/Q demodulation gives both the Coriolis signal and quadrature error signal, which can be used for an active error compensation using servo feedback that adjusts the electrical spring constants. This way the quadrature signal is nulled so it does not overload the amplifier.

In addition to the basic measurement functionality described above, the sense circuit usually includes a temperature sensor to compensate for the temperature variations in the resonator and the electronics. For example, the air damping in the Analog Devices gyroscope is temperature dependent and leads to 15% variations in the Coriolis signal over the temperature range [206, 207]. Calibration in multiple temperatures is often needed to reduce the temperature dependency, which increases the calibration cost.

Overall, the gyroscope measuring circuit complexity far surpasses what is needed for accelerometers or pressure sensors. This complexity is reflected in the gyroscope die size and total power consumption.

22.6 Noise in gyroscopes

As was seen in Chapter 9, the capacitive detection and synchronous demodulation can result in very good noise performance. Ideally, the gyroscope noise performance is limited by mechanical noise [206]. In this section we review the mechanical noise in gyroscopes with matched or separated drive and sense modes.

22.6.1 Noise in gyroscopes with matched modes

Figure 22.9(a) illustrates the noise in a mode matched gyroscope where the gyroscope vibration frequency $\omega = \omega_{0d}$ coincides with the sense mode resonance frequency ω_{0s}. To detect the Coriolis signal, the signal current is demodulated and filtered. As the noise is confined to a narrow band near the resonance, filtering after demodulation does not significantly reduce the mechanical noise seen at the filter output. Thus, most of the mechanical noise is captured by the sensor electronics, and the detected rms-noise displacement is $x_{\mathrm{rms}} = \sqrt{k_B T / k_s}$.

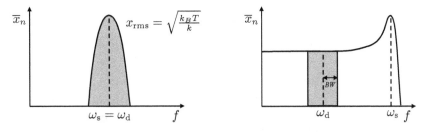

(a) Due to the high Q in matched mode gyroscopes, most of the mechanical noise is near the sense frequency and is seen at the demodulated output.

(b) With separated modes, the vibration frequency $\omega = \omega_{0d}$ is below the sense mode resonance frequency ω_{0s}. The rms-noise is set by the sensor electronics bandwidth.

Figure 22.9: Mechanical noise in gyroscopes.

The input referred noise is obtained with the help of Equation (22.9). The sense mode displacement due to a angular rotation rate Ω is

$$x_s = -2\frac{m_d}{m_s}\frac{Q_s}{\omega}\Omega x_d. \tag{22.14}$$

Equating Equation (22.14) with the rms-noise displacement gives the input referred rms-noise

$$\Omega_{\mathrm{rms}} = \frac{m_s\omega}{2m_d Q_s x_d}x_{\mathrm{rms}} = \frac{m_s}{2m_d Q_s x_d}\sqrt{\frac{k_B T}{m_s}}. \tag{22.15}$$

Usually the drive and sense mode masses are equal ($m_s = m_d = m$) and Equation (22.15) simplifies to

$$\Omega_{\mathrm{rms}} = \frac{1}{2Q_s x_d}\sqrt{\frac{k_B T}{m}}. \tag{22.16}$$

As expected, a large quality factor, vibration amplitude, and mass reduce the gyroscope noise.

22.6.2   Noise in gyroscopes with separated modes

Figure 22.9(b) illustrates the mechanical noise in a gyroscope where drive frequency $\omega = \omega_{0d}$ is much lower than the sense mode resonance frequency ω_{0s}. The Coriolis signal is detected via demodulation and low pass filtering. This filtering will limit the gyroscope bandwidth and reduce the amount of noise seen at the filter output.

In demodulation, noise above and below sense frequency is aliased to signal band as is illustrated in Figure 22.10. Given the filter bandwidth BW, the rms-noise due to mechanical vibrations is approximately

$$x_{\mathrm{rms}} = \sqrt{2}\overline{x}_{n,s}\sqrt{BW}, \tag{22.17}$$

where $\overline{x}_{n,s} = \overline{F}_{n,s}/k_s$ is the spectral density for the sense mode noise displacement and BW is the gyroscope bandwidth limited by the sensor electronics. The factor $\sqrt{2}$ in Equation (22.17) accounts for the noise on both sides of the vibration frequency $\omega = \omega_{0d}$ as shown in Figures 22.9(b) and 22.10.

The input referred noise is obtained by equating the noise force $\sqrt{2}\overline{F}_{n,s} = \sqrt{2}\sqrt{4k_B T\gamma_s}$ with the Coriolis force $|F_c| = 2m_d\omega x_d\Omega$. Again, the factor $\sqrt{2}$ in the noise force accounts for the aliasing of noise in demodulation. Solving $|F_c| = \sqrt{2}\overline{F}_{n,s}$ for the angular rotation rate gives the input referred noise spectral density

$$\overline{\Omega} = \frac{\sqrt{2}\overline{F}_{n,s}}{2m_d\omega x_d}. \tag{22.18}$$

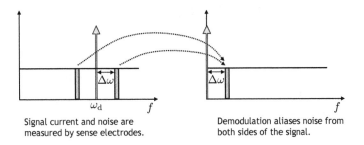

Signal current and noise are
measured by sense electrodes.

Demodulation aliases noise from
both sides of the signal.

Figure 22.10: When the signal is demodulated, noise at distance $\Delta\omega$ above or below the sense frequency $\omega = \omega_d$ is transferred to $\Delta\omega$.

The input referred rms-noise is

$$\Omega_{\mathrm{rms}} \approx \overline{\Omega}\sqrt{BW}. \qquad (22.19)$$

As expected, a large mass and drive mode vibration amplitude reduce the input referred noise.

Example 22.4: Surface micromachined gyroscope noise
Problem: Consider the capacitive gyroscope in Example 22.3. Assume that the sensor electronics bandwidth is $BW = 50$ Hz and calculate:
(1) the input referred noise spectral density,
(2) the input referred rms-noise,
(3) the sense current noise spectral density, and
(4) the sense current rms-noise.
Solution: (1) The noise force is $\sqrt{2}\,\overline{F}_{n,s} = \sqrt{2}\sqrt{4k_BT\gamma_s} \approx 3.3 \cdot 10^{-13}$ N/$\sqrt{\mathrm{Hz}}$. The input referred noise spectral density from Equation (22.18) is

$$\overline{\Omega} = \frac{\overline{F}_{n,s}}{2m_d\omega x_d} \approx 6.6 \cdot 10^{-4} \text{ rad/s/}\sqrt{\mathrm{Hz}}. \qquad (22.20)$$

(2) The input referred rms-noise is $\Omega_{\mathrm{rms}} = \overline{\Omega}\sqrt{BW} \approx 0.005$ rad/s.
(3) The mass displacement spectral density is $\sqrt{2}\overline{x}_s = \sqrt{2}\,\overline{F}_n/k_s$ and the noise current spectral density around the oscillation frequency $\omega = \omega_{0d}$ is $\overline{i}_n = 2\eta_s\omega_{0d}\overline{x}_s = 0.8$ fm/$\sqrt{\mathrm{Hz}}$.
(4) Given the filter bandwidth BW, the rms-noise current is $i_{\mathrm{rms}} = \overline{i}_n\sqrt{BW} \approx$ 6 fA.

22.7   Commercial gyroscopes

The gyroscope market is growing fast. The automotive market demands both high performance sensors for stability control and lower performance parts for rollover detection. In the consumer market, high-end digital cameras utilize gyroscopes for image stabilization. Gyroscopes can also be used to enhance motion based computer user interfaces.

In this section, we take a closer look at several miniature gyroscopes: a quartz tuning fork, piezoelectric triangle bar, and surface micromachined silicon resonator. The wide selection of sensors illustrates that silicon MEMS is not the only way to make micromachined gyroscopes.

22.7.1   Case study: Quartz tuning fork gyroscope

Piezoelectric materials are well suited for vibrating gyroscopes as the drive- and the sense-mode can be excited and sensed without the need for the dc bias voltage. This simplifies the interface electronics especially in comparison to the electrostatic excitation that may require voltages greater than are available in standard CMOS IC processes. Quartz is especially interesting material as it is available in wafer form and can be micromachined, although the feature sizes are much larger than for silicon. Crystalline quartz is piezoelectric, has very predictable material properties, and high internal quality factor. As we saw in Chapter 21, more than a billion quartz tuning fork resonators are sold every year demonstrating that large scale, low cost manufacturing is possible with quartz.

Figure 22.11 illustrates the operation principle of a double tuning fork gyroscope that is commercially available [208, 209]. The principle of operation is the following: One side of the tuning fork is excited in the normal tuning fork motion where the two tines oscillate in anti-phase. This drive-mode does not exert net force on the anchor as the moments of two tuning fork tines cancel. When the gyroscope rotates around its axis, however, the Coriolis force on the two tines is in the opposite directions and there is net torque on the anchor. This torque excites the sense-mode oscillations of the second tuning fork that is connected to the same anchor.

A low cost micromachined angular rate sensor LCG50 from Systron Donner has full scale range $\pm 100°$/s, noise $< 0.005°$/s/$\sqrt{Hz}$ and temperature stability $8°$/s (8% of the full scale) in a 2.94 cm$\times$2.94 cm$\times$1.07 cm PC board package. The inertial grade versions offer stabilities better than $0.4°$/s for the temperature range from -40 °C to 80 °C.

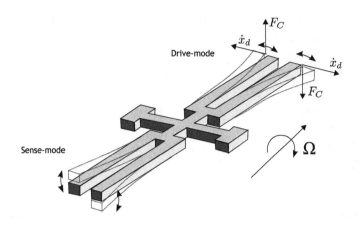

Figure 22.11: Piezoelectric quartz tuning fork gyroscope.

22.7.2 Case study: Piezoelectric metal/ceramic gyroscope

A low cost metal/ceramic gyroscope by Murata shown in Figure 22.12 has a single triangle bar with orthogonal drive and sense modes [210]. The bar is made of "Elinvar" nickel steel alloy (59% iron, 36% nickel, and 5% chromium) that is a unique metal alloy as its Youngs modulus is relatively constant over temperature. The temperature of frequency is smaller than 3 ppm/K, which is about ten times better than for single-crystal silicon resonators. The resonator quality factor is more than 2,000 at 20 kHz.

The drive-mode in X_1-direction is excited with the bottom piezoelectric plate attached to the triangle bar. The motion is sensed differentially with two

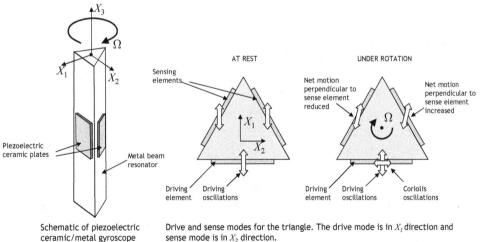

Figure 22.12: Piezoelectric bar gyroscope.

plates. If the bar vibrates only in the drive-mode, the signal from the two sense plates cancel. When the bar rotates around its length axis, the Coriolis force excites the sense-mode that is orthogonal to the drive-mode. The sense-mode vibrations in the X_2-axis direction results in a differential signal from the two sense plates.

Murata also makes even smaller ceramic resonators that are based on patterned piezoelectric ceramic beams. These devices have package dimensions of 8 mm×4 mm×2 mm, which compares well to micromachined silicon gyroscopes [211].

22.7.3   Case study: Surface micromachined gyroscope

The surface micromachined gyroscope from Analog Devices is an engineering feat: The full-scale movement of the Coriolis sensor is less than 2 Å, about the size of a hydrogen atom, and the noise limit corresponds to approximately a 16-fm sense electrode movement movement. This amazing performance is partially due to the single-chip integration of the measurement electronics and the mechanical element. Figure 22.13 shows a simplified schematic of the gyroscope element together with the published gyroscope element parameters [206, 212].

The operation of the Analog Devices gyroscope is similar to the surface micromachined gyroscope in Example 22.3. The element is biased to 12 V and the drive-mode is excited with the comb actuators in the middle of the element. The comb drive enables a large $|x_d| = 7$ μm vibration amplitude. The Coriolis force excites the orthogonal sense-mode vibrations that are sensed with the differential parallel plate capacitors. The sense electrodes and sense amplifiers are biased at 1.5 V so the net dc bias for generating the sense current is $V_{\mathrm{dc}} = $ 12 V $-$ 1.5 V $=$ 10.5 V.

The drive- and sense-mode frequencies are not matched. To minimize the quadrature signal, the sense-mode frequency is significantly higher than the drive-mode. The sense-mode mass is a combination of both the drive-mode mass and the frame mass. The sense-mode frequency, the mass, and the quality factor are not published but with reasonable guesses (see Example 22.5), we can get a more complete view of the gyroscope operation.

The data sheet for the ADXRS150 gyroscope specify the full scale range as $\Omega = \pm 150°$/s, the noise as $0.05°$/s/$\sqrt{\mathrm{Hz}}$, and temperature stability as $23°$/s (15% of the full scale) [207]. The device dimensions are 7 mm×7 mm×3 mm. In comparing the surface micromachined silicon gyroscope to the commercial quartz and ceramic gyroscopes, we note that the stability performance is lower than for the significantly larger quartz tuning fork gyroscope in Section 22.7.1. The stability performance of ceramic/metal gyroscope analyzed in Section 22.7.2 is not known but it is interesting to note that the sizes for the two devices are comparable.

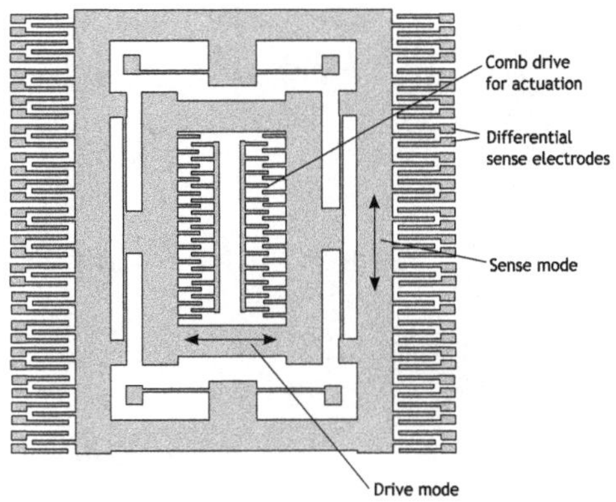

Parameter	Value
Drive-mode frequency, f_{0d}	15 kHz
Drive-mode amplitude, $|x_d|$	7 μm
Drive-mode Q, Q_d	45
Thickness, h	4 μm
Sense cap. sensitivity, $\frac{\Delta C}{\Delta x_s}$	0.7 μF/m
Sense-mode dc bias, V_s,	10.5 V
Mechanical noise floor, $\overline{i}_n$	12 fA/$\sqrt{\text{Hz}}$
Full scale range, Ω	$\pm$150 $^\circ$/s
Full scale cap. change, ΔC_{FS}	120 aF

Figure 22.13: Schematic view and published element parameters for a surface micro-machined gyroscope from Analog Devices [206, 212].

Example 22.5: Analog Devices gyroscope parameter estimation

Problem: Based on the published parameters and Figure 22.13 for the Analog Devices gyroscope, estimate the sense-mode resonance frequency and quality factor.

Solution: The drive-mode mass and the frame mass can be estimated from the published die photograph in reference [206]. Based on the published thickness and measured lateral dimensions, we estimate the drive-mode mass as $m_d = 2.5$ nkg and the frame mass as $m_f = 1.6$ nkg. The sense-mode mass consists of both drive-mode and frame masses and is estimated to be $m_s = m_d + m_f = 4.1$ nkg. The full scale sense-mode displacement from the full scale capacitance change and sensitivity is $x_{FS} = \Delta C_{FS}/|\Delta C/x| = 1.7$ Å.

The sense-mode frequency is not known. Here, we estimate it by calculating the known noise current and full scale displacements as a function of sense-mode resonant frequency and comparing the results to the published values in Figure 22.13.

The thermal noise force acting on sense-mode is $\overline{F}_n = \sqrt{4 k_B T \gamma_s}$ where the damping coefficient depends on the sense-mode resonance frequency as $\gamma_s = m_s \omega_{0s}/Q_s$. Here ω_{0s} and Q_s are the sense-mode resonance frequency and quality factor, respectively. The sense-mode spring constant is $k_s = \omega_{s0}^2 m_s$.

The thermal noise induced displacement is

$$\overline{x}_n = \frac{\overline{F}_n}{k_s}$$

and the corresponding vibration velocity near the sensing frequency $\omega = \omega_{0d}$ is

$$\overline{\dot{x}}_n = \omega \overline{x}_n.$$

The sensed noise current from the capacitance sensitivity and sense bias voltage is

$$\bar{i}_n = 2\bar{\bar{x}}_n \frac{\Delta C}{\Delta x_s} V_s,$$

where the factor of two is due to the differential measurement with the two electrodes.

The magnitude of the Coriolis force from Equation (22.2) is

$$|F_c| = 2m\Omega\dot{x}_d.$$

The sense-mode vibration amplitude from Equation (22.8) is

$$x_s(\omega) = H_s(\omega)F_c = -j\omega 2m_d H_s(\omega)\Omega x_d,$$

where $H_s(\omega)$ is the sense-mode transfer function at the drive frequency ω given by Equation (22.6)

$$H_s(\omega) = \frac{1}{m_s} \frac{1}{\omega_{0s}^2 - \omega^2 + j\omega\omega_{0s}/Q_s}.$$

The unknown parameters in these equations are the sense-mode resonant frequency ω_{s0} and quality factor Q_s. By trial and error, we find that $f_{s0} = 23$ kHz and $Q_s = 18$ give thermal noise floor

$$\bar{i}_n = 2\bar{\bar{x}}_n \frac{\Delta C}{\Delta x_s} V_s = 12 \text{ fA}/\sqrt{\text{Hz}}$$

and the full scale displacement corresponding to $\Omega = 150°/\text{s}$ is

$$x_s(\omega_d) = H_s(\omega_d)F_c = -j\omega_d 2m_d H_s(\omega_d)\Omega x_d = 1.7 \text{ Å}.$$

These compare well to reported noise current of $\bar{i}_n = 12$ fA/$\sqrt{\text{Hz}}$ at 15 kHz in Figure 22.13 and the calculated full scale displacement $x_{FS} = 1.7$ Å.

Key concepts

- Gyroscopes measure the angular rotation rate.

- Coriolis force is proportional to the product of the proof mass velocity and the rotation rate.

- Micromechanical gyroscopes are based on two orthogonal vibration modes. The resonator is excited in the drive-mode and the Coriolis force induces the sense-mode vibrations.

- Quadrature error is unwanted movement in the sense-mode direction due to manufacturing non-idealities. The quadrature error signal is in phase with the drive-mode oscillations.

- Piezoelectric gyroscopes present a strong competition to the silicon MEMS gyroscopes.

- Gyroscopes measure very small signals. The measurement accuracy is heavily affected for example by temperature induced errors.

Exercises

Exercise 22.1
By visiting Analog Device's web-site, compare the package size and current consumption for the Analog Devices gyroscopes and accelerometers. How much larger are the package size and current consumption for the gyroscopes?

Exercise 22.2
Explain why choosing the sense mode resonance frequency to be lower than the drive mode resonance frequency $w_s \ll w_d$ is not a good gyroscope design.

Exercise 22.3
A microresonator is vibrating at the resonant frequency $f_0 = 22$ kHz with a vibration amplitude $x_0 = 1.5$ μm along X_1-axis. The resonator mass is $m = 0.35$ nkg. Calculate the Coriolis force acting on the resonator mass that is on a rotating platform with a rotation rate $\Omega = 2\pi$ rad/s around the X_3-axis.

Exercise 22.4
A silicon micromechanical gyroscope has $f_d = 10$ kHz, $f_s = 25$ kHz, $Q_s = Q_d = 20$ and masses $m_s = m_d = 50$ nkg. The drive mode vibration amplitude is 1 μm. Calculate (1) the gyroscope bandwidth if it is limited by the mechanical element response time and (2) the gyroscope resolution limited by the mechanical noise. Calculate both the input referred rms-noise and noise spectral densities assuming that the gyroscope bandwidth is limited by electronics to $BW = 100$ Hz.

Exercise 22.5
A piezoelectric gyroscope has matched resonant frequencies $f_s = f_d = 15$ kHz, quality factors $Q_s = Q_d = 3,000$, and masses $m_s = m_d = 0.2$ nkg. The drive mode vibration amplitude is 1 μm. Calculate (1) the gyroscope bandwidth if it is limited by the mechanical element response time and (2) the gyroscope resolution limited by the mechanical noise.

Exercise 22.6
Consider the Analog Devices gyroscope in Figure 22.13. How would you change

the design to reduce the noise by $5\times$ if the performance is limited by mechanical noise?

Exercise 22.7

Explain the quadrature error and why it is so significant. List methods to reduce it.

Exercise 22.8

Figure 22.14 shows a differential transimpedance amplifier that is used in the Analog Devices gyroscopes to convert the motional current from the two sense electrodes into voltage. Ideally, the capacitor gives a perfect $-90°$ phase shift between the voltage and the current. Due to biasing resistors, however, the phase shift is less than the perfect $-90°$. If $C = 2$ pF and $R = 2.5$ GΩ, how large is the phase shift? What would happen if the resistance was "only" 25 MΩ? Note that the operation frequency for the Analog Devices gyroscope is 15 kHz.

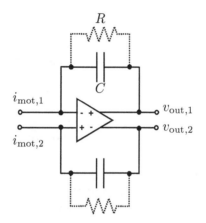

Figure 22.14: Differential transimpedance amplifier.

Exercise 22.9

Follow the phase shift in the Analog Devices qyroscope shown in Figure 22.13. What is the phase shift between (1) drive signal and drive mode vibrations, (2) drive mode and Coriolis force, (3) Coriolis force and sense mode vibrations, (4) sense mode vibrations and motional current, and (5) motional current and Coriolis amplifier output voltage (see also Exercise 22.8)? Calculate also the the total phase shift between the drive signal and the output voltage. Hint: if you did all the steps correctly, the total phase shift is zero and the drive signal can directly demodulate the output voltage as is shown in Figure 22.13.

Exercise 22.10

The Analog devices gyroscope can measure the sense electrode movement down to 16 fm. With 10.5 V bias, what is the sense electrode charge change due to the 16-fm movement and how many electrons does this correspond to?

Exercise 22.11

How does the gyroscope sense mode displacement given by Equation (22.10) scale if all dimensions and the drive mode displacement x_d are reduced by an equal factor?

Exercise 22.12

Follow the phase shift in a quartz double tuning fork gyroscope that has matched drive and sense modes. What is the phase shift between (1) drive signal and drive mode vibrations, (2) drive mode and Coriolis force, (3) Coriolis force and sense mode vibrations, (4) sense mode vibrations and motional current.

Exercise 22.13

Estimate the input referred noise $\overline{\Omega}_n$ due to mechanical noise for the Analog Devices gyroscope in Example 22.5. Compare your result to the noise density of $0.05°/s/\sqrt{Hz}$ in manufacturer's datasheet.

23

Microfluidics

Microfluidic devices have been in development since 1970s. The first devices include gas chromatographs [213] and inkjet print head nozzles [214]. While micromachined gas chromatographs have not had wide commercial success, the inkjet printers have been hugely successful with the market volume which measured in hundreds of millions of dollars.

The follow-up to the early devices has taken a long time, but since the late 1980s significant research effort has been put into the development of microfluidic devices including cooling for integrated circuits [215], implantable drug delivery devices [216], and biochemical analysis [217–221]. The state of the art total analysis system (TAS) strives to automatically carry out all aspects of chemical and biological sample processing including the sampling, the sample transport, chemical reactions, separations, and detection. Micro total analysis system (μTAS) take this one step further by miniaturizing the desktop-sized analysis systems into hand held instruments or smaller. A typical μTAS has a disposable microfluidic analysis chip ("lab-on-chip") and macroscopic nondisposable components such as pumps, valves, light sources, detectors, control electronics, and a display.

The advantages of the μTAS are:

1. The amount of used essays and solutions is reduced. Especially biological essays such as specific DNA sequences are very expensive and reduction of dead volumes can translate into significant cost savings. The reduction in the sample volume can also enhance the user experience; for example, less blood may be required for the analysis.

2. The batch fabrication can lower the analysis cost. Currently, laboratory analysis is done by trained personnel in expensive laboratories. The μTAS

could enable low cost analysis systems so that every doctor's and nurse's office could be equipped with the analysis system. The financial barrier to carry out the needed test would be reduced which would improve the quality of the care.

3. In addition to lowering the cost, the analysis at the point of care would eliminate the sample transport which is both a cost and a time issue. With μTAS at the point of care, the time from sampling to the analysis is short.

4. Microfluidic systems can be used for parallel analysis with different reagents. For example, the same sample could be checked for multiple illnesses.

Currently, μTAS market is in its infancy. Several lab-on-a-chip manufacturers, including Aclara (Mountain View, CA), Caliper (Newton, MA), Cepheid (Sunnyvale, CA), Micronics Inc. (Redmond, WA), and Orchid Biosciences (Princeton, NJ) have developed technologies that are just entering the market. The market size, however, is expected to grow rapidly.

In this chapter we review the microfluidic devices. The focus is on physical characteristics of the microflows and challenges in miniaturing microfluidic components and pumps.

23.1   Flow in microchannels

The liquid and gas flow is characterized by the Reynold's number

$$\mathrm{Re} = \frac{\rho \dot{x} L}{\mu}, \tag{23.1}$$

where ρ is the fluid density, $\dot{x}$ the velocity characteristic to the flow, L the length scale characteristic to the flow, and μ is the fluid viscosity [222]. The Reynold's number is the ratio of inertial forces to viscous forces and categorizes the flow:

- When $\mathrm{Re} \ll 1$, viscous effects dominate, inertial effects can be neglected, and the flow is completely laminar.

- When $\mathrm{Re} \sim 1$, viscous effects are comparable to inertial effects and vortices began to appear.

- When $\mathrm{Re} > 2,000 - 3,000$, the inertia effects dominate and flow is turbulent.

In general, both the velocity $\dot{x}$ and the length scale L are small in microchannels and the flow is laminar. For example, typical microflow parameters $\dot{x} = 1$ mm/s and $L = 100$ μm give $\mathrm{Re} = 0.1$ for water ($\rho = 1,000$ kg/m^3 and $\mu = 1 \cdot 10^{-3}$ Pas).

Figure 23.1 illustrates the laminar flow in microfluidic channels. The volume flow rate is given by

$$q_V = \dot{V} = \frac{\Delta p}{R}, \tag{23.2}$$

where Δp is the pressure drop across the channel length and R is the flow resistance. For low Reynold's numbers (laminar flow), the flow resistance depends on the the viscosity μ, the channel length L, and the channel radius r as

$$R = \frac{8\mu L}{\pi r^4}. \tag{23.3}$$

As Equation (23.3) shows, the flow resistance is very sensitive to the channel radius r.

As with the electrical networks, more complex channels can be analyzed as parallel and series combinations of simple channels. For series connected channels, the flow resistances are added ($R_{\text{tot}} = R_1 + R_2 + \ldots + R_N$). For parallel connected channels, total flow resistance is obtained by adding the reciprocals of the resistances and then taking the reciprocal of the total ($1/R_{\text{tot}} = 1/R_1 + 1/R_2 + \ldots + 1/R_N$).

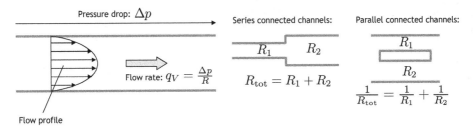

Figure 23.1: Flow rate in microfluidic channels is generally laminar and the flow rate is controlled by the flow resistance R and the pressure drop Δp across the channel length. Fluidic "circuits" can be analyzed as parallel and series combination of resistors.

Example 23.1: Blood flow through capillaries
Problem: Consider blood ($\mu = 3$ mPas) flow in human capillary veins (diameter 8 μm, length 1 mm, pressure drop 10 kPa). What is the volume flow rate and how long does it take for the blood entering the capillary to flow through it? How high should the flow velocity be for the flow to be turbulent?
Solution: The flow resistance is

$$R = \frac{8\mu L}{\pi r^4} \approx 3.0 \cdot 10^{16} \text{ Pas/m}^3.$$

The volume flow rate is

$$q_v = \frac{p}{R} \approx 4.4 \cdot 10^{-14} \text{ m}^3/\text{s}.$$

The flow velocity is $\dot{x} = q/A$ and the time to flow through the capillary is

$$t = \frac{L}{\dot{x}} \approx 1.2 \text{ s}.$$

The Reynolds number is $\text{Re} = \rho \dot{x} r / \mu = 0.001$ so the flow is clearly laminar. For the flow to be tubulent, the velocity should be greater than $1,000-2,000$ m/s!

Example 23.2: Flow rate in microchannels
Problem: A 1-mm long microchannel should have a minimum flow rate of 5 nl/min with $\Delta p = 10$ kPa. What is the minimum channel diameter and how long does it take for a flow to go through the channel?
Solution: The flow resistance is

$$R = \frac{8\mu L}{\pi r^4}$$

and the volume flow rate is

$$q_v = \frac{\Delta p}{R}.$$

Solving for the channel radius gives $r = 16$ μm.
    The flow velocity is $\dot{x} = q_V/A$ and the time for the flow to travel through the capillary length is

$$t = \frac{L}{\dot{x}} \approx 10 \text{ s}.$$

Bubbles in microchannels can be very problematic. If the bubble is smaller than the channel diameter, the bubble will restrict the flow as is illustrated in Figure 23.2. A large bubble can completely block the channel. The minimum pressure needed to move a bubble is

$$p_c = \frac{2\gamma}{r}, \tag{23.4}$$

where γ is the surface tension which for water is $\gamma = 0.07$ N/m and r is the channel radius [223]. For example, the pressure needed to move a bubble in a

10-μm diameter channel filled with water is $p_c = 28$ kPa, which is not trivial for a micropump. Even if gas bubbles can be totally avoided in the operational use of microfluidic devices, the surface tension effects still have practical relevance to the initial filling of these devices.

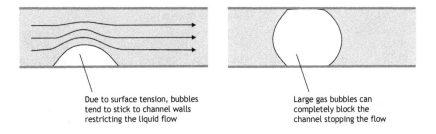

Figure 23.2: Bubbles in microfluidic channels can restrict or even block the flow.

Example 23.3: Bubble in a channel
Problem: Estimate the minimum channel size that is not blocked by a bubble for a microfluidic system that has the pumping pressure $p = 4$ kPa.
Solution: Using $\gamma = 0.07$ N/m and $p_c = 4$ kPa, Equation (23.4) gives the smallest radius

$$r = \frac{2\gamma}{p_c} = 35 \ \mu\text{m}$$

that does not block the flow. The channel diameter should be more than 70 μm.

23.2 Mixing

Since the flow in microchannels is generally laminar, there is very little mixing in microchannels. The lack of mixing has surprising consequences: As is illustrated in Figure 23.3, just combining two laminar flows in a microchannel is not sufficient to mix them. Due to the lack of turbulence, the only mixing mechanism is the diffusion of the molecules and particles. The average distance a particle diffuses in time t is

$$d = \sqrt{2Dt}, \tag{23.5}$$

where D is the diffusion coefficient of the particle. As Table 23.1 illustrates, the diffusion coefficients vary significantly for biological molecules with larger diffusing more slowly. For example, hemoglobin ($D = 70 \ \mu\text{m}^2/\text{s}$) takes one minute to diffuse a typical channel diameter of 100 μm [222]. The mixing by diffusion is further illustrated in Example 23.4.

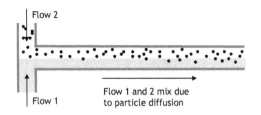

Figure 23.3: Due to laminar flow, passive mixing happens only due to diffusion. Long channel length is needed for passive mixing.

Table 23.1: Diffusion coefficients for typical biological ions or particles [9].

Molecule	Diffusion coefficient in water $[\mu m^2/s]$
H^+	9,000
Na^+	2,000
O_2	1,000
Glycine	1,000
Hemoglobin	70
Myosin	10
Virus	5

To increase the mixing speed, several active and passive mixers have been developed. One approach to improve the passive mixer in Figure 23.3 is to introduce sharp bends in the channel to generate turbulent flow. In macroscopic channels, this can drastically cut the length of the channel needed for mixing. The beneficial effect, however, can be small in microchannels where even sharp bends may not generate turbulence [9]. Active mixers generate turbulent flow by inducing large particle velocities. Typical examples include piezoelectric mixers and electrokinetic mixers.

Example 23.4: Mixing in a microchannel
Problem: Estimate the time for myosin to mix in a 50-μm diameter channel. If the flow rate is $q_V = 50$ nl/min, how long should the channel be to enable the mixing?
Solution: Equation (23.5) gives the diffusion time across the channel

$$t = \frac{d^2}{2D} \approx 125 \text{ s}.$$

From the channel cross sectional area and the flow rate, the average flow velocity is

$$\dot{x} = \frac{q_V}{A} \approx 0.4 \text{ mm/s}.$$

The channel length should be at least $L = \dot{x}t = 5$ cm to enable the mixing.

23.3 Microfluidic systems: valves and pumps

Initiating and controlling fluid flow requires valves and pumps. In a typical μTAS application, the biological samples are moved through miniature assay systems. Other applications include miniature coolant systems for microelectronics and dispensing therapeutic agents into the body. For example, implanted micropumps could be used for releasing insulin required by diabetics. Currently, these implanted pumps are large systems $V > 50$ cm^3 that use static pressure reservoirs and solenoid-driven valves. Miniaturization of the pumps and valves could drastically reduce the system size.

In this section, the mechanical valves and pumps are reviewed. The design requirements and challenges of microfluidic systems are considered with the focus on micromachined structures.

23.3.1 Microvalves

The small size of the micromechanical valves brings benefits in terms of response time, small dead volume, and improved fatigue properties [224]. Unfortunately, because of their small size, the micromechanical valves are prone to clogging. Also, the demonstrated valves often have had significant leakage flow. The leakage has prevented the use of micromechanical valves for example in insulin dispensing applications where very accurate dosage is required.

A typical passive valve is shown in Figure 23.4(a). The valve passes flow in one direction but is closed for reverse pressure. Typically, cantilevers and diaphragms are used as moving parts, as these shapes can be easily defined with planar microfabrication technologies. The passive valves are common as check valves are in microfluidic pumps.

The passive valve design is a compromise between leakage and a threshold pressure difference Δp_c across the valve needed to actuate it. For low leakage, the valve should compress tightly against the inlet/outlet orifice when the valve is in the closed position. This calls for a stiff design that provides large closing forces; however, this also means that a large pressure difference across the valve is needed to actuate it. This is further illustrated in Example 23.5.

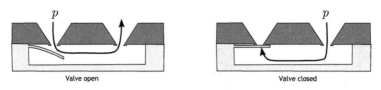

(a) A passive valve passes fluid in one direction but blocks the flow in the other direction.

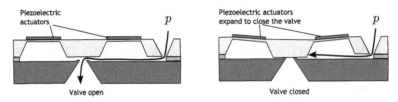

(b) Actuated valve made with microfacbrication techniques [225].

Figure 23.4: Microvalve operation principles.

Figure 23.4(b) shows an actively actuated valve design [225]. The active valve can pass the fluid in both directions and does not require a threshold pressure for fluid to flow through. An ideal valve actuator should provide large forces so that the valve can be closed tightly for low leakage. In addition, the valve displacement should be large to fully open the valve. These requirements are similar to the ideal micropump actuator. The different actuation schemes including capacitive, piezoelectric, and thermal actuation are further discussed in the next section where micropumps are introduced.

Example 23.5: A passive microvalve
Problem: A passive microvalve is used in an implantable insulin pump. A 10-kPa pressure difference across the valve is required for opening it and the back pressure in the body is 25 kPa. How large is the pumping pressure needed to open the valve?
Solution: Since the pressure outside the valve is 25 kPa and valve requires a 10-kPa pressure difference across it, the pump pressure should be 35 kPa.

23.3.2 Micropumps

The basic design for a mechanical two-stroke microfluidic pump is illustrated in Figure 23.5. The pump has a single chamber and two passive valves [216]. During the suction stroke, the pump chamber is expanded, drawing working fluid in through the inlet valve. During the discharge stroke, the driver acts to reduce the pump chamber volume, expelling working fluid through the outlet valve.

The first reported micromachined pump comprised of two silicon wafers bonded between two glass wafers [226]. The pump chamber diameter was 12.5 mm diameter and the chamber height was 130 μm. The glass diaphragm thickness was 190 μm and it was actuated with a piezoelectric disk. Reported performance was $q_{max} = 8$ μl/min for the flow rate and $p_{max} = 10$ kPa for the pumping pressure at $f = 1$ Hz actuation frequency and $V = 125$ V actuation voltage.

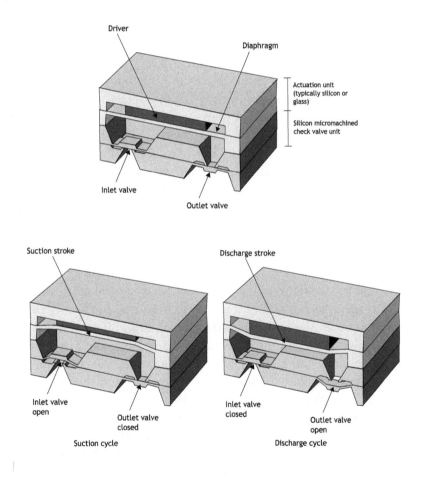

Figure 23.5: Structure and operation principle of a micromachined pump with passive valves (After [216]).

Example 23.6: Bubble in a microfluidic pump
Problem: A micropump is based on an actuated diaphragm and passive valves
that require a 35-kPa pressure to open. The pump chamber diameter is $r =$
5 mm and the height is $h = 250$ μm. When the chamber is fully filled with an
incompressible liquid, the diaphragm will directly actuate the valve and only
small displacement is needed. When the pump is filled or partially filled with
gas, for example by an air bubble, the gas will compress before the valve opens.
Calculate the diaphragm displacement needed to open the valve if there is a
bubble in the chamber that fills one third of the chamber volume.
Solution: The ideal gas law gives the isothermal pressure change as

$$\Delta p = \frac{V_1 - V_2}{V_2} p_0,$$

where V_1 is the initial bubble volume, V_2 is the final volume, and $p_0 = 101.3$ kPa
is the initial atmospheric pressure in the chamber. The initial bubble volume is
$V_1 = \frac{1}{3}\pi a^2 h$, where a is the chamber diameter and h is the chamber height. The
final bubble volume is $V_1 = \frac{1}{3}\pi a^2 (h - 3x_a)$, where x_a is the average diaphragm
displacement. Requiring that $\Delta p = 35$ kPa gives $x_a \approx 22\mu$m. Displacement this
large is not easily obtained with microactuators.

This example shows that that ratio of the chamber volume in the discharge
and the suction cycle should be small. In other words, the dead volume in the
pump should be small.

As Example 23.6 illustrates, micropump actuator should be capable of both
a large force and a large stroke. To address this need, several actuators have been
proposed. The piezoelectric actuator shown in Figure 23.6(a) has a piezoelectric
disc mounted between the pump diaphragm and a rigid frame. When voltage is
applied to the disk, it pushes the diaphragm. The piezoelectric actuator is capable of very large forces but the displacement is not large. Another piezoelectric
actuator is shown in Figure 23.6(b) has a piezoelectric disk mounted directly
onto the diaphragm. When voltage is applied to the disk electrodes, the lateral
dimensions change and this lateral strain causes the diaphragm to deflect. Due
to the bi-morph effect, the displacement is amplified but the pumping force
generated is smaller than for the disk actuator in Figure 23.6(a).

The thermopneumatic pump shown in Figure 23.6(c) is based on heating
liquid or gas in a closed chamber. The heated fluid expands, exerting pressure on
the pump diaphragm. This type of pump is capable of large force and displacement but requires a significant amount of power to operate. The capacitively
actuated pump in Figure 23.6(d) operates by electrostatic forces expanding the
pump chamber. Thus, the capacitive pump has a powered suction stroke and
an unpowered discharge stroke. The electrostatic pumps are capable of large

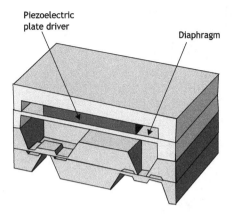

(a) A piezoelectric disk is mounted between the pump diaphragm and rigid support.

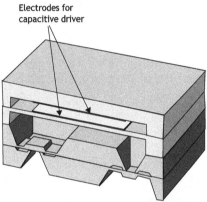

(b) Piezoelectric driver in the lateral-strain configuration.

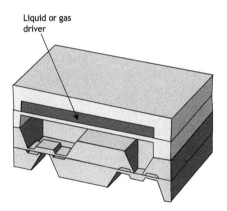

(c) Thermopneumatic pump is based on heat expansion of fluid in a closed chamber.

(d) Electrostatically driven reciprocating micropump.

Figure 23.6: Different actuation methods for micropumps (After [216]).

displacements but requires large actuation voltages and do not provide very large actuation forces.

23.4 Nonmechanical pumps

Given the challenges in mechanical pumping at a small scale, several nonmechanical pumping mechanisms have been investigated. Figure 23.7 illustrates the popular nonmechanical methods to actuate fluid at micro-scale:

Electrophoretic pumping shown in Figure 23.7(a) relies on the presence of ions in the solution [227, 228]. When the electric field $\mathcal{E}$ is applied across the channel length, the ions experience a force $F = z_i q \mathcal{E}$, where z_i is the ion charge and $q = 1.602 \cdot 10^{-23}$ is the electron charge. This electrical force is countered by the frictional drag force, which for ions with the radius r_i moving in a fluid with the viscosity μ at velocity $\dot{x}$ is $F = 6\pi r_i \mu \dot{x}$. Thus, the ion velocity is

$$\dot{x} = \frac{z_i q}{6\pi r_i \mu} \mathcal{E}. \tag{23.6}$$

For example, ion with the charge $z_i = +1$ and the radius $r_i = 10$ nm has the drift velocity of $\dot{x} \approx 500$ μm/min in water under electric field of $\mathcal{E} = 10$ V/mm.

The electrophoretic pumping is commonly used for the separation of ions with the different charge and size. For example, DNA is routinely separated with gel electrophoresis [228]. In this method, the negatively charged DNA molecules move with electrical field through a gel. Shorter molecules move faster and migrate farther than longer ones which allow separation of DNA strands by their length.

The challenge in electrophoretic pumping is the generation of the high voltage needed for pumping. The typical voltages are $1 - 10$ kV, which is not a problem for a desktop power supply but is more difficult to handle in microscale. As a result, electrophoretic pump instruments are usually measured in tens of centimeters. Also, electrophoretic pumping acts only on the charged ions and does not, in general, generate significant net fluid flow.

Electroosmotic pump shown in Figure 23.7(b) is also based on the ionic actuation [215]. But unlike electrophoretic pumping that mostly moves just the ionic particles and molecules, the electroosmotic pumping generates significant liquid flow. Electroosmotic pumping works due to the surface charge that spontaneously develops when a liquid comes in contact with a solid. For example, silica based ceramics such as glass become negatively charged in liquid with pH>4. The charged surface then attracts positive ions that accumulate near the surface. With applied electric fields, the positive ions near the channel walls move in the direction of the electric field, which induces net liquid flow.

The electroosmotic pump can generate significant pressures. For example, an electroosmotic pump with silica capillary packed with silica beads has demonstrated pressures up to 20 MPa at an applied voltage of 7 kV [216]. Electroosmotic effect has been used in laboratory analysis for several decades and several μTAS utilize the electroosmotic pumping.

Ultrasonic pumping is based on nanometer-scale mechanical vibrations that induce mechanical flow. The generated static pressure is small but the flow rates can be significant. Since there are no moving parts apart from the

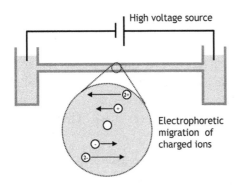

(a) Electrophoretic pumping is based on Coulomb's force on ionized particles [227].

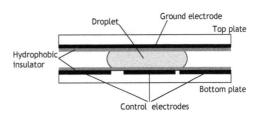
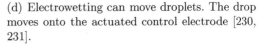

(b) Electroosmotic pump is based on electric charge layer forming near liquid solid interface [215].

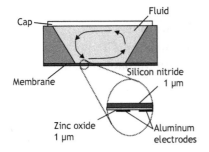

(c) Ultrasonic vibrations can be used to induce liquid flow. For example, a traveling wave can be excited on a thin membrane [229].

(d) Electrowetting can move droplets. The drop moves onto the actuated control electrode [230, 231].

Figure 23.7: Selected non-mechanical pumps.

vibration, the ultrasonic pumps are robust and relatively easy to manufacture. In comparison to the electrophoretic and electroosmotic pumping, the required actuation voltages are usually low, measured in tens of volts.

Figure 23.7(c) illustrates an ultrasonic thin-film liquid actuator [229]. A traveling wave is exited on a thin piezoelectric membrane. The wave amplitude is small, only a few nanometers; however, the high vibration frequency translates into a high vibration velocity that generates significant fluid motion.

Electrowetting can be used to manipulate discrete microdroplets. As is illustrated in Figure 23.7(d), the actuation is accomplished by direct electrical control of the surface tension using patterned electrodes covered with a thin dielectric [230, 231]. The voltage dependency of the surface tension is described by

$$\gamma = \gamma(0) - \frac{\epsilon}{2d}V^2,\qquad(23.7)$$

where ϵ is the permittivity of the insulator, d is the thickness of the insulator, and V is the applied potential. The applied electric potential tends to pull the droplet down, onto the electrode, lowering the macroscopic contact angle and increasing the droplet contact area. This causes the droplet to move into the actuated electrode.

The electrowetting can operate at relatively low voltages $V < 50$ V which makes it attractive for microactuation. The method, however, is mainly limited to fluoropolymer surfaces and works only for droplets – continuous fluid manipulation is not possible as the actuation method relies on the surface tension which requires a liquid interface.

While all micropumping methods have their own advantages, no actuation method is perfect for μTAS. Most nonmechanical pumps require large voltages or do not generate significant pressures. Micropumping remains an active research topic.

23.5   Minimum sample volume

The microfluidic devices can be used to process and analyze small sample volumes. This can save the essays and reduce the analysis time. The small sample volume, however, is not guaranteed to contain analyte that is to be measured. For example, virus concentration in human blood may be very diluted in the early stages of a disease. Thus, a small blood sample may not contain a single virus even if the human has been infected.

The minimum volume of sample required to detect a given analyte concentration is

$$V_1 = \frac{1}{\beta C_a},\qquad(23.8)$$

where β is the sensor efficiency that ranges from 0 to 1 and C_a is the analyte concentration (units/m^3) [221,232]. Equation (23.8) gives the minimum volume that is likely to have one molecule that is detected – some samples may have less, some have more molecules. To obtain statistical confidence in the measurement, the sample volume should be ten times bigger than the minimum volume.

Figure 23.8 shows the scaling of the sample volume graphically together with common biochemical analyte concentrations. The classical clinical chemistry works in the region of high sample concentrations ($C_a > 10^{14}$ mol/ml). This analysis region is expected to work well also for μTAS. A low concentration does not preclude the use of microfluidic systems, but the microfluidic device should be able to handle a large sample volume with a reasonable through-put. For example, to detect DNA fingerprints, the microfluidic system should continuously process, concentrate, and measure the sample.

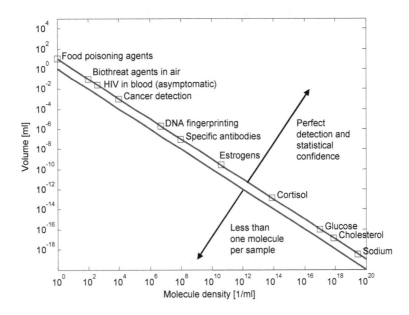

Figure 23.8: Scaling of the sample volume and common biochemical analyte concentrations (After [232]). For example, food poisoning agents have a concentration of $C_a \sim 1$ 1/ml and the minimum volume that has on average of one agent is 1 ml.

Example 23.7: Minimum sample volume to detect HIV

Problem: Estimate the volume needed to detect the HIV virus in a blood sample ($C_a < 400$ 1/ml). If the sample is analyzed with a microfluidic analyzer that has a flow rate of 5 nl/min, how long does processing of the sample take? Assume that the detector efficiency is $\beta = 0.3$.

Solution: From Equation (23.8), the minimum volume that is likely to have one HIV virus is

$$V_1 = \frac{1}{\beta C_a} = \frac{1}{0.3 \cdot 400 1/1} \approx 8.3 \ \mu l. \tag{23.9}$$

For statistical confidence, the minumim sample volume should be ten times bigger $V_{min} = 10 V_1 = 83 \ \mu l$. To process this sample with a $q_V = 5$ nl/flow rate would take $t = V_{min}/q_V = 1.2$ days!

Key concepts

- Micro total analysis system (μTAS) is a hand held or smaller instrument that carries out biological or chemical sample processing including the sampling, the sample transport, chemical reactions, separations, and detection.

- Microfluidic systems can save cost as the amount of assays needed for the analysis is reduced.

- The Reynold's number in microfluidic systems is small and the flow is therefore laminar. The flow rate can be calculated from the flow resistance and the pressure drop over the channel length.

- Due to the lack of turbulence, mixing in microchannels happens only by diffusion.

- Valves and pumps are needed for flow control. Good pumps should have both large strokes and large force. This is difficult with microactuators.

- Nonmechanical pumps do not have moving parts which makes them reliable. Large actuation voltages, however, may be needed.

Exercises

Exercise 23.1
Estimate the practical limits for the smallest possible microfluidic channel size (diameter) that can still be filled with fluid even with a bubble in the channel (Murphy: there always is a bubble). Assume that you are trying to force the liquid into the channel with a suction pump so one end of the channel is in atmospheric pressure and the other is in vacuum so that the maximum pressure difference acting on the bubble is 1 bar.

Exercise 23.2
A micromechanical diaphragm pump has a circular pump chamber with the radius $r = 800$ μm and the height $h = 60$ μm. The average diaphragm displacement is $\Delta h = 5$ μm. What is the pump flow rate with the actuation frequency $f = 4$ Hz?

Exercise 23.3
Calculate the pressure needed to move the pump diaphragm in Exercise 23.2 by an average of $\Delta h = 5$ μm (the diaphragm center displacement will be more) if the diaphragm thickness is $t = 10$ μm. Use the diaphragm equations in Chapter 13 and ignore the pressure needed to open the valves.

Exercise 23.4

A 2-mm long microchannel should have a minimum flow rate of 1 μl/min with a 200-Pa pressure drop along it. What is the minimum channel diameter and how long does it take for a flow to go through the channel?

Exercise 23.5

Estimate the time for oxygen to mix in a 300-μm diameter channel. If the flow rate is $q_V = 0.5$ μl/min, how long should the channel be to enable the mixing?

Exercise 23.6

Estimate the minimum channel size that is not blocked by a bubble for a microfluidic system that has the pumping pressure $p = 500$ Pa.

Exercise 23.7

Estimate the volume needed to measure glucose in a blood sample ($C_a < 10^{18}$ 1/ml). Assume that the detector efficiency is $\beta = 10^{-5}$ (ten parts per million) and that to measure the glucose level, the concentration should be 10^4 times larger than the minimum detection level. Compare your results to actual sample volume of 300 nl for a commercial glucose meter.

Exercise 23.8

Find, and review the paper: C.G. Wilson, Y.B. Gianchandani, "Spectral detection of metal contaminants in water using an on-chip microglow discharge," *IEEE Trans. Electron Devices*, 49(12), Dec 2002, pp. 2317-2322. The LEd-SpEC device described in this utilizes two electrified microfluidic channels to create an arc discharge. Water, and impurities are sputtered into the created discharge, and the impurities are measured with optical spectroscopy. To do this, the following must happen: 1) Water gets delivered to the electrode area. 2) Water gets sputtered into the discharge 3) Some fraction of the impurity atoms ionize, and 4) some fraction of the light from the ionized impurity atoms is collected and measured.

Assuming the water reservoir is 3 mm in diameter and 1 mm deep, filled with water, and loaded with 1 ppm sodium, how many sodium atoms are in the reservoir? What would dictate the fraction of light collected from the arc? Estimate this fraction of light. If the entire reservoir is delivered to the arc discharge (absolute best case), and 1/100th of the atoms ionize, how many photons (one per ionization) would be collected. What would you estimate as the lowest possible impurity concentrations in water that could be measured with this system? What is the detector efficiency β?

24

Economics of microfabrication

Making microdevices is relatively easy but manufacturing devices for profit is surprisingly difficult. The MEMS components are low cost, often less than $10 per device, so large sales volume is necessary to support the expensive manufacturing infrastructure. The parallels to the IC industry are apparent. A typical IC cost is in the dollar range but the production line can cost over $1B. The IC manufacturers can stay in business only by selling devices by millions. The competition between the IC manufacturers is fierce as everyone seeks to capture as much of the market share as possible. Similarly, the competition between MEMS manufacturers is cutthroat. For example, there are more than a dozen players in the accelerometer market and everyone is trying to capture as much of the market share as possible.

While the MEMS business appears similar to the IC manufacturing, there are few important differences. The ICs are complex systems made of standard components. Even a low cost circuit can contain over a million transistors and there is significant value in the design itself. Many "fabless" design houses make a living just by developing new designs that are manufactured by foundries – IC manufacturers that are open to making components based on customers proprietary designs. Some design houses make a living just by selling designs to others to use in their IC designs. In comparison, the MEMS devices usually have only a few moving parts and the device complexity pales in comparison to modern microcircuits.

Another key difference with the IC industry is that the MEMS components are strikingly dissimilar. Whereas IC manufacturers can recoup the large investments by manufacturing a large number of circuits using the same standard process, the new MEMS devices often require new process development. Even "fabless" MEMS companies need to work very closely with the foundry to develop a process for their design. This can prevent otherwise good devices from entering the market, as developing a new device can require several years of work before the investment starts to yield returns.

In this chapter, the economics of microfabrication is investigated. The effects of manufacturing yield, wafer size, and integration on device costs are explored. The cost and profit of MEMS business is illustrated with several examples and case studies.

24.1 Yield analysis

Batch manufacturing is based on fabrication of a large number of devices on a single wafer. Manufacturing yield is the ratio of good devices divided by the device potential. For example, if the wafer contains 100 devices and 80 of them yield good components, then the total yield is 80%. Obviously, a high yield is desirable which calls for a robust device design. However, an aggressive design might be able to fit more devices on a wafer. The optimal design will maximize the total number of yielded dies per wafer. The main yields affecting MEMS manufacturing are the wafer yield, the die yield, and the assembly yield:

Wafer yield gives the fraction of wafers that are finished successfully. Wafers are occasionally lost, for example, due to careless handling (wafer breaks), equipment malfunction, or drift in process parameters that results in all parts being out of specifications. During the learning stages as a new product is introduced to the market, the wafer yield may be low; in a mature process, the wafer yield should be high ($> 90\%$).

Die yield is the fraction of working dies on the wafer. Some devices may not work for example due to dirt particles at a critical location. The yield may also vary across the wafer. The center of the wafer is typically the optimal location for the process, and the yield is good, but drops toward the edge of the wafer. For example, if the etch process is optimized for the wafer center where majority of the dies are, the devices at the wafer edge may be over- or under-etched.

The simplest model for the die yield is

$$Y_{\text{die}} = e^{-DA}, \tag{24.1}$$

where D is the critical defect density and A is the die area. Typical defect sources are mask defects, pinholes or bubbles in the photoresist, photoresist residues, and airborne particles that land onto the wafer. For small dies, the yield is high as there is only a small probability that a defect is within the die area. As the die area increases, the yield decreases exponentially. The critical defect density D increases with decreasing minimum dimensions as even smaller defects, for example in the form of particles, can cause the device to fail. More complex models have been developed to account for the variations across the wafer [13].

The die yield decreases with increasing process complexity. The yield for a process with multiple process steps is a product of yield for individual steps

$$Y_{\text{die}} = e^{-D_1 A} e^{-D_2 A} e^{-D_3 A} \cdots e^{-D_N A} = e^{-\sum D_i A}, \qquad (24.2)$$

where is D_i:s are the critical defect densities for the process steps.

Assembly yield is the fraction successfully assembled devices. Several factors may lower the assembly yield: package interconnects may fail, the package itself may be defective, or the bonding to the package may fail. The assembly yield is typically high but can be problematic with small "flip-chip" packages.

Total yield is a product of the individual yields

$$Y_{\text{total}} = Y_{\text{wafer}} Y_{\text{die}} Y_{\text{assembly}}. \qquad (24.3)$$

Due to the multiplying nature of yields, the yield of individual steps should be high or the total yield can become unacceptably low.

Example 24.1: MEMS yield
Problem: A surface micromachining process has 95% wafer yield and 90% assembly yield. The fabrication consists of three lithography steps each having defect density of 0.5 cm^{-2}. The die size is 16 mm^2. What is the total yield?
Solution: The die yield is

$$Y_{\text{die}} = e^{-3DA} = e^{-3 \cdot 0.5 \cdot 0.16} = 0.79.$$

The total yield is

$$Y_{\text{total}} = Y_{\text{wafer}} Y_{\text{die}} Y_{\text{assembly}} = 0.67 = 67\%.$$

This example is illustrative as it shows that due to multiplication of the yields, one third of the dies are lost even though the yield for the individual steps is high.

24.2   Cost analysis

The cost per device is a function of manufacturing costs and non-recurring engineering cost (NRE). The NRE is the one time product development cost for designing and prototyping the device. We are mainly interested in the cost of yielded or working devices that can be sold. For example, if the raw device manufacturing cost is $C_{raw} = \$1.00$ and the yield is $Y = 0.5$, then the yielded cost is $C_{yielded} = C_{raw}/Y = \$2.00$.

Dividing the manufacturing into wafer processing, testing, and assembly, the total cost per packaged device with a MEMS and IC die is

$$C_{total} = \frac{1}{Y_{assembly}} (C_{ICdie} + C_{MEMSdie} + C_{test} + C_{package}) + \frac{C_{NRE}}{n} \qquad (24.4)$$

where C_{ICdie} is the cost of yielded IC die, $C_{MEMSdie}$ is the cost of yielded MEMS die, C_{test} is the testing costs, $C_{package}$ is the cost of assembly and packaging, $Y_{assembly}$ is the assembly yield, C_{NRE} is the non-recurring engineering expenses (design and prototyping costs), and n is the number of devices sold. In Equation (24.4), the NRE costs are divided by the total number of devices made to obtain engineering cost per device.

The yielded die cost is

$$C_{die} = \frac{W}{Y_{die} Y_{wafer} N}, \qquad (24.5)$$

where W is the wafer cost, Y_{die} is the die yield, Y_{wafer} is the wafer yield, and N is the number of dies per wafer. The wafer cost includes all direct costs (labor, starting materials, and consumables) required for the process. In addition, capital cost tied to the manufacturing equipment may also be included per usage basis. For example, a process equipment that costs $500k might be up-to-date for four years. Assuming reasonable down time, the machine can be used for a total of 30,000 hours over the four-year period so cost due to capital expenses for using this machine would be $\$500k/3{\cdot}10^4 = \$17$/hour. For simplicity, we have ignored the interest in this capital cost example.

The wafer cost is dependent on wafer size and manufacturing process. The typical wafer cost for an older 0.25-0.35 μm CMOS process suitable for MEMS devices requiring good analog performance is $1-2k. Compared to the CMOS wafers, the MEMS wafer cost might be expected to be significantly lower as the process is simple in comparison to CMOS. For example, the MEMS processes typically require less than ten masks whereas the CMOS processes require over twenty masks. However, the manufacturing volumes for MEMS are also lower, and the MEMS wafer cost may vary from $300-2k depending on the complexity of the MEMS process.

The number of dies per wafer depends on the die area and the wafer radius. As dies are squares and wafers are circles, a portion of the wafer cannot be used for the dies. The number of dies on the wafer can be estimated from

$$N = \pi \left[\frac{r^2}{A} - \frac{2r}{\sqrt{2A}} \right], \qquad (24.6)$$

where r is the wafer radius and A is the die area [233].

Figure 24.1 illustrates the relationship between the die area and the yield. With a large die area shown in Figure 24.1(a), the probability that there is a defect on the die is high and a large area at the die edge is unusable. With smaller dies shown in Figure 24.1(b), the yield will be higher than for Figure 24.1(a) even though the defect density is the same. Thus, reducing the die size will result in more dies and the number of yielded dies will increase even more.

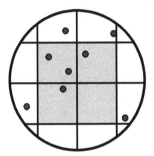
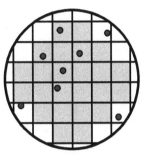

(a) Large die area results in a small yield and very high cost per working die. The wafer has total of four dies out of which only one is defect free.

(b) Small die area results in more dies and better yield. The wafer has 24 dies out which 19 are working.

Figure 24.1: Relation ship between the die area, number of dies, and yield. The circles denote point defects on the wafer.

The cost vs. die area is further illustrated in Figure 24.2 where the die cost and yield are plotted as a function of the die area for two different defect densities. The number of dies is estimated based on Equation (24.6) and the yield follows Equation (24.2). As the yield drops exponentially with increasing area, the cost given by Equation (24.5) becomes prohibitively high with large die sizes unless the defect density is reduced.

Example 24.2: MEMS die cost
Problem: How does the die cost change if process line is updated from 6" to 8". Assume that the finished 6" wafer costs $800.00 and the 8" wafer costs $1000.00, the die size is 10 mm^2, and the yield is 80% for both processes.

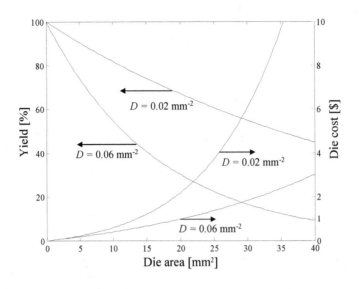

Figure 24.2: The yield and die cost as a function of die area for a 8 inch wafer. The wafer cost is assumed to be $1k.

Solution: The approximate number of dies is

$$N = \pi \left[\frac{r^2}{A} - \frac{2r}{\sqrt{2A}} \right] = 1717$$

for the 6" process and

$$N = \pi \left[\frac{r^2}{A} - \frac{2r}{\sqrt{2A}} \right] = 3100$$

for the 8" process. The die costs are

$$C_{\text{die}} = \frac{W}{Y_{\text{die}} N} = \$0.58$$

for the 6" process and

$$C_{\text{die}} = \frac{W}{Y_{\text{die}} N} = \$0.40$$

for the 8" process. With the given assumptions, the number of dies increase by 80% and the die cost decreases by 31%.

24.2.1   Cost case study: MEMS integration

To illustrate the cost analysis, we will look at the benefits of two chip vs. single chip integration from the cost perspective. The actual costs incurred are strongly dependent on the device, process, and packaging options. Thus, the costs presented should be considered as an illustrative example.

First, we will analyze the cost structure for a product where the MEMS and IC die are almost equal in size. Table 24.1 shows the example yields, die sizes, and costs associated with the products. For simplicity, the assembly yield is assumed high and is ignored in the analysis. In all cases, we assume that the wafer diameter is 8 inches.

As shown in Table 24.1, the integrated MEMS and IC wafer is expensive, almost equal to the cost of separate MEMS and IC wafers, as there is no synergy between the processes. The single-chip integrated MEMS+IC solution will require a die area that is the sum of the MEMS and IC areas. This assumption is valid for current commercial polycrystalline silicon surface micromachining processes where the MEMS and IC areas cannot overlap. In addition to the usable die size, we assume that 200 μm margin is needed for each die for dicing.

For the two-chip solutions, the MEMS and IC dies are tested separately and the packaging is slightly more expensive than the single-chip solution, as larger number interconnects are required. Thus, the testing and assembly costs for the two-chip approach are more than for the single-chip approach. Still, the cost for the integrated approach is three times higher than the cost of combining separate MEMS and IC dies.

There are two reasons for the large cost differential. First, there is little savings in the wafer cost, as the integrated solution requires completion of both MEMS and IC processes. As the combined MEMS+IC die is large, there are

Table 24.1: Comparison of two chip and integrated MEMS+IC product when MEMS and IC have about the same die size.

	IC	MEMS	MEMS+IC	Comments
Wafer cost, W	$1,000	$600	$1,450	MEMS+IC reduces wafer cost.
Die yield, Y_{die}	0.95	0.8	0.760	$Y_{MEMS+IC} = Y_{IC} Y_{MEMS}$
Device area, $A[\text{mm}^2]$	4.0	6.0	10.0	$A_{MEMS+IC} = A_{IC} + A_{MEMS}$
Edge area, $A_e[\text{mm}^2]$	1.6	2.0	2.5	200 μm rim around the die
Die area, $A_{tot}[\text{mm}^2]$	5.6	8.0	12.5	Total area (die and rim)
Dies, N	12341	8644	5453	Dies for 8" wafer (Eq (24.6))
Yielded dies, $N \cdot Y$	11724	6915	4144	Total number of yielded dies
Die cost	$0.34	$0.52	$3.50	
Test cost	$0.30	$0.30	$0.30	Fixed test cost per die
Packaging	$0.40		$0.30	Price is almost the same
Total Cost	**$1.86**		**$4.10**	

fewer dies available from the wafer. Secondly, the large cost of finished die is compounded by a lower die yield for the single chip MEMS+IC. As both the MEMS and IC have to work on the same die, the yields multiply. For the two chips approach, the MEMS and IC dies can be tested first, and only the known good dies are assembled for the final products. Thus, the overall yield for the two chip approach is larger.

Table 24.2 shows the cost structure when the MEMS device is small compared to the IC. As the MEMS device takes only a small area, the number of finished dies with IC is almost same for the single-chip and two chip approaches. Unfortunately, the yields still multiply and the number of yielded dies is lower for the MEMS+IC. The yield hit is offset by the lower cost for testing and packaging, and the final costs are almost equal for the two approaches. This conclusion could easily tip to either direction if the yields or testing/packaging costs were to change.

To summarize the results of this case study, the single-chip approach that combines MEMS and IC on the same die is not attractive if the MEMS requires a large area that is not available for the IC. The two-chip approach won with a small margin even in the case of a small MEMS device but this conclusion could change and the single-chip integration is attractive if:

1. The same performance could not be obtained with the two-chip approach. For example, micromirror arrays would not be feasible without the single-chip integration. To a lesser extent, the integration of sensors on a same chip with the circuit offers performance benefits as parasitic capacitances are reduced.

2. MEMS could be integrated on top of the IC so that no additional die area is required on the expensive IC wafer. This approach has been taken in the digital micromirror device (DMD) and is one of the benefits of SiGe surface micromachining that can be completed after the IC processing [234].

3. The cost of MEMS processing was lower and the yield was higher. In this case, the added cost of MEMS on IC is small in comparison to the two-chip packaging costs.

4. The assembly yield for the two-chip approach was low. This would be likely if a large number of interconnects were needed. Again, the micromirror arrays are a perfect example of a product where the interconnect requirements mandate the single-chip integration.

5. There was synergy in MEMS and IC fabrication. If the MEMS could be made using the structures available in IC process, the additional cost of integrating MEMS on IC is low. This is the rational with the "CMOS MEMS" where the MEMS structures are defined in a standard CMOS IC as a post-processing step [235].

As the manufacturing technologies evolve, companies are constantly evaluating whether to do single-chip or two-chip integration. In few cases, one approach is clearly better than the other, but in many cases both approaches can result in a viable business.

Table 24.2: Comparison of two chip and integrated MEMS+IC product when MEMS device is small.

	IC	MEMS	MEMS+IC	
Wafer cost, W	$1,000	$600	$1,450	MEMS+IC reduces wafer cost
Die yield, Y_{die}	0.95	0.8	0.760	$Y_{MEMS+IC} = Y_{IC}Y_{MEMS}$
Device area, $A[\text{mm}^2]$	4.0	0.5	4.5	$A_{MEMS+IC} = A_{IC} + A_{MEMS}$
Edge area, $A_e[\text{mm}^2]$	1.6	0.6	1.7	200 μ rim around the die
Die area, A_{tot}	5.6	1.1	6.2	Total area (die and rim)
Dies, N	12341	65683	11139	Dies for 8" wafer (Eq (24.6))
Yielded dies, $N \cdot Y$	11724	52547	8465	Total number of yielded dies
Die cost	$0.34	$0.01	$0.77	
Test cost	$0.30	$0.30	$0.30	Fixed test cost per die
Packaging	$0.40		$0.30	Price is almost the same
Total Cost	**$1.35**		**$1.37**	

24.3 Profit analysis

Business is about making profits. The sales profit will be used to cover indirect expenses (administration salaries, sales expenses, buildings, and other costs not directly related to manufacturing). The operating profit, the income after all the expenses, can be used to grow the business or may be transferred to the owners of the company.

Profit margin p indicates the portion of the sales that contribute to the income of a company. Given the manufacturing costs and desired profit margin, the sales price for the finished device is

$$S = \frac{C_{total}}{1 - p} \tag{24.7}$$

where C_{total} is the total cost for manufacturing the yielded device. Given the sales price S and profit margin p, the profit per device is

$$P = pS. \tag{24.8}$$

To cover all indirect costs, it is desirable to have a profit margin of around 30-50% or more if the manufacturing volume is low. The reader familiar with profit margins for consumer goods will note that the desired profit margin for

MEMS is quite high in comparison. The high margin, however, is needed to cover the large indirect costs due to the expensive cleanroom infrastructure.

Example 24.3: MEMS profit
Problem: An accelerometer that costs \$2.00 to manufacture is sold for \$3.50. What is the profit margin, profit per device, and how large is the sales volume necessary to reach \$10M total profit?
Solution: From Equation (24.7), the profit margin is

$$p = 1 - C/S \approx 43\%$$

and the profit per device is

$$P = pS \approx \$1.50.$$

The required sales volume is \$10M$/p \approx$\$23.3M or 6.7M units.

24.3.1 Profit case study: Fabless start-up

As an example of MEMS profits, we will consider a fabless MEMS start-up that makes MEMS gizmos with a target sales price of \$3.00 and sales volume of 20 million devices over four years. As the manufacturing is contracted to external foundries, the initial investment is relatively small. To develop the product, however, the company needs a team of good engineers and must cover the prototyping costs. Again, the costs are for illustration only and the actual spending in a MEMS start-up may be higher or lower than in this example.

The yielded die costs for the MEMS and IC combined is assumed to be \$1.80 and the assembly and test cost is assumed to be \$0.30. The assembly yield is presumed to be good enough as not to influence the cost significantly. To estimate non-recurring engineering costs, we assume that the product development will take two years and will require a team of two MEMS designers and two IC designers. Two MEMS and IC prototype runs are completed in the two-year development period. We are optimistic that the second prototype run will result in a device meeting all the specs and the device is production ready.

Table 24.3 shows the expenditure breakdown per year. Most cost items speak for themselves. Even a small start-up requires a CEO, as someone needs to make decisions, talk to the investors, and meet future customers. The designer salaries are the biggest expenditure. Perhaps less obvious is that the designers will require expensive CAD and simulation software in order to be effective. In comparison to designer salaries, the prototyping costs are small in our example. The IC prototyping round is roughly $60k. The MEMS prototype is more expensive as the process is not standard. In our example, the MEMS prototype cost is $150k. This low cost can be realistic if the device is made with an almost standard process. If new process development is needed, the prototyping costs can be much higher.

Table 24.3: Yearly costs for a fabless MEMS start-up.

Senior IC designer		Senior MEMS designer		Test engineer	
Salary	$100k	Salary	$100k	Salary	$65k
Overhead	$30k	Overhead	$30k	Overhead	$18k
Computer	$10k	Computer	$10k	Computer	$10k
CAD tools	$40k	CAD tools	$30k	CAD tools	$30k
Total	$180k	Total	$170k	Total	$123k
IC designer		MEMS designer		CEO	
Salary	$70k	Salary	$80k	Salary	$120k
Overhead	$21k	Overhead	$24k	Overhead	$40k
Computer	$10k	Computer	$10k	Computer	$2k
CAD tools	$40k	CAD tools	$30k	Travel	$40k
Total	$141k	Total	$144k	Total	$202k
Prototyping cost		Support staff		Miscallenius	
IC prototype	$60k	Salary	$45k	Lawyer fees	$30k
MEMS prototype	$150k	Overhead	$13k	Office rent	$40k
Testing cost	$20k	Computer	$2k	Patenting costs	$40k
Total	$230k	Total	$60k	Total	$190k
		Total cost	**$1.36M**		

While the mileage may vary, the expenditure per year or the capital "burn-rate" for a fabless start-up is around $1-2M/year in the initial stage. In our specific example, the total cost for a two-year development is about $3M. The targeted sales volume is 20M units over four years, after which time the gizmo will be obsolete. The NRE cost per chip is $3M/20M=$0.15. The total device cost is $2.25, and with the sales price of $3.00, the profit per device is $0.75 (profit margin is 25%). Over the four-year life span, the total sales profit from the product is $15M.

This example paints a rosy picture for our MEMS start-up. Reality is often much more harsh. The total cost for the initial start-up phase is realistic if everything goes as planned, but as the product is introduced to the market, additional costs will occur. These include cost of ramping up production, hiring sales staff, making marketing material such as datasheets, and hiring engineers to support the customer adoption. So even with the perfect design execution, additional capital is needed. Most recent MEMS start-ups have required two or three rounds of financing totaling in excess of $10-20M before being able to introduce their product to the market. Several start-ups have failed in the product launch and have not made any profit. Given this reality, the business model should either target a higher sales price or higher volume to make the risks worthwhile.

24.3.2   Profit case study: VTI Technologies

VTI Technologies is the leading manufacturer of low-G accelerometers for the automotive and medical markets. It is also one of the few manufacturers that focus solely in MEMS and gives out financial information in quarterly reports. Table 24.4 shows VTI's revenue and expenses for one year [236]. The net sales revenue (income) was €73.5 million and the cost of sales (manufacturing costs) was €50.9 million. This left VTI with a healthy gross profit margin of 31%. Before the owner can pocket the profits, however, indirect costs (R&D, sales and marketing, and administrative expenses) need to be factored in. These expenses add up to total of €17.6 million or 24% of the revenue, and the operating profit was €7.7 million (profit margin of 7%).

Table 24.4: Financial statement for VTI Technologies.

	M€	% of revenue
Net sales revenue	73.5	
Cost of sales	-50.9	69%
Gross profit	22.6	31%
R&D expenses	-8.6	12%
Sales and marketing expenses	-4.2	6%
Administrative expenses	-4.8	7%
Operating profit/loss	**5.0**	**7%**

Healthy profit is important for a component manufacturer as long investment cycles in manufacturing infrastructure can lead to overcapacity problems. For example, overcapacity in the IC memory manufacturing periodically leads to a glut of chips and the sales prices collapse. If this happens in the MEMS market, the companies will need to have a strong balance sheet to carry them through the bad years stressing the importance of racking up the profit when the business is good.

It is interesting to compare the operating profit margin to IC manufacturing. Analog IC companies typically have gross profit margins in excess of 50% which is needed to build a buffer against hard times and to invest in next generation of manufacturing facilities. The key difference with MEMS and IC manufacturing is the scale: IC manufacturer can use standard process for wide number of different devices. Due to the divergence of MEMS components, it is difficult to support large number of niche devices with a single manufacturing line. This leads to lower profits but can also provide more stable revenue.

A panacea answer to increase the MEMS profits is to focus on high volume applications such as consumer market as opposed to niche applications where the volumes are low. The problem with this approach is that if everyone makes the same conclusion, the market will be saturated with MEMS devices – a frequent occurrence in memory manufacturing but something that has not yet happened with MEMS.

24.4   Beyond the high cost manufacturing

The case studies in this chapter showed that the low cost of batch fabrication is only partially true. Just the manufacturing costs can be in the dollar range per device. A visit to a "dollar store" or the Japanese equivalent "100 yen store" reveals that cost for many other goods is much lower. In markets where MEMS devices compete with established technologies, the cost advantage from batch fabrication is often small. For example, in the timing market, the MEMS resonators compete with quartz crystals that have typical manufacturing cost of $0.05-$0.5. For MEMS to be successful in the timing market, the die size needs to be very small to drive down the cost and/or there must be other advantages such as better integration with IC or smaller package size.

The relative high manufacturing cost for MEMS is compounded by the high NRE costs. Often a good MEMS solution can be found for a given problem but the cost of design, mask layout and prototyping is prohibitively high if projected manufacturing volume is low. In such cases, traditional manufacturing methods are more cost competitive.

To lower the cost of MEMS devices, we can reduce either the manufacturing or NRE costs. The first approach of reducing the manufacturing cost has been taken with microfluidic devices that quickly moved from expensive silicon substrate to plastic or glass [10]. The microfluidic channels may not even be lithographically defined but can be done with inexpensive molding techniques. As the conventional manufacturing methods continue to improve, we will see increasing number of microdevices that do not follow the traditional clean room and silicon fabrication paradigm.

The reduction of NRE is a natural consequence of the industry becoming more mature. As more experience is gained for example in pressure sensor design, it becomes easier to do a variation of the product for another application. Also, the design tools are getting easier and cheaper to use. This development parallels the IC design where the time for a new designer to become productive has shortened from more than a year to a few months.

The big challenge in reducing the manufacturing and NRE costs is standardizing the manufacturing processes. Currently, the MEMS paradigm is "one device, one process" which means that each new device requires a new manufacturing process. The success of IC industry is based in large part on being able to make different IC designs using the same standard manufacturing process. Due to divergence of MEMS devices, this has not been possible in micromanufacturing. New processing techniques such as maskless laser micromachining that resemble series manufacturing may change the MEMS cost structure in the future. For the time being MEMS, remains expensive business.

Key concepts

- MEMS is a volume business. The patch fabrication makes sense only if sales volume justifies the high nonrecurring engineering (NRE) costs and the expensive manufacturing infrastructure.

- The die cost depends on the die size and manufacturing yield.

- Due to point defects, the yield decreases with increasing die size. Therefore the die cost increases exponentially with larger die sizes.

- Single-chip integration can be attractive if the similar performance cannot be obtained with two-chip approach and/or the MEMS is small. For large MEMS elements, the two-chip approach is often more economical.

- The price for the MEMS chip is in the dollar range and large sales volumes are needed to support the expensive manufacturing infrastructure.

Exercises

Exercise 24.1
A micromachining process has 90% wafer yield, 80% die yield, and 95% assembly yield. What is the total yield?

Exercise 24.2
A MEMS display product has 1920×1080 mirrors on each die. Calculate the required yield for each pixel if the desired die yield is 90%.

Exercise 24.3
Pressure sensor process has 95% assebly yield, 95% IC die yield, 70% MEMS die yield (includes the wafer yield). The raw cost not accounting for the yield losses are $0.40 for the MEMS die and $0.20 for the IC die. The testing and packaging cost is $0.15. The NRE cost is $1.2M and 50M pressure sensors are

sold. Calculate the total sensor cost and the sales price if the desired profit margin is 40%.

Exercise 24.4

Estimate the number of dies on a 8-inch wafer if the die size is 9 mm^2 (accelerometer) and 0.5 mm^2 (microresonator).

Exercise 24.5

Calculate the yielded die cost for a pressure sensor element in a 6-inch wafer process. Assume that the wafer cost is $500, the die size is 4 mm^2, and the wafer yield is 90% and the die yield is 85%.

Exercise 24.6

Find the latest financial information for a Nasdaq traded MEMS company MEMSIC, Inc. (sticker symbol MEMS). What is the gross profit margin and operating profit margin for the company. In the operating profit margin, do not count the income for non-MEMS related revenues (for example, exclude "other income: interest earnings").

Exercise 24.7

A company is considering an upgrade from 150 mm wafers to 200 mm wafers. The processing cost ($650/wafer) and the die size (25 mm^2) is about the same for both processes but the 200 mm wafer costs more than 150 mm wafer ($120 vs. $75) and the yield for the new process is initially lower (80% vs. 90%). The upgrade is estimated to cost $20M. If your company's production volume is 5M units a year and your selling price for the dies is $0.80, how long will it take for the investment to pay off? You may assume 100% assembly yield and you don't have to consider interest for your investment. Repeat the problem for an annual production volume of 50M units.

Exercise 24.8

A MEMS device requires a rim around the rectangular element area for dicing and packaging. Show that a square element shape minimizes the rim area.

A

Laplace transform

The Laplace transform is used to transform a differential equation into an easily solvable algebraic equation. In this book, the Laplace transform is used for solving electrical circuits and mechanical systems. The Laplace transform can be viewed as a transformation from the time-domain, in which inputs and outputs are functions of time to the frequency-domain, where the same inputs and outputs are functions of complex frequency.

The transform is denoted $\mathcal{L}\{f(t)\}$, where the time domain function $f(t)$ is transformed to a function $F(s)$. Mathematically, the transformation is

$$F(s) = \mathcal{L}\{f(t)\} = \int_0^\infty e^{-st} f(t)\mathrm{d}t, \tag{A.1}$$

where the integration is carried out over all positive times. The Laplace transform is not valid times $t < 0$. Table A.1 lists several frequently encountered transforms.

The inverse of the Laplace can also be obtained by integration, but due to complex s, the integration is seldom carried out in practice. Instead, the transformation and inverse transformation are obtained using standard tables as shown in Table A.1. The use of Table A.1 to obtain inverse transform is illustrated in Examples A.1 and A.2.

The continuous Fourier transform is equivalent to the Laplace transform with complex argument $s = j\omega$

$$F(\omega) = \int_{-\infty}^{+\infty} e^{-j\omega t} f(t)\mathrm{d}t = \mathcal{F}\{f(t)\} = F(s)|_{s=j\omega}. \tag{A.2}$$

This is often used to determine the frequency response function $H(\omega)$ of the analyzed system.

Table A.1: Table of Laplace transforms

$f(t)$ for $t \geq 0$	$\mathcal{L}(f)$
1	$\frac{1}{s}$
e^{at}	$\frac{1}{s-a}$
t^n	$\frac{n!}{s^{n+1}}$ $(n = 0, 1, \ldots)$
$\sin at$	$\frac{a}{s^2+a^2}$
$\cos at$	$\frac{s}{s^2+a^2}$
$\sinh at$	$\frac{a}{s^2-a^2}$
$\cosh at$	$\frac{s}{s^2-a^2}$
$u(t-a) = \begin{cases} 0 & t \leq a \\ 1 & t > a \end{cases}$	$\frac{e^{-as}}{s}$
$\delta(t-a)$	e^{-as}
$f'(t)$	$s\mathcal{L}(f) - f(0)$
$f''(t)$	$s^2\mathcal{L}(f) - sf(0) - f'(0)$

Example A.1: RC-circuit

Problem: Using Laplace transform, calculate the $v_{\text{out}}/v_{\text{in}}$ in s-domain for the circuit in Figure A.1. What is the frequency response $H(\omega) = v_{\text{out}}(\omega)/v_{\text{in}}(\omega)$ and the time domain step response for a unit step input at $t = 0$?

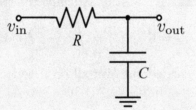

Figure A.1: RC-circuit

Solution: The current through the resistor is $i = v/R$. Taking the Laplace transform of the current through the capacitor $i = C\frac{\partial v}{\partial t}$ gives

$$i = \mathcal{L}\left\{C\frac{\partial v}{\partial t}\right\} = sCv \equiv \frac{1}{Z}v, \tag{A.3}$$

where we have defined the complex impedance $Z = 1/sC$.

Using the complex impedance, the output voltage from voltage division is

$$v_{\text{out}}(s) = \frac{Z}{R+Z}v_{\text{in}}(s) = \frac{1}{1+sRC}v_{\text{in}}(s) \tag{A.4}$$

and the response function $H(s)$ is

$$H(s) = \frac{v_{\text{out}}(s)}{v_{\text{in}}(s)} = \frac{Z}{R + Z} = \frac{1}{1 + sRC}. \tag{A.5}$$

The frequency response is obtained by substituting $s = j\omega$ to give

$$H(\omega) = \frac{1}{1 + j\omega RC}. \tag{A.6}$$

The step response is

$$v_{\text{out}}(s) = H(s)v_{\text{in}}(s) = \frac{1}{1 + sRC} \frac{1}{s}, \tag{A.7}$$

where we have used $\mathcal{L}\{1\} = 1/s$ for the unit step input. Solution to Equation (A.7) is obtained by partial fraction expansion. We seek to write Equation (A.7) as

$$\frac{A}{s} + \frac{B}{1 + sRC} = \frac{A(1 + sRC) + Bs}{s(1 + sRC)}. \tag{A.8}$$

Equations (A.7) and (A.8) are equal if $A = 1$ and $B = -RC$. Thus, the step response is

$$v_{\text{out}}(s) = \frac{1}{s} + \frac{-RC}{1 + sRC} = \frac{1}{s} - \frac{1}{s + 1/RC}. \tag{A.9}$$

From Table A.1, $\mathcal{L}^{-1}\{1/s\} = 1$ and $\mathcal{L}^{-1}\{1/(s+a)\} = e^{-at}$, and the inverse of Equation (A.9) is

$$v_{\text{out}}(t) = 1 - e^{-t/RC}. \tag{A.10}$$

Example A.2: Free fall velocity

Problem: The free fall of an object is governed by the equation of motion

$$m\frac{\partial \dot{x}(t)}{\partial t} + \gamma \dot{x}(t) = F, \tag{A.11}$$

where m is the object mass, $\dot{x}$ is the velocity, γ is the air drag coefficient, and F is a constant gravity force. Find the object velocity as a function of time.

Solution: Taking the Laplace-transformation of Equation (A.11) gives

$$sm\dot{x}(s) + \gamma \dot{x}(s) = \frac{F}{s}, \tag{A.12}$$

where we have used $\mathcal{L}\{\dot{x}'(t)\} = s\dot{x}(s)$, $\mathcal{L}\{\dot{x}(t)\} = \dot{x}(s)$, and $\mathcal{L}\{F\} = F/s$. Solving Equation (A.12) for the velocity gives

$$\dot{x}(s) = \frac{F}{s}\frac{1}{ms+\gamma}. \tag{A.13}$$

Using the partial fraction expansion (see Example A.1), Equation (A.13) can be written as

$$\dot{x}(s) = \frac{F}{\gamma}\left(\frac{1}{s} - \frac{1}{s+\gamma/m}\right). \tag{A.14}$$

From Table A.1, $\mathcal{L}^{-1}\{1/s\} = 1$ and $\mathcal{L}^{-1}\{1/(s+a)\} = e^{-at}$, and the inverse of Equation (A.14) is

$$\dot{x}(t) = \frac{F}{\gamma}\left(1 - e^{-\frac{\gamma}{m}t}\right). \tag{A.15}$$

Exercises

Exercise A.1

Using Laplace transformation, calculate the $v_{\text{out}}/v_{\text{in}}$ in s-domain for circuits in Figure A.2. What is the frequency and step responses?

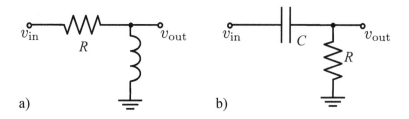

a)   b)

Figure A.2: A RL-circuit and a RC-circuit.

Exercise A.2

A damped spring is governed by the equation of motion

$$\gamma\frac{\partial x(t)}{\partial t} + kx(t) = F,$$

where γ is the damping coefficient, $x(t)$ is the spring displacement, k is the spring constant, and F is the unit step force acting on the spring. If the initial spring displacement is zero ($x = 0$ at $t = 0$), derive an expression for the spring displacement as a function of time using the Laplace transformation.

B

Mechanical harmonic resonators

The harmonic resonator is a good basis for modeling most engineering and physical systems. Harmonic resonator examples include atomic vibrations, interaction of electrons and electromagnetic waves, plasma oscillations, operational amplifier response, and car suspension. Micromachined structures that can be modeled as a harmonic resonator include accelerometers, gyroscopes, microphones, and RF filters.

There are two reasons for the prevalence of the harmonic (or second-order) response in engineering analysis. First, as the list above suggests, second-order system accurately models wide range of phenomena. Second, the second-order system is easy to understand. By starting with a simple model and adding complexity only when required gives intuitive understanding of the underlying physics and design trade-offs.

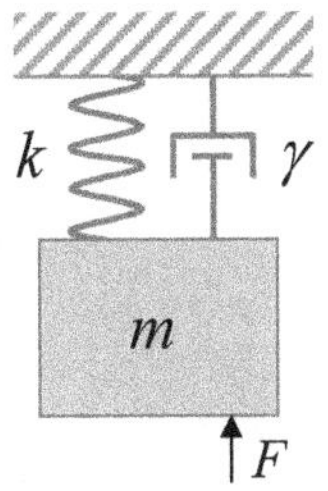

Equation of motion:

$$m\frac{\partial^2 x}{\partial t^2} + \gamma\frac{\partial x}{\partial t} + kx = F(t)$$

Substitute $x = x_0 e^{st}$ and $F = F_0 e^{st}$

$$x_0 = \frac{F_0}{s^2 m + s\gamma + k} \equiv H(s)F_0$$

Figure B.1: Harmonic resonator with mass m, spring k, and dash-pot γ. A fast way to obtain the frequency response and the response function $H(s)$ is to substitute $F = F_0 e^{st}$ for the excitation force, $x = x_0 e^{st}$ for the displacement, and solving for x_0.

Here we will review the frequency and time domain characteristics of the harmonic oscillator using phasors and Laplace transform that is reviewed in Appendix A. The equation of motion for the harmonic resonator in Figure B.1 is

$$m\frac{\partial^2 x}{\partial t^2} + \gamma\frac{\partial x}{\partial t} + kx = F, \tag{B.1}$$

where x is the displacement of the mass m, γ is the damping coefficient, k is the spring constant and F is the excitation force. The solution to Equation (B.1) is obtained, for example, using Laplace method. By using the $\dot{x} = sx - x(0)$ and $\ddot{x} = s^2 x - s\dot{x}(0) - x(0)$, Equation (B.1) gives

$$x = \frac{F + (\gamma + m)x(0) + sm\dot{x}(0)}{s^2 m + s\gamma + k}. \tag{B.2}$$

For the sake of simplicity, we assume that the initial conditions are zero ($x(0) = 0$ and $\dot{x}(0) = 0$) and define the response function $H(s)$ as

$$H(s) \equiv x/F = \frac{1}{s^2 m + s\gamma + k}. \tag{B.3}$$

to write Equation (B.3) as

$$x = \frac{F}{s^2 m + s\gamma + k} \equiv H(s)F. \tag{B.4}$$

It is useful to define two additional quantities: the natural or resonance frequency $\omega_0 = \sqrt{\frac{k}{m}}$ and the quality factor $Q = \frac{\omega_0 m}{\gamma}$ to rewrite Equation (B.4) as

$$x = \frac{F/m}{s^2 + s\omega_0/Q + \omega_0^2} \equiv H(s)F. \tag{B.5}$$

In what follows, we will study the harmonic resonator response function given by Equation (B.4) or (B.5). First, we will study the resonator frequency response that describes the resonator response to a sinusoidal force. Next, the time domain impulse and step responses are reviewed.

B.1 Frequency response

Assuming time harmonic excitation $F = F_0 e^{st}$ where F_0 is the complex amplitude of excitation force, the frequency response is obtained by substituting $x = x_0 e^{st}$ and $F = F_0 e^{st}$ into Equation (B.1). This gives the complex vibration amplitude

$$x_0 = \frac{F_0/m}{s^2 + s\omega_0/Q + \omega_0^2} \equiv H(s)F_0. \tag{B.6}$$

We note that the frequency response given by Equation (B.6) is identical to the response function $H(s)$ obtained with Laplace method and given by Equation (B.5).

By substituting $s = j\omega$ into Equation (B.6), we obtain the complex frequency response function

$$H(\omega) = \frac{1}{m} \frac{1}{\omega_0^2 - \omega^2 + j\omega\omega_0/Q} \equiv H_R(\omega) + jH_I(\omega), \tag{B.7}$$

where the real and imaginary parts of the response function are

$$H_R(\omega) = \frac{1}{m} \frac{\omega_0^2 - \omega^2}{(\omega_0^2 - \omega^2)^2 + (\omega\omega_0/Q)^2} \tag{B.8}$$

and

$$H_I(\omega) = \frac{1}{m} \frac{-\omega\omega_0/Q}{(\omega_0^2 - \omega^2)^2 + (\omega\omega_0/Q)^2}. \tag{B.9}$$

Using this short hand notation, we can write the frequency response given by Equation (B.6) as

$$x_0 = H_R(\omega)F_0 + jH_I(\omega)F_0. \tag{B.10}$$

Two important frequencies are the dc response with $H_R(0) = 1/m\omega_0^2 = 1/k$ and $H_I(0) = 0$ and resonance response with $H_R(\omega_0) = 0$ and $H_I(\omega_0) = -Q/\omega_0^2 m = -Q/k$.

B.1.1   Amplitude response

Taking the absolute value of Equation (B.6) gives the amplitude of the displacement for the time harmonic excitation

$$|x_0| = \frac{|F_0|/m}{\sqrt{\left(\omega^2 - \omega_0^2\right)^2 + (\omega\omega_0/Q)^2}} = \frac{|F_0|/k}{\sqrt{\left(1 - \frac{\omega^2}{\omega_0^2}\right)^2 + \left(\frac{\omega}{Q\omega_0}\right)^2}}. \tag{B.11}$$

The amplitude at dc is $|x_0(0)| = |F|/k$ and amplitude at resonance is $|x_0(\omega_0)| = Q|F|/k$. If $Q > 1$ we see peaking at the resonance and the resonance response is higher than the dc response. Above resonance, the response falls as $1/\omega^2$.

The maximum response amplitude can be found by taking a derivative of Equation (B.11) and solving for zero. From $\frac{\partial|x_0|}{\partial\omega} = 0$ we get $\omega_{max} = \omega_0\sqrt{1 - 1/2Q^2}$ which is real only if $Q > 1/\sqrt{2}$. Thus, for $Q > 1/\sqrt{2}$ there is a maximum at $\omega > 0$ and for $Q \leq 1/\sqrt{2}$ no maximum is found and the amplitude $|x|$ decreases with frequency as is illustrated in Figure B.2(a). The amplitude at ω_{max} is

$$|x|_{max} = \frac{Q^2 F_0/k}{\sqrt{Q^2 - 1/4}}. \tag{B.12}$$

In the limit of high quality factor ($Q \gg 1$) we find that $\omega_{max} \approx \omega_0$ and $|x|_{max} \approx QF_0/k$.

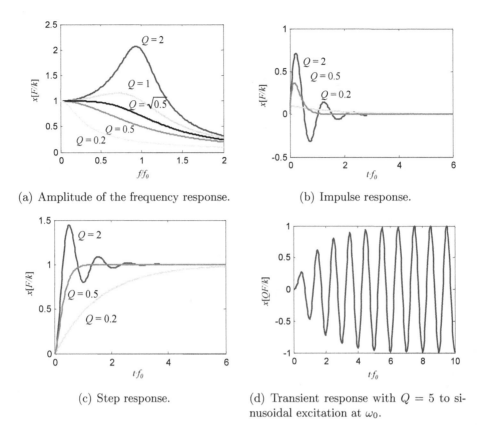

(a) Amplitude of the frequency response.

(b) Impulse response.

(c) Step response.

(d) Transient response with $Q = 5$ to sinusoidal excitation at ω_0.

Figure B.2: Harmonic resonator responses with several different quality factors.

B.1.2   Phase response

The phase of the displacement x_0 relative to the excitation force F_0 given by Equation (B.10) is

$$\theta = \tan^{-1} \frac{H_I}{H_R} = \tan^{-1} \frac{-\omega\omega_0/Q}{\omega_0^2 - \omega^2}. \tag{B.13}$$

The phase at low frequencies ($\omega \ll \omega_0$) is $\theta \approx 0$ which means that there is no lag between the excitation force F_0 and displacement x_0. At resonance the phase is $\theta(\omega_0) = -\pi/2$ and the displacement lags the excitation force exactly by $90°$. Above resonance frequency ($\omega \gg \omega_0$) the displacement lags the excitation force by $\approx 180°$.

B.2   Impulse response

The time domain response of the harmonic resonator is important as real applications operate mostly in the time domain. For example, time domain response will tell how quickly an accelerometer that is used to deploy air bags gives out signal after the crash has occurred. The impulse response gives the resonator displacement after a very short duration impact.

To obtain the impulse response, we start with the Laplace transform of the unit impulse $F(t) = F_0 \delta(t)$ which is $F(s) = F_0$. Substituting $F(s) = F_0$ to Equation (B.5) gives the impulse response

$$x = \frac{F_0/m}{s^2 + s\omega_0/Q + \omega_0^2} = \frac{F_0/m}{(s - r_p)(s - r_n)}. \tag{B.14}$$

where r_p and r_n are the positive and negative roots of the denominator given by

$$r_{p,n} = -\frac{\omega_0}{2Q} \pm \omega_0 \sqrt{\left(\frac{1}{2Q}\right)^2 - 1} = -\frac{\omega_0}{2Q} \pm j\omega_d. \tag{B.15}$$

The $\omega_d = \omega_0 \sqrt{1 - \left(\frac{1}{2Q}\right)^2}$ is the damped resonance frequency and as we will see shortly, sets the frequency of unforced vibrations.

The inverse Laplace transform of the impulse response given by Equation (B.14) is obtained by partial fraction expansion that gives

$$x = \frac{F_0}{m}\left(\frac{c_1}{s - r_p} + \frac{c_2}{s - r_n}\right). \tag{B.16}$$

where $c_1 = 1/(r_p - r_n)$ and $c_2 = -1/(r_p - r_n)$. Using Table A.1 on page 406, the inverse of (B.16) is

$$x = F_0/m \left(c_1 \exp(r_p t) + c_2 \exp(r_n t)\right) \tag{B.17}$$

$$= j\frac{F_0}{2m\omega_d} \exp\left(-\frac{\omega_0 t}{2Q}\right) \left(\exp\left(-j\omega_d t\right) - \exp\left(j\omega_d t\right)\right) \tag{B.18}$$

While Equation (B.18) is the desired solution, the complex numbers make it difficult to interpret. To visualize the results, we will rewrite the solution in more intuitive forms for under damped, critically damped, and over damped systems.

B.2.1   Under damped

When $Q > 0.5$, the resonator is under damped. The damped resonance frequency $\omega_d = \omega_0 \sqrt{1 - \left(\frac{1}{2Q}\right)^2}$ is positive real number and Equation (B.18) can be rewritten as

$$x = \frac{F_0}{m\omega_d} \exp\left(-\frac{\omega_0 t}{2Q}\right) \sin\left(\omega_d t\right). \tag{B.19}$$

As is clearly seen from Equation (B.19) and Figure B.2(b), the resonator goes through periodic vibrations at frequency ω_d before settling with the time constant $2Q/\omega_0$. We note that the damped vibration frequency ω_d for the freely vibrating resonator is lower than the undamped resonance frequency ω_0.

B.2.2 Critically damped

When $Q \to 0.5$, the damped resonance frequency approaches zero and system is critically damped. Taking the limit $Q \to 0.5$ of Equation (B.18) gives

$$x = \frac{F_0 t}{m} \exp\left(-\omega_0 t\right). \tag{B.20}$$

As Equation (B.20) and Figure B.2(b) show, critically damped system settles with the time constant $\tau_c = 1/\omega_0$ without vibrations. As the time constant is inversely proportional to the resonance frequency ω_0, a high resonance frequency results in fast time domain response (high bandwidth).

B.2.3 Over damped

For $Q < 0.5$ the damped resonance frequency $\omega_d = \omega_0\sqrt{1 - \left(\frac{1}{2Q}\right)^2}$ is imaginary. Equation (B.18) can be written as

$$x = \frac{F_0}{m|\omega_d|} \exp\left(-\frac{\omega_0 t}{2Q}\right) \sinh\left(|\omega_d|t\right) \tag{B.21}$$

$$\approx \frac{F_0}{2m|\omega_d|} \exp\left(-\left(\frac{\omega_0}{2Q} - |\omega_d|\right)t\right) \text{ for } t \gg 1/|\omega_d|. \tag{B.22}$$

Equation (B.21) behaves much like the critically damped system but as shown in Figure B.2(b), the magnitude of the initial response smaller and the decay time constant is longer given by $1/(\frac{\omega_0}{2Q} - |\omega_d|)$.

B.3 Step response

The step response describes how the system behaves after applying constant force. The Laplace transform of the step force $F(t) = F_0 u(t)$ is $F(s) = F_0/s$. Here F_0 is the amplitude of the force and $u(t)$ is the step function ($u(t) = 1$ for $t \geq 0$ and $u(t) = 0$ for $t < 0$). Substituting $F(s) = F_0/s$ to Equation (B.5) gives the step response

$$x = \frac{F_0/m}{s(s^2 + s\omega_0/Q + \omega_0^2)} = \frac{F_0/m}{s(s - r_p)(s - r_n)}. \tag{B.23}$$

where r_p and r_n are the positive and negative roots of the denominator given by Equation (B.15). By carrying out the partial fraction expansion of Equation (B.23) we get

$$x = \frac{F}{m}\left[\frac{c_1}{s} + \frac{c_2}{s - r_p} + \frac{c_3}{s - r_n}\right],$$
(B.24)

where $c_1 = 1/(r_p r_n)$, $c_2 = 1/r_p/(r_p - r_n)$, and $c_3 = -1/r_n/(r_p - r_n)$. As with the impulse response, the inverse of (B.24) is obtained using standard tables for Laplace transform

$$x = F_0/m(c_1 + c_2 \exp(r_p t) + c_3 \exp(r_n t)).$$
(B.25)

Again, it is desirable to write Equation (B.25) in more intuitive forms for under damped, critically damped, and over damped cases.

B.3.1   Under damped

When $Q > 0.5$, the resonator is underdamped and $\omega_d = \omega_0\sqrt{1 - \left(\frac{1}{2Q}\right)^2}$ is a positive real number. Equation (B.25) can be written as

$$x = \frac{F_0}{m\omega_0^2}\left[1 - \frac{1}{\omega_d}\exp\left(-\frac{\omega_0 t}{2Q}\right)\left(\frac{\omega_0}{2Q}\sin\left(\omega_d t\right) + \omega_d\cos\left(\omega_d t\right)\right)\right].$$
(B.26)

As Figure B.2(c) and Equation (B.26) shows, the under damped resonator overshoots and exhibits periodic oscillations at frequency ω_d before settling to the final value $x = F_0/m\omega_0^2 = F_0/k$ with the time constant $\tau_u = 2Q/\omega_0$. Increasing the quality factor will make the initial response faster but the system will have larger overshoot and exhibit more ringing. The maximum overshoot is 100% which is obtained when Q is high.

B.3.2   Critically damped

When $Q = 0.5$, the damped resonance frequency ω_d is zero. Taking the limit $Q \to 0.5$, Equation (B.25) gives

$$x = \frac{F_0}{m\omega_0^2}\left[1 - \exp\left(-\omega_0 t\right)\left(\omega_0 t + 1\right)\right].$$
(B.27)

As Figure B.2(c) and Equation (B.27) shows, the critically damped system does not overshoot. The final value is reached with the approximate time constant $\tau_c \approx 1/\omega_0$ but it is not possible to obtain a closed form expression for the exact settling time. To decrease the response time to the external step force $F = F_0 u(t)$, the resonant frequency should be increased to increase system bandwidth. Table B.1 shows selected calculated settling times. For example, if $f_0 = 100$ Hz, the time to settle within 10% is $0.62/(100$ Hz$) = 6.2$ ms.

Table B.1: Settling times for a critically damped system.

Percentage error	Settling time
20%	$0.48 \cdot f_0$
10%	$0.62 \cdot f_0$
5%	$0.76 \cdot f_0$
2%	$0.93 \cdot f_0$
1%	$1.06 \cdot f_0$

B.3.3 Over damped

For over damped systems $Q < 0.5$ and $\omega_d = \omega_0\sqrt{1 - (1/2Q)^2}$ is imaginary. Equation (B.25) can be written as

$$x = \frac{F_0}{m\omega_0^2}\left[1 - \frac{1}{|\omega_d|}\exp\left(-\frac{\omega_0 t}{2Q}\right)\left(\frac{\omega_0}{2Q}\sinh\left(|\omega_d|t\right) + |\omega_d|\cosh\left(|\omega_d|t\right)\right)\right] \quad \text{(B.28)}$$

$$\approx \frac{F_0}{m\omega_0^2}\left[1 - \exp\left(-\omega_0 Q t\right)\right] \text{ for } Q \leq 0.3. \quad \text{(B.29)}$$

As seen from Equation (B.28) and Figure B.2(c), the over damped system behaves much like critically damped system but settles more slowly with the approximate time constant given by $\tau \approx 1/\omega_0 Q$.

B.4 Transient response of forced vibrations

For oscillators and resonant sensors such gyroscopes, it is of interest to know how quickly the resonator reaches steady state of oscillations if the resonator is subjected to a time harmonic excitation at the resonant frequency ω_0. The Laplace transform of $F = F_0\sin(\omega_0 t)u(t)$ is $F(s) = \omega_0/(s^2 + \omega_0^2)$. Substituting this into Equation (B.5) and taking the inverse Laplace transform gives

$$x = \frac{QF_0}{m\omega_0^2}\left[-\cos(\omega_0 t) + \exp\left(-\frac{\omega_0 t}{2Q}\right)\left(\cos(\omega_d t) + \frac{\omega_0\sin(\omega_d t)}{2\omega_d Q}\right)\right]. \quad \text{(B.30)}$$

Figure B.2(d) shows the transient response for resonator with $Q = 5$. We see that the forced resonator response approaches the steady state motion with time constant $\tau = 2Q/\omega_0$. Increasing the quality factor increases the system response time to the excitation. This is significant for example in mode matched gyroscopes where increasing the quality factor increases the sensitivity but decreases the bandwidth.

Exercises

Exercise B.1
Explain how the resonance frequency affects the harmonic resonator time domain response.

Exercise B.2
Explain why high quality factor may not be desirable.

Exercise B.3
Derive Equation (B.30). Hint: A fair amount of algebra may be required and computers are pretty good at this. You may want to look at *ilaplace*-command in Matlab to shorten the time to calculate the inverse Laplace-transformation.

Exercise B.4
Calculate the time that it takes for a resonator with $\omega_0 = 100$ Hz to settle within 2% of the final value when excited with a unit force step in case of i. high-Q ($Q = 20$), ii. critically damped ($Q = 0.5$), and iii. overdamped ($Q = 0.1$) resonator.

C

Nonlinear vibrations in resonators

Many MEMS devices are strongly nonlinear. For example, the nonlinearity in the parallel plate capacitive actuators is so strong that the large displacement modeling requires numerical simulations. In this chapter, we focus on weakly nonlinear resonators that can still be analyzed analytically. The nonlinear spring forces in (micro)resonators cause the resonator response to deviate from the linear harmonic response covered in Appendix B. The nonlinearity is most apparent around the resonant frequency as the nonlinear springs effectively make the resonant frequency amplitude dependent.

Our motivation for the analysis is: to understand the nonlinear effects observed with real resonators and to estimate the maximum vibration amplitude where vibrations are still almost linear. Knowing the range for linear vibrations sets the dynamic range for resonators. On the lower end, the smallest usable vibration amplitude is limited by the noise. The upper limit is set by the resonator power handling capacity, which is limited by nonlinear effects. The analysis in this chapter follows closely to reference [237], but the presentation has been slightly modernized.

C.1 Nonlinear spring forces

In general, the nonlinear spring force can be written as

$$F = -kx - k_1 x^2 - k_2 x^3 + O(x^4), \tag{C.1}$$

where k is the normal linear spring constant, k_1 and k_2 are the first and second order corrections, respectively, x is the spring displacement, and $O(x^4)$ indicates the higher order terms that are ignored in this analysis. With the nonlinear spring, the equation of motion for the mass-spring-dash pot resonator becomes

$$m\frac{\partial^2 x}{\partial t^2} + \gamma\frac{\partial x}{\partial t} + kx + k_1 x^2 + k_2 x^3 = F_\omega \cos\omega t, \tag{C.2}$$

where F_ω is the amplitude of the actuation force at the frequency ω. Equation (C.2) is solved in two parts: First, the unforced and undamped vibrations are analyzed. Next, the obtained solution is used to approximate forced vibrations with damping.

C.2   Unforced vibrations

Setting $\gamma = 0$ and $F_\omega = 0$ in Equation C.2, we obtain

$$m\frac{\partial^2 x}{\partial t^2} + kx + k_1 x^2 + k_2 x^3 = 0. \tag{C.3}$$

From physics, we expect the unforced, undamped harmonic oscillator to oscillate infinitely with constant amplitude at the resonance frequency ω_0. The nonlinear terms will change the oscillation frequency to ω_0'. Moreover, the oscillation frequency ω_0' will depend on oscillation amplitude due to $k_1 x^2$ and $k_2 x^3$ terms in Equation (C.3). To obtain this relationship, we will carry out perturbation analysis around the linear oscillation frequency ω_0. After all, we are interested in almost linear systems!

Dividing Equation (C.3) by m and making the change of variables as

$$\begin{aligned} k_1 &= \epsilon\alpha m \\ k_2 &= \epsilon^2\beta m \\ \omega_0' t &= \tau \end{aligned} \tag{C.4}$$

gives

$$\omega_0'^2\frac{\partial^2 x}{\partial \tau^2} + \omega_0^2 x + \epsilon\alpha x^2 + \epsilon^2\beta x^3 = 0. \tag{C.5}$$

In Equation (C.4) the first-order correction to spring constant k_1 is proportional to the perturbation parameter ϵ while the second-order correction k_2 is proportional to ϵ^2. This approach provides a convenient way to group the perturbation terms by their order. Next, we substitute our trial solution $\omega_0' = \omega_0 + \epsilon\omega_1 + \epsilon^2\omega_2$ and $x = x_0 + \epsilon x_1 + \epsilon^2 x_2$ into Equation (C.5) and group the terms in powers of ϵ:

$$\begin{aligned} &\omega_0^2\frac{\partial^2 x_0}{\partial \tau^2} + \omega_0^2 x_0 \\ &+\epsilon\left[\omega_0^2\frac{\partial^2 x_1}{\partial \tau^2} + \omega_0^2 x_1 + \alpha x_0^2 + 2\omega_0\omega_1\frac{\partial^2 x_0}{\partial \tau^2}\right] \\ &+\epsilon^2\left[\omega_0^2\frac{\partial^2 x_2}{\partial \tau^2} + \omega_0^2 x_2 + 2\alpha x_0 x_1 + \beta x_0^3 + (\omega_1^2 + 2\omega_0\omega_2)\frac{\partial^2 x_0}{\partial \tau^2} + 2\omega_0\omega_1\frac{\partial^2 x_1}{\partial \tau^2}\right] \\ &+O(\epsilon^3) = 0. \end{aligned} \tag{C.6}$$

For Equation (C.6) to be satisfied for $\epsilon \neq 0$, the following Equations must be satisfied:

$$\omega_0^2 \frac{\partial^2 x_0}{\partial \tau^2} + \omega_0^2 x_0 = 0, \tag{C.7}$$

$$\omega_0^2 \frac{\partial^2 x_1}{\partial \tau^2} + \omega_0^2 x_1 + \alpha x_0^2 + 2\omega_0 \omega_1 \frac{\partial^2 x_0}{\partial \tau^2} = 0, \tag{C.8}$$

and

$$\omega_0^2 \frac{\partial^2 x_2}{\partial \tau^2} + \omega_0^2 x_2 + 2\alpha x_0 x_1 + \beta x_0^3 + (\omega_1^2 + 2\omega_0 \omega_2)\frac{\partial^2 x_0}{\partial \tau^2} + 2\omega_0 \omega_1 \frac{\partial^2 x_1}{\partial \tau^2} = 0. \tag{C.9}$$

Our next step is to solve the Equations (C.7) to (C.9) for the unknown x_0, x_1, x_2, ω_0, ω_1, and ω_2.

Equation (C.7) is just the harmonic resonator and has the solution

$$x_0 = X_0 \cos \tau, \tag{C.10}$$

where X_0 is the vibration amplitude. Substituting Equation (C.10) to (C.8) gives

$$\omega_0^2 \frac{\partial^2 x_1}{\partial \tau^2} + \omega_0^2 x_1 = -\frac{1}{2}\alpha X_0^2(1 + \cos 2\tau) + 2\omega_0 \omega_1 X_0 \cos \tau. \tag{C.11}$$

The resonant term[1] $2\omega_0 \omega_1 X_0 \cos \tau$ on the right side of Equation (C.11) would result in x_1 growing infinitely. This is physically not possible as no energy is pumped into the resonator system. Thus, we require the resonant term to be zero leading to $\omega_1 = 0$. Solving Equation (C.11) with the resonant term set to zero leads to

$$x_1 = -\frac{3\alpha}{6\omega_0^2}X_0^2 + \frac{\alpha}{6\omega_0^2}X_0^2 \cos 2\tau. \tag{C.12}$$

We thus have two additional frequency components due to the first-order corrections: a dc-term and a higher harmonic at twice the oscillation frequency. Substituting Equations (C.10) and (C.12) to (C.9) gives

$$\omega_0^2 \frac{\partial^2 x_2}{\partial \tau^2} + \omega_0^2 x_2 = -\left[-\frac{5\alpha^2}{6\omega_0^2}X_0^3 + \frac{3\beta}{4}X_0^3 - 2\omega_0\omega_2 X_0\right]\cos\tau - \left[\frac{\alpha^2}{6\omega_0^2}X_0^3 + \frac{\beta}{4}X_0^3\right]\cos 3\tau. \tag{C.13}$$

We again require that the resonant term is zero giving

$$\omega_2 = \left[-\frac{5\alpha^2}{12\omega_0^3} + \frac{3\beta}{8\omega_0}\right]X_0^2. \tag{C.14}$$

Solving Equation (C.13) with the resonant term set to zero leads to

$$x_2 = \left[\frac{\alpha^2}{24\omega_0^4} + \frac{\beta}{16\omega_0^2}\right]X_0^3 \cos\tau + \left[\frac{\alpha^2}{48\omega_0^4} + \frac{\beta}{32\omega_0^2}\right]X_0^3 \cos 3\tau. \tag{C.15}$$

---

[1]The resonant excitation terms are sometimes referred to as the secular terms.

Due to the second-order correction, the displacement has two additional components: a term at oscillation frequency and a term at three times the oscillation frequency. The additional term at oscillation frequency is a fingerprint of odd-order nonlinearity. It is very detrimental to communication systems as it deteriorates the signal at the signal frequency.

To summarize this section, the nonlinear spring constant changes the resonance frequency of the resonator as

$$\omega_0' = \omega_0 + \epsilon^2 \omega_2 = \omega_0 + \kappa X_0^2, \tag{C.16}$$

where

$$\kappa = \frac{3k_2}{8k}\omega_0 - \frac{5k_1^2}{12k^2}\omega_0. \tag{C.17}$$

Equation (C.16) will be used in the next section to analyze forced vibrations of damped resonator.

C.3   Forced vibrations

As we saw in the previous section, the main effect nonlinear springs is to change the resonant frequency of the resonator. Therefore, we can analyze weakly nonlinear forced vibrations by using of Equation (C.16) and substituting $\omega_0 \to \omega_0'$ into Equation (B.1). The time harmonic vibration amplitude near the resonance is then given by

$$X_0 = \frac{F_\omega/m}{\sqrt{(\omega^2 - \omega_0'^2)^2 + (\omega\omega_0'/Q)^2}}, \tag{C.18}$$

where from the previous section we have $\omega_0' = \omega_0 + \kappa X_0^2$. Equations (C.16), (C.17) and (C.18) show that due to a positive or negative k_1, the peak-frequency given by $\omega = \omega_0'$ shifts to a lower frequency with an increasing vibration amplitude X_0 as illustrated in Figure C.1. Similarly, a negative k_2 results in the peak-frequency shifting to a lower frequency while a positive k_2 results in a higher peak-frequency.

A useful measure of the maximum vibration amplitude is obtained by calculating the bifurcation point x_b shown in Figure C.2. At higher excitation levels, the amplitude-frequency relationship is no longer a single valued function and shows hysteresis. Thus, the maximum vibration amplitude before hysteresis x_c can be used to estimate the limit for linear operation.

To obtain an analytical estimate for the bifurcation we write $\Delta\omega = \omega - \omega_0$ and make use of the approximations $(\omega_0 + \Delta\omega)^2 - \omega_0'^2 = (\omega_0 + \Delta\omega + \omega_0')(\omega_0 + \Delta\omega - \omega_0') \approx 2\omega_0(\Delta\omega - \kappa X_0^2)$ and $(\omega_0 + \Delta\omega)\omega_0' \approx \omega_0^2$ to write Equation (C.18) as

$$X_0^2 = \frac{F^2/m^2}{4\omega_0^2\left[(\Delta\omega_0 - \kappa X_0^2)^2 + \omega_0^2/4Q^2\right]} \tag{C.19}$$

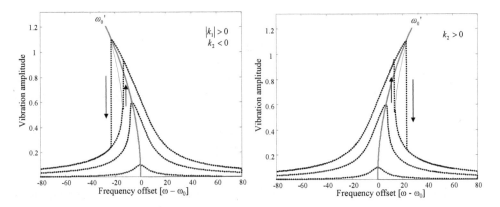

Figure C.1: Simulated (dotted lines) and analytical (solid lines) resonator amplitude-frequency response curves around the resonance frequency. Depending on the spring constant, the peak-frequency can shift to either higher or lower frequency. At large vibration amplitudes, the response shows hysteresis.

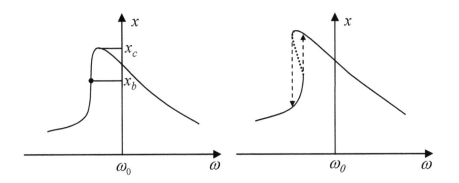

Figure C.2: At bifurcation point, the slope of the amplitude-frequency curve becomes infinite. After bifurcation, the amplitude-frequency curve has an unstable region (dotted line) resulting in frequency hysteresis. Notice also that after bifurcation, the amplitude-frequency curve is no longer a single valued function.

or

$$X_0^2 4\omega_0^2 \left[((\Delta\omega_0 - \kappa X_0^2)^2 + \omega_0^2/4Q^2 \right] = F^2/m^2. \tag{C.20}$$

As Figure C.2 indicates, at bifurcation point the slope $\partial X_0/\partial\Delta\omega = \infty$. Deriving Equation (C.20), solving for $\partial X_0/\partial\Delta\omega$, and requiring that denominator is zero to give $\partial X_0/\partial\Delta\omega = \infty$ leads to

$$X_0^2 = \frac{4Q^2\Delta\omega\kappa \pm \sqrt{4Q^4\Delta\omega^2\kappa^2 - 3Q^2\kappa^2\omega_0^2}}{6Q^2\kappa^2}. \tag{C.21}$$

For Equation (C.21) to be the bifurcation point, it has to be single valued which means that the square root term has to be zero. Requiring that $4Q^4\Delta\omega^2\kappa^2 -$

$3Q^2\kappa^2\omega_0^2 = 0$ gives

$$\Delta\omega = \pm\frac{\sqrt{3}\omega_0}{2Q} \tag{C.22}$$

at bifurcation, where the positive and negative sign are for the positive and negative κ, respectively. Substituting Equation (C.22) back into Equation (C.18) gives

$$x_b = \sqrt{\frac{\omega_0}{\sqrt{3}Q|\kappa|}} \tag{C.23}$$

for the bifurcation point.

As indicated in Figure C.2, the critical vibration amplitude (or the greatest vibration amplitude) x_c is slightly higher than the vibration amplitude at the bifurcation point. It is obtained by substituting Equations (C.22) and (C.23) into Equation (C.18) and solving for the force. The amplitude of vibrations at resonance due to this force is

$$x_c = \sqrt{\frac{4\omega_0}{3\sqrt{3}Q|\kappa|}}. \tag{C.24}$$

As Equation (C.24) shows, increasing the quality factor *lowers* the critical vibration amplitude x_c as the resonator is made more susceptible to nonlinearities.

D

Thermal noise generator

In Chapter 2 we learned that on average there is $\frac{1}{2}k_BT$ of thermal energy associated with each degree of freedom. While this allows the calculation of the rms-noise, the frequency spectrum of the noise remains to be determined. Here we will derive an expression for the thermal noise generator to calculate the frequency characteristics of the noise. In electrical circuits, the thermal noise generator is a voltage source associated with each resistor. In mechanical devices, the noise generator is a force associated with each damper. In both cases, the noise generator is associated with dissipation and it should be no surprise that the resulting expressions for the noise generator are very similar.

D.1 Derivation of noise voltage generator

We will first derive an expression for the noise voltage generator as most readers are already familiar with the thermal noise in resistors. Once we have derived the correct expression for the resistor noise voltage generator, $\overline{v_n^2} = 4k_BTR$, a similar derivation for the mechanical dissipation will be easier to follow. The derivation presented here is a bit mathematical. Those who enjoy good argumentation without many equations should obtain a copy of the original article on the resistor noise by Nyquist [24], as it is still one of the most readable derivations of the resistor noise.

The noise voltage generator can be derived by considering the series RLC-resonator shown in Figure D.1(a). The circuit is fully described by two variables: current and voltage. From the equipartition theorem we know that there is an average of $\frac{1}{2}k_BT$ of thermal energy stored in each inductor and capacitor. Associated with the resistor is a thermal noise generator $\overline{v}_n$. This noise generator is just another representation for the thermal noise, and the energy stored in the inductor and capacitor due to $\bar{v}_n$ should equal $\frac{1}{2}k_BT$. Thus, to obtain an expression for the noise generator, we can calculate the energy stored in the inductor or the capacitor due to the noise voltage generator $\bar{v}_n$ and equate this energy with $\frac{1}{2}k_BT$.

The current through the RLC-resonator due to a noise voltage v_n is

$$i_n = \frac{v_n}{R + sL + 1/sC},\tag{D.1}$$

where $s = j\omega$ as usual. From Equation (D.1), the magnitude of the mean square current is

$$\overline{i_n^2} = \frac{\overline{v_n^2}}{R^2 + (\omega L - 1/\omega C)^2}.\tag{D.2}$$

Equation (D.2) can be written in terms of resonance frequency $\omega_0 = 1/\sqrt{LC}$ and quality factor $Q = \omega_0 L/R$ as

$$\overline{i_n^2} = \frac{1}{R^2}\frac{\overline{v_n^2}}{1 + Q^2(\omega/\omega_0 - \omega_0/\omega)^2}.\tag{D.3}$$

The energy stored in the inductor in a frequency interval df is $dW = \frac{1}{2}L\overline{i_n^2}df$ and the total energy stored in the inductor is

$$W = \frac{1}{2}L\int_0^\infty \overline{i_n^2}df.\tag{D.4}$$

Substituting Equation (D.3) into Equation (D.4) gives

$$W = \frac{1}{4\pi R}\int_0^\infty \frac{\overline{v_n^2}Qd(f/f_0)}{1 + Q^2(f/f_0 - f_0/f)^2}.\tag{D.5}$$

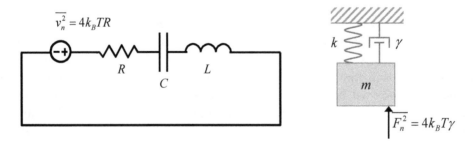

(a) Electrical resonator showing the resistor R, the inductance L, and the capacitance C and the noise voltage generator $\bar{v}_n$.

(b) Mechanical resonator represented with a mass m, a spring k, a dash-pot damper γ, and a thermal noise generator $\bar{F}_n$.

Figure D.1: Noise generators in second-order systems.

In order to evaluate the integral in Equation (D.5), we assume that the quality factor is high so that essentially all energy is confined near the resonance. Within the small frequency range around the resonance the voltage generator is approximately constant $\overline{v_n^2} \approx \overline{v_n^2(f_0)}$ giving

$$W = \frac{\overline{v_n^2(f_0)}}{4\pi R} \int_0^\infty \frac{Q\mathrm{d}(f/f_0)}{1 + Q^2(f/f_0 - f_0/f)^2}. \tag{D.6}$$

To evaluate the integral, we do a change of variables as $\frac{f}{f_0} = e^x$ to write the integral in Equation (D.6) as

$$Int = \int_{-\infty}^\infty \frac{Qe^x\mathrm{d}x}{4Q^2\sinh^2 x + 1} \tag{D.7}$$

The integral equation is unchanged if e^x is replaced with e^{-x} so the integral can be further rewritten as

$$Int = \frac{1}{2}\int_{-\infty}^\infty \frac{Q(e^x + e^{-x})\mathrm{d}x}{4Q^2\sinh^2 x + 1} = \int_{-\infty}^\infty \frac{Q\cosh x\mathrm{d}x}{4Q^2\sinh^2 x + 1} = \frac{1}{2}\int_{-\infty}^\infty \frac{\frac{1}{2}\mathrm{d}u}{u^2 + (\frac{1}{2})^2} = \frac{\pi}{2}. \tag{D.8}$$

From Equations (D.6) and (D.8), the energy stored in the inductor is

$$W = \frac{\overline{v_n^2(f_0)}}{4\pi R}Int = \frac{\overline{v_n^2(f_0)}}{8R}. \tag{D.9}$$

Equating (D.9) with the thermal energy $\frac{1}{2}k_B T$ gives

$$\overline{v_n^2(f_0)} = 4k_B T R. \tag{D.10}$$

Since the result does not depend on frequency f_0, it is valid for all frequencies. We have

$$\overline{v_n^2} = 4k_B T R, \tag{D.11}$$

which is the desired noise voltage generator for the thermal noise in resistors.

D.2 Derivation of mechanical noise force generator

We'll now proceed to the problem of noise in a mechanical resonator with the mass m, the spring k, and the damper γ shown in Figure D.1(b). We note that the system is fully described by two variables: velocity and position. Associated with the damping, there is a noise force generator $\overline{F}_n$. In what follows, we will derive the expression for the noise force generator. The analysis follows closely our steps for the voltage noise in the previous section: We will first integrate the total noise energy due to the noise generator. By equating this energy with the total thermal noise energy $\frac{1}{2}k_BT$ for each degree of freedom, we obtain the magnitude of $\overline{F}_n$.

The equation of motion for the system is

$$m\frac{\partial^2 x}{\partial t^2} + \gamma\frac{\partial x}{\partial t} + kx = F_n. \tag{D.12}$$

Equation (D.12) can be written in terms of velocity $\dot{x}$ by noting that $\dot{x} = \frac{\partial x}{\partial t} = sx$ giving

$$sm\dot{x} + \gamma\dot{x} + \frac{k}{s}\dot{x} = F_n. \tag{D.13}$$

From (D.13), the mean square velocity due to noise generator $\overline{F_n^2}$ is

$$\overline{\dot{x}_n^2} = \frac{\overline{F_n^2}}{\gamma^2 + (\omega m - k/\omega)^2}. \tag{D.14}$$

Equation (D.14) can be rewritten in terms of resonance frequency $\omega_0 = \sqrt{k/m}$ and quality factor $Q = \omega_0 m/\gamma$ as

$$\overline{\dot{x}_n^2} = \frac{1}{\gamma^2}\frac{\overline{F_n^2}}{1 + Q^2(\omega/\omega_0 - \omega_0/\omega)^2}. \tag{D.15}$$

From the integration over all frequencies, the average kinetic energy stored in the resonator is

$$W_k = \int_0^\infty \frac{1}{2}m\overline{\dot{x}_n^2}\,\mathrm{d}f = \frac{1}{4\pi\gamma}\int_0^\infty \frac{\overline{F_n^2}Q\mathrm{d}(f/f_0)}{1 + Q^2(f/f_0 - f_0/f)^2}, \tag{D.16}$$

which is the same as Equation (D.5) with the variable change $\overline{F_n^2} \rightarrow \overline{v_n^2}$ and $\gamma = R$. Following our previous analysis for Equation (D.5), Equation (D.16) gives

$$W_k = \frac{\overline{F_n^2}}{8\gamma}. \tag{D.17}$$

Equating this with the total kinetic energy due to the thermal energy $\frac{1}{2}k_BT$ gives the desired noise force generator

$$\overline{F_n^2} = 4k_BT\gamma. \tag{D.18}$$

The similarity to the electrical noise voltage generator $\overline{v_n^2} = 4k_BTR$ is evident.

E

Anisotropic elasticity of silicon

For an anisotropic material such as silicon, the Young's modulus, Poisson's ratio, and shear modulus depend on which crystal direction the material is being deformed. Looking at Figure E.1, this should be no surprise as the silicon crystal is highly structured. In some directions, the covalent bond density is higher and we expect silicon to be stiffer in those directions.

Figure E.1 is also a quick introduction to the crystallographic notation: Different directions are indicated with respect to crystal basis using Miller indexes. In a cubic crystal such as silicon the [100]-, [010]- and [001]-direction can be chosen to coincide with X, Y, and Z-axes. The Miller indexes can be thought as vectors. For example, [210] means two in [100]-direction and one in [010]-direction. Another example is shown in Figure E.1 where [101]-direction is shown. Also shown in the Figure E.1 is (100)-plane which is the plane orthogonal to the [100]-direction.

Here, we cover the calculation of the silicon Young's modulus, the Poisson's ratio, and the shear modulus in any crystalline direction. Our analysis follows reference [238] that derives the Young's modulus and Poisson's ratio from the elastic constants for silicon and germanium. The algebra is the same for any elastic material with the cubic symmetry and the formalism in this chapter can be applied to other semiconductors as well.

To account for the silicon anisotropy, we need to use the tensor formalism. The general relationship between the stress and the strain is

$$T_{ij} = \sum_k \sum_l C_{ijkl} S_{kl}, \qquad (E.1)$$

where T_{ij} is the stress, C_{ijkl} is the second-order stiffness tensor, and S_{kl} is the strain. As discussed in Chapter 4, subscripts ij refer to direction of stress and

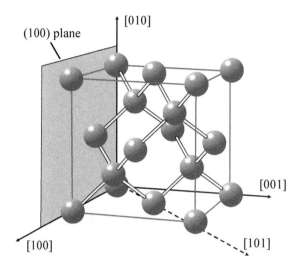

Figure E.1: Silicon crystal structure: Different crystal orientations are indicated with Miller indexes with [100] coinciding with X-axis. Also shown is (100)-plane (that is plane orthogonal to [100] direction) and crystal unit cell (bounding box).

the indexes k and l are looped through X, Y, and Z: For example, the normal stress T_{XX} stress can be written as

$$T_{XX} = C_{XXXX}S_{XX} + C_{XXXY}S_{XY} + C_{XXXZ}S_{XZ} + C_{XXYY}S_{YY}$$
$$+ C_{XXYX}S_{YX} + C_{XXYZ}S_{YZ} + C_{XXZZ}S_{ZZ} + C_{XXZX}S_{ZX} + C_{XXZY}S_{ZY}.$$

Equation (E.1) can be simplified with the short hand matrix notation that is convenient way to represent the stress-strain relationship in anisotropic materials. The notation takes use of symmetry relationships $S_{XY} = S_{YX}$, $S_{XZ} = S_{ZX}$, and $S_{YZ} = S_{ZY}$ between shear stresses to write $XX \rightarrow 1$, $YY \rightarrow 2$, $ZZ \rightarrow 3$, $ZY \rightarrow 4$, $ZX \rightarrow 5$, and $YZ \rightarrow 6$. For example C_{XXZY} then becomes C_{14} and S_{ZY} is simply S_4. With this notation, Equation (E.1) can be rewritten as

$$
\begin{bmatrix} T_1 \\ T_2 \\ T_3 \\ T_4 \\ T_5 \\ T_6 \end{bmatrix}
=
\begin{bmatrix}
C_{11} & C_{12} & C_{13} & C_{14} & C_{15} & C_{16} \\
C_{21} & C_{22} & C_{23} & C_{24} & C_{25} & C_{26} \\
C_{31} & C_{32} & C_{33} & C_{34} & C_{35} & C_{36} \\
C_{41} & C_{42} & C_{43} & C_{44} & C_{45} & C_{46} \\
C_{51} & C_{52} & C_{53} & C_{54} & C_{55} & C_{56} \\
C_{61} & C_{62} & C_{63} & C_{64} & C_{65} & C_{66}
\end{bmatrix}
\begin{bmatrix} S_1 \\ S_2 \\ S_3 \\ S_4 \\ S_5 \\ S_6 \end{bmatrix}.
\tag{E.2}
$$

The main advantage of the short hand notation is that we have gotten rid of the nasty tensor double sum in Equation (E.1) and many calculations can be solved with matrix algebra.

The silicon stiffness matrix in [100]-crystal axes is

$$[C] = \begin{bmatrix} 1.66 & 0.64 & 0.64 & 0 & 0 & 0 \\ 0.64 & 1.66 & 0.64 & 0 & 0 & 0 \\ 0.64 & 0.64 & 1.66 & 0 & 0 & 0 \\ 0 & 0 & 0 & 0.80 & 0 & 0 \\ 0 & 0 & 0 & 0 & 0.80 & 0 \\ 0 & 0 & 0 & 0 & 0 & 0.80 \end{bmatrix} \cdot 10^{11} \text{ Pa.} \qquad (E.3)$$

Equation (E.2) allows direct calculation of the Young's modulus. To obtain Young's modulus in [100] direction (or X-direction), we set all other stresses to zero and solve for $E_{[100]} = T_1/S_1$. This gives

$$E_{[100]} = C_{11} - 2\frac{C_{12}}{C_{11} + C_{12}}C_{12} = 130 \text{ GPa.} \qquad (E.4)$$

The Poisson's ratio can be obtained similarly as

$$\nu_{[100]} = \frac{C_{12}}{C_{11} + C_{12}} = 0.28. \qquad (E.5)$$

The expressions and values for Young's modulus to [110]- and [111]-direction are given in Table E.1.

Unfortunately as Table E.1 illustrates, calculations become complicated very quickly and a computerized method is desirable. Since the stiffness matrix is known, the Young's modulus can be numerically computed by taking inverse of $[C]$. This is called the compliance matrix $[C]^{-1}$. The compliance matrix can be used to calculate strains due to applied stress in desired directions as $[S] = [C]^{-1}[T]$.

From the compliance matrix, the Young's modulus in X is obtained by calculating the strain due to T_1 with all the other stresses set to zero. From $[S] = [C]^{-1}[T]$, the Young's modulus is then simply

$$E = \frac{T_1}{S_1} = 1/C_{11}^{-1}, \qquad (E.6)$$

Table E.1: Silicon Young's modulus in different crystalline directions

Direction	Expression	Value [GPa]
[100]	$C_{11} - 2\frac{C_{12}}{C_{11}+C_{12}}C_{12}$	130
[110]	$4\frac{(C_{11}^2+C_{12}C_{11}-2C_{12}^2)C_{44}}{2C_{44}C_{11}+C_{11}^2+C_{12}C_{11}-2C_{12}^2}$	170
[111]	$3\frac{C_{44}(C_{11}+2C_{12})}{C_{11}+2C_{12}+C_{44}}$	189

where C_{11}^{-1} is the $(1,1)$-component of the compliance matrix $[C]^{-1}$. Similarly, Poisson's ratio is

$$\nu_{ij} = -\frac{S_j}{S_i} = -\frac{C_{ij}^{-1}}{C_{ii}^{-1}}. \tag{E.7}$$

The compliance matrix is useful in other situations too. Figure E.2 shows 2D stretching of a plate. In this case $T_1 = T_2$ and the other stresses are zero. This gives $E_{2D} = T_1/S_1 = 181$ GPa. Interestingly this does not depend on axis orientation.

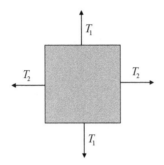

Figure E.2: 2D stretching of a silicon plate.

The Young's modulus in any direction can be obtained by calculating the stiffness matrix in rotated coordinates. An example of a coordinate axis rotation is shown in Figure E.3, where the coordinates are rotated 45° around the [001]-axis to obtain Young's modulus in [110]-direction. In calculating the rotated stiffness matrix, it has to be remembered that the stiffness matrix $[C]$ is based on a second-order tensor, and that the tensor rotation is slightly more complicated than for matrix rotation. The rotated $[C^R]$ can be found in literature [238] but due to complex algebra, it may be easier to calculate it numerically. The C^R-matrix is obtained from

$$C_{ijkl}^R = \sum_p \sum_q \sum_r \sum_s Q_{pi} Q_{qj} Q_{rk} Q_{sl} C_{pqrs}, \tag{E.8}$$

where $[Q]$ and the indexes p, q, r, s are looped through X, Y, and Z. Calculating the C^R-matrix is tedious for hand calculation but a breeze for a computer. Once stiffness matrix is known in the new coordinates, the Young's modulus is obtained from Equation (E.6) as before.

This is all that is needed to calculate the silicon Young's modulus and Poisson's ratio in any direction. As an example, they are shown in (100)-plane in Figure E.4. The Matlab-script used in calculating the images can be downloaded from `http://www.kaajakari.net/PracticalMEMS`. The script can be adapted to calculate the Young's modulus in other planes. The (111)-plane gives an interesting result of constant Young's modulus and Poisson's ratio. This makes

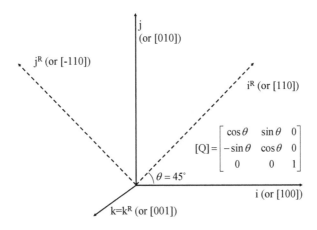

Figure E.3: Axis rotation using rotation matrix $[Q]$. The new axes is obtained with $[x^R] = [Q][x]$

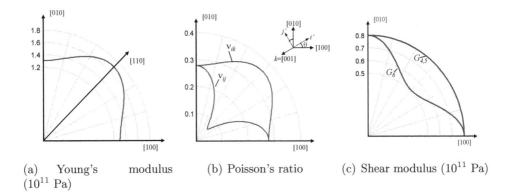

(a)   Young's   modulus
$(10^{11}$ Pa$)$

(b) Poisson's ratio

(c) Shear modulus $(10^{11}$ Pa$)$

Figure E.4: Calculated silicon Young's modulus and Poisson's ratio in (100)-plane.

(111) wafers an interesting material for micromachining as the spring constant does not depend on the orientation in the (111)-plane [37].

F

Anisotropic piezoresistivity of silicon

The voltage-current relationship in silicon with no applied stress is isotropic and electric field $\mathcal{E}$ is related to the current density J by $\mathcal{E} = \rho J$, where ρ is the resistivity. The change in electric field due to a change in resistivity in an isotropic material such as polycrystalline silicon is then simply $\delta \mathcal{E} = \delta \rho J$ or

$$\delta \mathcal{E} / \rho = \Delta J, \tag{F.1}$$

where we have defined the fractional change in resistivity as $\Delta = \delta \rho / \rho$.

For single-crystal silicon under stress, the $\mathcal{E} - J$-relationship in is anisotropic, and Equation (F.1) has to be written as

$$\begin{bmatrix} \delta \mathcal{E}_1 / \rho \\ \delta \mathcal{E}_2 / \rho \\ \delta \mathcal{E}_3 / \rho \end{bmatrix} = \begin{bmatrix} \Delta_1 & \Delta_6 & \Delta_5 \\ \Delta_6 & \Delta_2 & \Delta_4 \\ \Delta_5 & \Delta_4 & \Delta_3 \end{bmatrix} \begin{bmatrix} J_1 \\ J_2 \\ J_3 \end{bmatrix}, \tag{F.2}$$

where the short hand notation of Appendix E has been used. The six coefficients Δ_1, Δ_2, Δ_3, Δ_4, Δ_5, and Δ_6 are related to stress T by a matrix equation

$$\begin{bmatrix} \Delta_1 \\ \Delta_2 \\ \Delta_3 \\ \Delta_4 \\ \Delta_5 \\ \Delta_6 \end{bmatrix} = \begin{bmatrix} \pi_{11} & \pi_{12} & \pi_{13} & \pi_{14} & \pi_{15} & \pi_{16} \\ \pi_{21} & \pi_{22} & \pi_{23} & \pi_{24} & \pi_{25} & \pi_{26} \\ \pi_{31} & \pi_{32} & \pi_{33} & \pi_{34} & \pi_{35} & \pi_{36} \\ \pi_{41} & \pi_{42} & \pi_{43} & \pi_{44} & \pi_{45} & \pi_{46} \\ \pi_{51} & \pi_{52} & \pi_{53} & \pi_{54} & \pi_{55} & \pi_{56} \\ \pi_{61} & \pi_{62} & \pi_{63} & \pi_{64} & \pi_{65} & \pi_{66} \end{bmatrix} \begin{bmatrix} T_1 \\ T_2 \\ T_3 \\ T_4 \\ T_5 \\ T_6 \end{bmatrix}, \tag{F.3}$$

where π:s are the piezoresistive coefficients.

For silicon, there are three independent piezoresistive coefficients. If cubic axes [100], [010], and [001] are chosen as reference, Equation (F.3) simplifies to

$$
\begin{bmatrix} \Delta_1 \\ \Delta_2 \\ \Delta_3 \\ \Delta_4 \\ \Delta_5 \\ \Delta_6 \end{bmatrix} = \begin{bmatrix} \pi_{11} & \pi_{12} & \pi_{12} & 0 & 0 & 0 \\ \pi_{12} & \pi_{11} & \pi_{12} & 0 & 0 & 0 \\ \pi_{12} & \pi_{12} & \pi_{11} & 0 & 0 & 0 \\ 0 & 0 & 0 & \pi_{44} & 0 & 0 \\ 0 & 0 & 0 & 0 & \pi_{44} & 0 \\ 0 & 0 & 0 & 0 & 0 & \pi_{44} \end{bmatrix} \begin{bmatrix} T_1 \\ T_2 \\ T_3 \\ T_4 \\ T_5 \\ T_6 \end{bmatrix}. \tag{F.4}
$$

Table F.1 gives the three independent piezoresistive coefficients for p- and n-type silicon with low doping concentrations.

From Equation (F.4) and Table F.1, it is possible to calculate the piezoresistance for an arbitrary orientation. As an example, longitudinal piezoresistive coefficient(stress in the direction of the current) in [100]-direction is obtained from Equation (F.4) by setting $T_2 = T_3 = T_4 = T_5 = T_6 = 0$ to give $\Delta_1 = \pi_{11}T_1 \equiv \pi_l T_1$. The transverse piezoresistive coefficient (stress in orthogonal to the current) is obtained by setting $T_1 = T_3 = T_4 = T_5 = T_6 = 0$ to give $\Delta_1 = \pi_{12}T_2 \equiv \pi_t T_t$.

The piezoresistivity in any direction can be obtained by calculating the π-matrix in the rotated coordinates. In calculating the rotated piezoresistive coefficient matrix, it has the be remembered that π-matrix is based on a second-order tensor and the rotation is identical to stiffness tensor rotation in Appendix E. The rotated π^R-matrix is obtained from

$$
\pi^R_{ijkl} = \sum_p \sum_q \sum_r \sum_s Q_{pi}Q_{qj}Q_{rk}Q_{sl}\pi_{pqrs}, \tag{F.5}
$$

where $[Q]$ is the rotation matrix from Appendix E. Equation (F.5) has been used to plot Figure 5.1 in Chapter 5. You may download the Matlab-script used for plotting Figure 5.1 from http://www.kaajakari.net/PracticalMEMS.

Table F.1: Silicon piezoresistive coefficients at room temperature in units of 10^{-11} Pa^{-1} at low doping concentrations ($< 10^{17}$ cm^{-3}) [40].

	π_{11}	π_{12}	$2\pi_{44}$
Si(p)	6.6	−1.1	138.1
Si(n)	−102.2	53.4	−13.6

G

Constitutive equations for piezoelectric materials

The piezoelectric materials are characterized by the constitutive equations for the stress and electric displacement:

$$[T] = [C][S] - [e][\mathcal{E}] \tag{G.1}$$

$$[D] = [\epsilon][\mathcal{E}] + [e][S]. \tag{G.2}$$

Here $[T]$ is the stress vector, $[C]$ is the stiffness matrix, $[S]$ is the strain vector, $[e]$ is the piezoelectric coefficient matrix, $[D]$ is the electric displacement vector, $[\epsilon]$ is the permittivity matrix, and $[\mathcal{E}]$ is the electric field vector. In expanded form Equations (G.1) and (G.2) are

$$
\begin{bmatrix} T_1 \\ T_2 \\ T_3 \\ T_4 \\ T_5 \\ T_6 \end{bmatrix}
=
\begin{bmatrix}
C_{11} & C_{12} & C_{13} & C_{14} & C_{15} & C_{16} \\
C_{21} & C_{22} & C_{23} & C_{24} & C_{25} & C_{26} \\
C_{31} & C_{32} & C_{33} & C_{34} & C_{35} & C_{36} \\
C_{41} & C_{42} & C_{43} & C_{44} & C_{45} & C_{46} \\
C_{51} & C_{52} & C_{53} & C_{54} & C_{55} & C_{56} \\
C_{61} & C_{62} & C_{63} & C_{64} & C_{65} & C_{66}
\end{bmatrix}
\begin{bmatrix} S_1 \\ S_2 \\ S_3 \\ S_4 \\ S_5 \\ S_6 \end{bmatrix}
+
\begin{bmatrix}
e_{11} & e_{21} & e_{31} \\
e_{12} & e_{22} & e_{32} \\
e_{13} & e_{23} & e_{33} \\
e_{14} & e_{24} & e_{34} \\
e_{15} & e_{25} & e_{35} \\
e_{16} & e_{26} & e_{36}
\end{bmatrix}
\begin{bmatrix} \mathcal{E}_1 \\ \mathcal{E}_2 \\ \mathcal{E}_3 \end{bmatrix}
\tag{G.3}
$$

and

$$
\begin{bmatrix} D_1 \\ D_2 \\ D_3 \end{bmatrix}
=
\begin{bmatrix}
e_{11} & e_{12} & e_{13} & e_{14} & e_{15} & e_{16} \\
e_{21} & e_{22} & e_{23} & e_{24} & e_{25} & e_{26} \\
e_{31} & e_{32} & e_{33} & e_{34} & e_{35} & e_{36}
\end{bmatrix}
\begin{bmatrix} T_1 \\ T_2 \\ T_3 \\ T_4 \\ T_5 \\ T_6 \end{bmatrix}
+
\begin{bmatrix}
\epsilon_{11} & \epsilon_{12} & \epsilon_{13} \\
\epsilon_{21} & \epsilon_{22} & \epsilon_{23} \\
\epsilon_{31} & \epsilon_{32} & \epsilon_{33}
\end{bmatrix}
\begin{bmatrix} \mathcal{E}_1 \\ \mathcal{E}_2 \\ \mathcal{E}_3 \end{bmatrix}.
\tag{G.4}
$$

Equations (G.1) and (G.2) can also be written as

$$[S] = [C]^{-1}[T] + [d][\mathcal{E}] \tag{G.5}$$
$$[D] = [\epsilon^T][\mathcal{E}] + [d][T] \tag{G.6}$$

where $[C]^{-1}$ is the compliance matrix (inverse of $[C]$) and

$$[d] = [e][C]^{-1} \tag{G.7}$$
$$[\epsilon^T] = [\epsilon] + [e][C]^{-1}[e]' \tag{G.8}$$

are the piezoelectric constant and permittivity at constant stress [59].

The piezoelectric matrix equations are rather cumbersome to work with. The spirit of this book is to obtain simple approximate expressions and leave the accurate analysis for numerical simulations. In this spririt, we seek to simplify the matrix equations in the form of

$$T_3 = ES_3 - e_{33}\mathcal{E}_3 \tag{G.9}$$
$$D_3 = \epsilon_{33}\mathcal{E}_3 + e_{33}S_3 \tag{G.10}$$

for longitudinal actuator (see Equations (16.2) and (7.3) on pages 242 and 111) and in the form of

$$T_1 = ES_1 - e_{31}\mathcal{E}_3 \tag{G.11}$$
$$D_3 = \epsilon_{33}\mathcal{E}_3 + e_{31}S_1 \tag{G.12}$$

for the transverse actuator (see Equations (16.7) and (7.10) on pages 246 and 113).

Figure G.1(a) shows the simplest longitudinal actuator. The only non-zero strain and electric field components are S_3 and $\mathcal{E}_3$, respectively. These boundary conditions would be a realistic approximation for example for a thin film that is mounted on a substrate. The film is free to deform in the thickness direction but is prevented from deforming laterally. Using $[S] = [0 \; 0 \; S_3 \; 0 \; 0 \; 0]'$ and $[\mathcal{E}] = [0 \; 0 \; \mathcal{E}_3]'$ in Equations (G.1) and (G.2) gives

$$T_3 = C_{33}S_3 - e_{33}\mathcal{E}_3 \tag{G.13}$$
$$D_3 = \epsilon_{33}\mathcal{E}_3 + e_{33}S_3. \tag{G.14}$$

which are the desired Equations (G.9) and (G.10) for the longitudinal transducer if $E = C_{33}$.

Figure G.1(b) shows the boundary conditions for a longitudinal actuator that is allowed to freely deform laterally ($T_1 = T_2 = 0$). These boundary conditions are realistic for thick ceramic actuators used in bulk micromachined devices. As the lateral strains are not known but the stresses are, we take

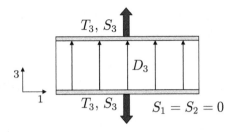

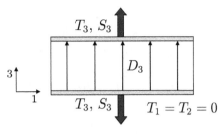

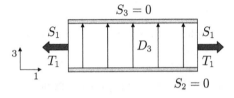

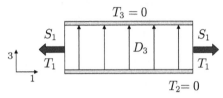

(a) Longitudinal configuration with no lateral deformation.

(b) Longitudinal configuration with no lateral stresses.

(c) Transverse configuration with no strain in X_3-direction.

(d) Transverse configuration with no stress in X_3-direction.

Figure G.1: Approximate boundary conditions for piezoelectric transducer.

Equations (G.5) and (G.6) as our starting point. Using $[T] = [0 \; 0 \; T_3 \; 0 \; 0 \; 0]'$ and $[\mathcal{E}] = [0 \; 0 \; \mathcal{E}_3]'$ gives

$$S_3 = C_{33}^{-1}T_3 + d_{33}\mathcal{E}_3 \Leftrightarrow T_3 = \frac{1}{C_{33}^{-1}}S_3 - \frac{d_{33}}{C_{33}^{-1}}\mathcal{E}_3 \tag{G.15}$$

$$D_3 = \epsilon_{33}^T\mathcal{E} + d_{33}T_3 \Leftrightarrow D_3 = \left(\epsilon_{33}^T - \frac{d_{33}^2}{C_{33}^{-1}}\right)\mathcal{E}_3 + \frac{d_{33}}{C_{33}^{-1}}S_3 \tag{G.16}$$

which are the desired Equations (G.9) and (G.10) if $E = \frac{1}{C_{33}^{-1}}$ and we define effective piezoelectric coupling as $e_{33}^{\text{eff}} = \frac{d_{33}}{C_{33}^{-1}}$ and effective permittivity as $\epsilon_{33}^{\text{eff}} = \epsilon_{33}^T - \frac{d_{33}^2}{C_{33}^{-1}}$.

Figure G.1(c) shows the simplest transverse transducer. The only non-zero strain and electric field components are S_1 and $\mathcal{E}_3$, respectively. These boundary conditions may not be very realistic but the low analysis effort makes them useful for approximate analysis. Using $[S] = [S_1 \; 0 \; 0 \; 0 \; 0 \; 0]'$ and $[\mathcal{E}] = [0 \; 0 \; \mathcal{E}_3]'$ in Equations (G.1) and (G.2) gives

$$T_1 = C_{11}S_1 - e_{13}\mathcal{E}_3 \tag{G.17}$$

$$D_3 = \epsilon_{33}\mathcal{E}_3 + e_{13}S_1. \tag{G.18}$$

which are the desired Equations (G.11) and (G.12) for the transverse transducer if $E = C_{11}$.

A more realistic transducer is shown in Figure G.1(d). The boundary conditions allow free deformation in X_2- and X_3-directions ($T_2 = T_3 = 0$). Using $[T] = [T_1\ 0\ 0\ 0\ 0\ 0]'$ and $[\mathcal{E}] = [0\ 0\ \mathcal{E}_3]'$ gives

$$S_1 = C_{11}^{-1}T_1 + d_{13}\mathcal{E}_3 \Leftrightarrow T_1 = \frac{1}{C_{11}^{-1}}S_1 - \frac{d_{13}}{C_{11}^{-1}}\mathcal{E}_3 \tag{G.19}$$

$$D_3 = \epsilon_{33}^T\mathcal{E}_3 + d_{13}T_1 \Leftrightarrow D_3 = \left(\epsilon_{33}^T - \frac{d_{13}^2}{C_{11}^{-1}}\right)\mathcal{E}_3 + \frac{d_{13}}{C_{11}^{-1}}S_1 \tag{G.20}$$

where we have used Equation (G.19) in Equation (G.20). Equations (G.19) and (G.20) are the desired Equations (G.11) and (G.12) for the transverse transducer if $E = \frac{1}{C_{11}^{-1}}$ and we define the effective piezoelectric coupling as $e_{13}^{\text{eff}} = \frac{d_{13}}{C_{11}^{-1}}$ and the effective permittivity as $\epsilon_{33}^{\text{eff}} = \epsilon_{33}^T - \frac{d_{13}^2}{C_{11}^{-1}}$.

Example G.1 illustrates the calculation of the effective piezoelectric coefficients for aluminum nitride. Depending on the boundary conditions, the effective coupling coefficients e_{33}^{eff} and e_{13}^{eff} for the longitudinal and transverse transducer are seen to vary by 20%. This variation is rather small as other effects such as process variations and electrode properties can have an equal or even bigger effect. This justifies our approximate approach in Chapters 7 and 16 where we just assumed $e_{33}^{\text{eff}} = e_{33}$ and $e_{13}^{\text{eff}} = e_{13}$ regardless of the exact boundary condotions.

Example G.1: Effective piezoelectric coupling for AlN
The material constants for AlN are [239]:

$$[C] = \begin{bmatrix} 3.45 & 1.25 & 1.20 & 0 & 0 & 0 \\ 1.25 & 3.45 & 1.20 & 0 & 0 & 0 \\ 1.20 & 1.20 & 3.95 & 0 & 0 & 0 \\ 0 & 0 & 0 & 1.18 & 0 & 0 \\ 0 & 0 & 0 & 0 & 1.18 & 0 \\ 0 & 0 & 0 & 0 & 0 & 1.10 \end{bmatrix} \cdot 10^{11}\ \text{N/m}^2$$

$$[e] = \begin{bmatrix} 0 & 0 & 0 & 0 & -0.48 & 0 \\ 0 & 0 & 0 & -0.48 & 0 & 0 \\ -0.58 & -0.58 & 1.55 & 0 & 0 & 0 \end{bmatrix}\ \text{C/m}^2$$

$$[\epsilon] = \begin{bmatrix} 8.0 & 0 & 0 \\ 0 & 8.0 & 0 \\ 0 & 0 & 9.5 \end{bmatrix} \cdot 10^{-11}\ \text{F/m}$$

Calculate the effective modulus E, the effective piezoelectric coupling cofficient and the effective permittivity for the four transducers in Figure G.1

Solution: The complience matrix is

$$[C]^{-1} = \begin{bmatrix} 0.35 & -0.10 & -0.08 & 0 & 0 & 0 \\ -0.10 & 0.35 & -0.08 & 0 & 0 & 0 \\ -0.08 & -0.08 & 0.30 & 0 & 0 & 0 \\ 0 & 0 & 0 & 0.85 & 0 & 0 \\ 0 & 0 & 0 & 0 & 0.85 & 0 \\ 0 & 0 & 0 & 0 & 0 & 0.91 \end{bmatrix} \cdot 10^{-11} \text{ m}^2/\text{N},$$

the d-matrix is

$$[d] = [e][C]^{-1} = \begin{bmatrix} 0 & 0 & 0 & 0 & -0.41 & 0 \\ 0 & 0 & 0 & -0.41 & 0 & 0 \\ -0.26 & -0.26 & 0.55 & 0 & 0 & 0 \end{bmatrix} \cdot 10^{-11} \text{ C/N},$$

and the permittivity matrix at constant stress is

$$[\epsilon^T] = [\epsilon] + [e][C]^{-1}[e]' = \begin{bmatrix} 8.2 & 0 & 0 \\ 0 & 8.2 & 0 \\ 0 & 0 & 10.6 \end{bmatrix} \cdot 10^{-11} \text{ F/m}.$$

For Figure G.1(a) we have

$$E = C_{33} = 3.95 \cdot 10^{11} \text{ N/m}^2,$$

$$\epsilon_{33} = 9.5 \cdot 10^{-11} \text{ F/m},$$

and

$$e_{33} = 1.55 \text{ C/m}^2.$$

For Figure G.1(b) we have

$$E = \frac{1}{C_{33}^{-1}} = 3.34 \cdot 10^{11} \text{ N/m}^2,$$

$$\epsilon_{33}^{\text{eff}} = 9.6 \cdot 10^{-11} \text{ F/m},$$

and

$$e_{33}^{\text{eff}} = \frac{d_{33}}{C_{33}^{-1}} = 1.85 \text{ C/m}^2.$$

For Figure G.1(c) we have

$$E = C_{11} = 3.45 \cdot 10^{11} \text{ N/m}^2,$$

$$\epsilon_{33}^{\text{eff}} = \epsilon_{33} = 9.5 \cdot 10^{-11} \text{ F/m},$$

and
$$e_{13}^{\text{eff}} = e_{13} = -0.58 \text{ C/m}^2.$$

For Figure G.1(d) we have

$$E = \frac{1}{C_{11}^{-1}} = 2.83 \cdot 10^{11} \text{ N/m}^2,$$

$$\epsilon_{33}^{\text{eff}} = \epsilon_{33}^{T} - \frac{d_{13}^2}{C_{11}^{-1}} = 1.05 \cdot 10^{-11} \text{ F/m},$$

and

$$e_{13}^{\text{eff}} = \frac{d_{13}}{C_{11}^{-1}} = -0.75 \text{ C/m}^2.$$

H

Often used constants and formulas

H.1 Constants

Table H.1: Physical constants.

Symbol	Name	Value
q_e	Electron charge	$1.6 \cdot 10^{-19}$ C
ϵ_0	Permittivity of free space	$8.85 \cdot 10^{-12}$ F/m
k_B	Boltzmann constant	$1.38 \cdot 10^{-23}$ J/K
$k_B T$	Thermal energy	$4.14 \cdot 10^{-21}$ J $(T = 300$ K$)$
$k_B T / q_e$	Thermal voltage	25.9 mV $(T = 300$ K$)$
N_A	Avogadro constant	$6.022 \cdot 10^{23}$ mol^{-1}
R	Gas constant	8.31 J mol^{-1} K^{-1}

Table H.2: Conversion factors.

1 atm = 760 Torr (exactly)
1 atm = $1.013 \cdot 10^5$ Pa
1 eV = $1.6 \cdot 10^{-19}$ J
T[K] = T[°C]+273.15

H.2 Decibel (dB) units

The decibel units are used to indicate magnitudes of the *relative powers*. For example, if signal power is $P_s = 1$ W and the noise power is $P_n = 1$ mW, the signal to noise ratio $SNR = P_s/Pn = 1,000$ in the decibel scale is

$$SNR_{\text{dB}} = 10\log_{10} P_s/P_n = 30 \text{ dB}. \tag{H.1}$$

To convert back to linear scale, Equation (H.1) is inverted to give

$$SNR = 10^{SNR_{\text{dB}}/10} = 1,000. \tag{H.2}$$

To use voltages instead of powers, we can substitute $P_s = 0.5V_s^2/R$ and $P_n = 0.5V_n^2/R$ to Equation (H.1) to give

$$SNR_{\text{dB}} = 10\log_{10} V_s^2/V_n^2 = 20\log_{10} V_s/V_n \tag{H.3}$$

where the arbitrary R cancels.

Sometimes absolute powers are used and the reference scale is indicated in the dB units. For example, dBm means that the power is referred to mW and $P_{\text{dBm}} = 23$ dBm equals

$$P = 10^{P_{\text{dBm}}/10} \cdot 1 \text{ mW} = 200 \text{ mW}.$$

H.3   Mode shapes and resonant frequencies for beams with different boundary conditions

The resonant frequency for a resonating beam is

$$\omega_0 = (\beta_n)^2 \sqrt{\frac{EI}{\rho AL^4}},$$

where β_n is the constant that depends on the resonator boundary conditions and mode number n, E is the Young's modulus, I is the second moment of inertia, and A is the cross-sectional area. Table H.3 gives the mode shape and β_n for several common boundary conditions. For example, for a beam with the width b and the height a, the second moment of inertia is $I = ab^3/12$ and the cross section area is $A = ab$. The lowest resonant frequency for a cantilever beam is $f_0 = \frac{\beta_n}{2\pi\sqrt{12}}\sqrt{\frac{E}{\rho}\frac{b}{L^2}} \approx 0.1615\sqrt{\frac{E}{\rho}\frac{b}{L^2}}$.

H.4   Second moment of inertias

Table H.4 shows the second moment of inertias about the central axis (the stress free plane in bending) for several possible MEMS beam cross sections.

Table H.3: Common boundary conditions for transverse beam vibrations.

End conditions	Mode shape	Value of β_n
Pinned-pinned 	$u_n(x) = C_n \sin \beta_n x/L$	$\beta_1 = \pi$ $\beta_2 = 2\pi$ $\beta_3 = 3\pi$ $\beta_4 = 4\pi$
Free-free 	$u_n(x) = C_n[\sin \beta_n x/L + \sinh \beta_n x/L$ $+\alpha_n(\cos \beta_n x/L + \cosh \beta_n x/L)]$ where $\alpha_n = \frac{\sin \beta_n - \sinh \beta_n}{\cosh \beta_n - \cos \beta_n}$	$\beta_1 = 4.730041$ $\beta_2 = 7.853205$ $\beta_3 = 10.995608$ $\beta_4 = 14.137165$
Fixed-fixed 	$u_n(x) = C_n[\sinh \beta_n x/L - \sin \beta_n x/L$ $+\alpha_n(\cosh \beta_n x/L - \cos \beta_n x/L)]$ where $\alpha_n = \frac{\sinh \beta_n - \sin \beta_n}{\cos \beta_n - \cosh \beta_n}$	$\beta_1 = 4.730041$ $\beta_2 = 7.853205$ $\beta_3 = 10.995608$ $\beta_4 = 14.137165$
Fixed-free 	$u_n(x) = C_n[\sin \beta_n x/L - \sinh \beta_n x/L$ $-\alpha_n(\cos \beta_n x/L - \cosh \beta_n x/L)]$ where $\alpha_n = \frac{\sin \beta_n + \sinh \beta_n}{\cos \beta_n + \cosh \beta_n}$	$\beta_1 = 1.875104$ $\beta_2 = 4.694091$ $\beta_3 = 7.854757$ $\beta_4 = 10.995541$

Table H.4: Second moment of inertias.

Form of section	Second moment of inertia

Circle

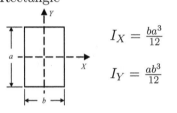

$$I_X = I_Y = \tfrac{\pi}{4} r^2$$

Rectangle

$$I_X = \tfrac{ba^3}{12}$$

$$I_Y = \tfrac{ab^3}{12}$$

Parallelogram

$$I_X = \tfrac{ba^3}{12}$$

$$I_Y = \tfrac{1}{12} ab(b^2 + c^2)$$

Trapezoid

$$I_X = \tfrac{a^3}{36} \tfrac{b_2^2 + 4b_1 b_2 + b_2^2}{b_1 + b_2}$$

$$I_Y = \tfrac{1}{48}(b_1 + b_2) \cdot (b_1^2 + b_2^2)a$$

Bibliography

[1] H. C. Nathansona, W. E. Newell, R. A. Wickstrom, and J. R. Davis, "The resonant gate transistor," *IEEE Trans. Electron Devices*, vol. 14, no. 2, pp. 117–133, Mar. 1967.

[2] K. Petersen, "Silicon as a mechanical material," *Proc. IEEE*, vol. 70, no. 5, pp. 420–457, May 1982.

[3] J. Bryzek, S. Roundy, B. Bircumshaw, C. Chung, K. Castellino, J. Stetter, and M. Vestel, "Marvelous MEMS," *IEEE Circuits Devices Mag.*, vol. 22, no. 2, pp. 8–28, Mar. 2006.

[4] S. Samaun, K. Wise, and J. Angell, "An IC piezoresistive pressure sensor for biomedical instrumentation," *IEEE Trans. Biomed. Eng.*, vol. 20, no. 2, pp. 101–109, Mar. 1973.

[5] C. Liu, *Foundations of MEMS*. New Jersey, NJ: Prentice Hall, 2005.

[6] S. Senturia, *Microsystem Design*. New York, NY: Springer, 2001.

[7] S. Franssila, *Introduction to Microfabrication*. Hoboken, NJ: John Wiley and Sons, 2004.

[8] M. Gad-el Hak, *MEMS: Introduction and Fundamentals*. Boca Raton, FL: CRC, 2006.

[9] G. Kovacs, *Micromachined Transducers Sourcebook*. Boston, MA: McGraw-Hill, 1998.

[10] M. Madou, *Fundamentals of Microfabrication*, 2nd ed. New York, NY: CRC Press, 2002.

[11] J. M. Bustillo, R. T. Howe, and R. S. Muller, "Surface micromachining for microelectromechanical systems," *Proc. IEEE*, vol. 86, no. 8, pp. 1552–1574, Aug. 1998.

[12] G. T. A. Kovacs, N. I. Maluf, and K. E. Petersen, "Bulk micromachining of silicon," *Proc. IEEE*, vol. 86, no. 8, pp. 1536–1551, Aug. 1998.

[13] W. R. Runyan and K. Bean, *Semiconductor Integrated Circuit Processing Technology*. New York, NY: Addison-Wesley, 1990.

[14] J. D. Plummer, M. D. Deal, and P. B. Griffin, *Silicon VLSI Technology: Fundamentals, Practice, and Modeling*. Upper Saddle River, NJ: Prentice Hall, 2000.

[15] M. S. Rodgers, J. J. Sniegowski, S. L. Miller, C. C. Barron, and P. J. McWhorter, "Advanced micromechanisms in a multi-level polysilicon technology," in *Proc. of SPIE*, vol. 3224, Austin,TX, Sep. 1997, pp. 120–130.

[16] K. E. Bean, "Anisotropic etching of silicon," *IEEE Trans. Electron Devices*, vol. 25, no. 10, pp. 1185–1193, Oct. 1978.

[17] F. Ayazi and K. Najafi, "High aspect-ratio combined poly and single-crystal silicon (HARPSS) MEMS technology," *J. Microelectromech. Syst.*, vol. 9, no. 3, pp. 288–294, Sep. 2000.

[18] M. Schmidt, "Wafer-to-wafer bonding for microstructure formation," *Proc. IEEE*, vol. 86, no. 8, pp. 1575–1585, Aug. 1998.

[19] T. J. Brosnihan, J. M. Bustillo, A. P. Pisano, and R. T. Howe, "Embedded interconnect and electrical isolation for high-aspect-ratio, SOI inertial instruments," in *Transducers'97. The 9th International Conference on Solid-State Sensors and Actuators*, Chicago, IL, 16-19 Jun. 1997, pp. 637–640.

[20] T. Gabrielson, "Mechanical-thermal noise in micromachined acoustic and vibration sensors," *IEEE Trans. Electron Devices*, vol. 40, no. 5, pp. 903–909, May 1993.

[21] A. N. Cleland and M. L. Roukes, "Noise processes in nanomechanical resonators," *J. App. Phys.*, vol. 92, no. 5, pp. 2758–2769, 2002.

[22] J. Gagnepain, J. Uebersfeld, G. Goujon, and P. Handel, "Relation between $1/f$ noise and Q-factor in quartz resonators at room and low temperatures, first theoretical interpretation," in *IEEE Annual Frequency Control Symposium*, Philadelphia, PA, 27-29 May 1981, pp. 476–483.

[23] M. Driscoll and W. Hanson, "Measured vs. volume model-predicted flicker-of-frequency instability in VHF quartz crystal resonators," in *IEEE International Frequency Control Symposium*, Salt Lake City, UT, 2-4 Jun. 1993, pp. 186–192.

[24] H. Nyquist, "Thermal agitation of electric charge in conductors," *Phys. Rev.*, vol. 32, pp. 110–113, Jul. 1928.

[25] S. W. Smith, *The Scientist and Engineer's Guide to Digital Signal Processing*, 2nd ed.   San Diego, California: California Technical Publishing, 1999.

[26] M. S. Keshner, "$1/f$ noise," *Proc. IEEE*, vol. 70, no. 3, pp. 212–218, Mar. 1982.

[27] S. Butefisch, A. Schoft, and S. Buttgenbach, "Three-axes monolithic silicon low-g accelerometer," *J. Microelectromech. Syst.*, vol. 9, no. 4, pp. 551–556, Dec. 2000.

[28] N. Yazdi, F. Ayazi, and K. Najafi, "Micromachined inertial sensors," *Proc. IEEE*, vol. 86, no. 8, pp. 1640–1659, Aug. 1998.

[29] L. Roylance and J. Angell, "A batch-fabricated silicon accelerometer," *IEEE Trans. Electron Devices*, vol. 26, no. 12, pp. 1911–1917, Dec. 1979.

[30] "ADXL50 monolithic accelerometer with signal conditioning," Analog Devices, One Technology Way, P.O. Box 9106, Norwood, MA 02062, USA, 1996. [Online]. Available: www.analog.com

[31] K.-L. Chau, S. Lewis, Y. Zhao, R. Howe, S. Bart, and R. Marcheselli, "An integrated force-balanced capacitive accelerometer for low-g applications," in *Transducers'95. The 8th International Conference on Solid-State Sensors and Actuators and Eurosensors IX*, Stockholm, Sweden, 25-29 Jun. 1995, pp. 593–596.

[32] S. Sherman, W. Tsang, T. Core, R. Payne, D. Quinn, K.-L. Chau, J. Farash, and S. Baum, "A low cost monolithic accelerometer; product/technology update," in *IEDM'92 Proceedings of the 1992 IEEE International Conference on Solid-State Circuits*, San Francisco, CA, Dec. 1992, pp. 501–504.

[33] "SCA620 series accelerometer," VTI Technologies, Myllynkivenkuja, P.O. Box 27, Vantaa, 01621, FINLAND, 2006.

[34] W. Young and R. Budynas, *Roark's Formulas for Stress and Strain*, 7th ed.   McGraw-Hill, 2002.

[35] J. W. Weaver, S. Timoshenko, and D. Young, *Vibration Problems in Engineering*, 5th ed.   New York, NY: Wiley, 1990.

[36] A. Selvakumar and K. Najafi, "A high-sensitivity z-axis capacitive silicon microaccelerometer with a torsional suspension," *J. Microelectromech. Syst.*, vol. 7, no. 2, pp. 192–200, Jun. 1998.

[37] J. Kim, D. Cho, and R. Muller, "Why is (111) silicon a better material for MEMS?" in *Transducers'03. The 12th International Conference on Solid-State Sensors, Actuators and Microsystems*, Boston, MA, 8-12 Jun. 2001, pp. 662–665.

[38] J. Cole, "A new sense element technology for accelerometer subsystems," in *Transducers'91. The 6th International Conference on Solid-State Sensors and Actuators*, San Francisco, CA, Jun. 24-27 1991, pp. 93–96.

[39] M. Chaparala and B. Shivaram, *Properties of Crystalline Silicon.* London, UK: Institution of Engineering and Technology, 1999, ch. 8.2 Piezoresistance of c-Si, pp. 421–429.

[40] C. S. Smith, "Piezoresistance effect in germanium and silicon," *Phys. Rev.*, vol. 94, no. 1, pp. 42–49, Apr. 1954.

[41] B. W. Chui, T. D. Stowe, T. W. Kenny, H. J. Mamin, B. D. Terris, and D. Rusgar, "Low-stiffness silicon cantilevers for thermal writing and piezoresistive readback with the atomic force microscope," *Appl. Phys. Lett.*, vol. 69, p. 2767, Oct. 1996.

[42] Y. Kanda, "A graphical representation of the piezoresistance coefficients in silicon," *IEEE Trans. Electron Devices*, vol. 29, no. 1, pp. 64–70, Jan. 1982.

[43] S. S. Li and W. R. Thurber, "The dopant density and temperature dependence of electron mobility and resistivity in n-type silicon," *Solid State Elect.*, vol. 20, no. 7, pp. 609–616, Jul. 1977.

[44] S. S. Li, "The dopant density and temperature dependence of hole mobility and resistivity in boron doped silicon," *Solid State Elect.*, vol. 21, no. 9, pp. 1109–1117, Sep. 1978.

[45] R. Pallás-Areny and J. G. Webster, *Sensors and Signal Conditioning*, 2nd ed. New York: John Wiley & Sons, 2001.

[46] W. Schottky, "Über spontane stromschwankungen in verschiedenen elektrizitätsleitern," *Ann. Phys.*, vol. 57, pp. 541–567, 1918.

[47] F. Hooge, "1/f noise sources," *IEEE Trans. Electron Devices*, vol. 41, no. 11, pp. 1926–1935, Nov. 1994.

[48] X. Yu, J. Thaysen, O. Hansen, and A. Boisen, "Optimization of sensitivity and noise in piezoresistive cantilevers," *J. App. Phys.*, vol. 92, no. 10, pp. 6296–6301, Jan. 2002.

[49] J. Harkey and T. Kenny, "1/f noise considerations for the design and process optimization of piezoresistive cantilevers," *J. Microelectromech. Syst.*, vol. 9, no. 2, pp. 226–235, Jun. 2000.

[50] J. Bergqvist, F. Rudolf, J. Maisano, F. Parodi, and M. Ross, "A silicon condenser microphone with a highly perforated backplate," in *Transducers'91. The 6th International Conference on Solid-State Sensors and Actuators*, San Francisco, CA, 24-27 June 1991, pp. 266–269.

[51] B. E. Boser and R. T. Howe, "Surface micromachined accelerometers," *IEEE J. Solid-State Circuits*, vol. 31, no. 3, pp. 366–375, Mar. 1996.

[52] H. Seidel and L. Csepregi, "Design optimization for cantilever-type accelerometers," *Sens. Actuators*, vol. 6, no. 2, pp. 81–92, Oct. 1984.

[53] L. K. Baxter, *Capacitive Sensors: Design and Applications.* New York, NY: IEEE Press, 1997.

[54] J. Molarius, A. Nurmela, T. Pensala, M. Ylilammi, and A. Dommann, "ZnO for thin film BAW devices," in *IEEE International Ultrasonic Symposium*, Rotterdam, Netherlands, 18-21 Sept. 2005, pp. 1816–1819.

[55] R. Ruby, P. Bradley, J. Larson III, Y. Oshmyansky, and D. Figueredo, "Ultra-miniature high-Q filters and duplexers using FBAR technology," in *ISSCC'01 Proceedings of the 2001 IEEE International Solid-State Circuits Conference*, San Francisco, CA, 5-7 Feb. 2001, pp. 120–121.

[56] J. Gualtieri, J. Kosinski, and A. Ballato, "Piezoelectric materials for acoustic wave applications," *IEEE Trans. Ultrason., Ferroelect., Freq. Contr.*, vol. 41, no. 1, pp. 53–59, Jan. 1994.

[57] H. Jaffe and D. A. Berlincourt, "Piezoelectric transducer materials," *Proc. IEEE*, vol. 53, no. 10, pp. 1372–1386, Oct. 1965.

[58] V. Bottom, *Introduction to Quartz Crystal Unit Design.* New York, NY: Van Nostrand Reinhold, 1982.

[59] B. A. Auld, *Acoustic fields and waves in solids.* Wiley, 1973.

[60] R. H. Bishop, Ed., *The Mechatronics Handbook.* CRC Press, 2002.

[61] D. DeVoe and A. Pisano, "A fully surface-micromachined piezoelectric accelerometer," in *Transducers'97. The 9th International Conference on Solid-State Sensors and Actuators*, Chicago, IL, 16-19 Jun. 1997, pp. 1205–1208.

[62] E. S. Kim and R. S. Muller, "IC-processed piezoelectric microphone," *IEEE Electron Device Lett.*, vol. 8, no. 10, pp. 467–468, Oct. 1987.

[63] R. P. Ried, E. S. Kim, D. M. Hong, and R. S. Muller, "Piezoelectric microphone with on-chip CMOS circuits," *J. Microelectromech. Syst.*, vol. 2, no. 3, pp. 111–120, Sep. 1993.

[64] S. S. Lee, R. P. Ried, and R. M. White, "Piezoelectric cantilever microphone and microspeaker," *J. Microelectromech. Syst.*, vol. 5, no. 4, pp. 238–242, Dec. 1996.

[65] N. P. Albaugh, *The Instrumentation Amplifier Handbook*.   Tucson, AZ: Burr- Brown Corporation, 2000.

[66] P. Horowitz and W. Hill, *The Art of Electronics*.   Cambridge, UK: Cambridge University Press, 1989.

[67] A. Sedra and K. Smith, *Microelectronic Circuits*, 5th ed.   New York, NY: Oxford University Press, 2003.

[68] B. Razavi, *Design of Analog CMOS Integrated Circuits*.   New York, NY: McGraw-Hill, 2001.

[69] P. Gray, P. Hurst, S. Lewis, and R. Meyer, *Analysis and Design of Analog Integrated Circuits*.   New Yotk, NY: Wiley, 2001.

[70] P. R. Gray and R. G. Meyer, "MOS operational amplifier design – a tutorial overview," *IEEE J. Solid-State Circuits*, vol. 17, no. 6, pp. 969–982, Dec. 1982.

[71] J. Tsai and G. Fedder, "Mechanical noise-limited CMOS-MEMS accelerometers," in *MEMS'05. Proceedings of the 18th IEEE International Conference on Micro Electro Mechanical Systems*, Miami, FL, 30 Jan.-3 Feb. 2005, pp. 630–633.

[72] J. Fischer, "Noise sources and calculation techniques for switched capacitor filters," *IEEE J. Solid-State Circuits*, vol. 17, no. 4, pp. 742–752, Aug. 1982.

[73] "Colibrys," 2005. [Online]. Available: http://www.colibrys.com/

[74] M. W. Judy and H. R. Samuels, "Inertial sensor," US Patent 2007/0180912 A1, Aug. 2007.

[75] V. Kaajakari, J. Kiihamäki, A. Oja, S. Pietikäinen, V. Kokkala, and H. Kuisma, "Stability of wafer level vacuum encapsulated single-crystal silicon resonators," *Sens. Actuators, A*, vol. 130-131, pp. 42–47, Aug. 2006.

[76] "SCA810-D01 single axis accelerometer with digital SPI interface," VTI Technologies, Myllynkivenkuja, P.O. Box 27, Vantaa, 01621, FINLAND, 2006. [Online]. Available: www.vti.fi

[77] "ADXL322 small and thin ±2 g accelerometer," Analog Devices, One Technology Way, P.O. Box 9106, Norwood, MA 02062, USA, 2007. [Online]. Available: www.analog.com

[78] K. Y. Yasumura, T. D. Stowe, E. M. Chow, T. Pfafman, T. W. Kenny, B. C. Stipe, and D. Rugar, "Quality factors in micron- and submicron-thick cantilevers," *J. Microelectromech. Syst.*, vol. 9, no. 1, pp. 117–125, Mar. 2000.

[79] J. Yang, T. Ono, and M. Esashi, "Surface effects and high quality factors in ultrathin single-crystal silicon cantilevers," *Appl. Phys. Lett.*, vol. 77, no. 23, pp. 3860–3862, Dec. 2000.

[80] X. Li, T. Ono, and Y. W. M. Esashi, "Study on ultra-thin NEMS cantilevers - high yield fabrication and size-effect on Young's modulus of silicon," in *MEMS'02. Proceedings of the 15th IEEE International Conference on Micro Electro Mechanical Systems*, Las Vegas, NV, 1-24 Jan. 2002, pp. 427–430.

[81] S. Pourkamali, Z. Hao, and F. Ayazi, "VHF single crystal silicon capacitive elliptic bulk-mode disk resonators – part II: Implementation and characterization," *J. Microelectromech. Syst.*, vol. 13, no. 6, pp. 1054–1062, Dec. 2004.

[82] L. Khine, M. Palaniapan, and W.-K. Wong, "12.9 MHz Lame-mode differential SOI bulk resonators," in *Transducers'07. The 14th International Conference on Solid-State Sensors, Actuators and Microsystems*, Lyon, France, 10-14 Jun. 2007, pp. 1753–1756.

[83] R. E. Mihailovich and N. C. MacDonald, "Dissipation measurements of vacuum-operated single-crystal silicon microresonators," *Sens. Actuators, A*, vol. 50, no. 3, pp. 199–207, Sep. 1995.

[84] R. Buser, "Very high Q-factor resonators in monocrystalline silicon," *Sens. Actuators*, vol. 21, no. 1, pp. 323–327, Feb. 1990.

[85] K. Numata, G. B. Bianc, M. Tanaka, S. Otsuka, K. Kawabe, M. Ando, and K. Tsubono, "Measurement of the mechanical loss of crystalline samples using a nodal support," *Phys. Lett. A*, vol. 284, no. 4-5, pp. 162–171, Jun. 2001.

[86] R. N. Candler, H. Li, M. Lutz, W. T. Park, A. Partridge, G. Yama, and T. W. Kenny, "Investigation of energy loss mechanisms in micromechanical resonators," in *Transducers'03. The 12th International Conference on Solid-State Sensors, Actuators and Microsystems*, Boston, MA, 8-12 Jun. 2003, pp. 332–335.

[87] H. Hosaka, K. Itao, and S. Kuroda, "Damping characteristics of beam-shaped micro-oscillators," *Sens. Actuators, A*, vol. 49, no. 1, pp. 87–95, Jun. 1995.

[88] Y. H. Park and K. C. Park, "High-fidelity modeling of MEMS resonators. Part I. Anchor loss mechanisms through substrate," *J. Microelectromech. Syst.*, vol. 13, no. 2, pp. 238–247, Apr. 2004.

[89] D. S. Binder, E. Quevy, T. Koyama, S. Govindjee, J. W. Demmel, and R. T. Howe, "Anchor loss simulation in resonators," in *MEMS'05. Proceedings of the 18th IEEE International Conference on Micro Electro Mechanical Systems*, Miami, FL, 30 Jan.-3 Feb. 2005, pp. 133–136.

[90] T. Mattila, J. Kiihamäki, T. Lamminmäki, O. Jaakkola, P. R. A. Oja, H. Seppä, H. Kattelus, and I. Tittonen, "12 MHz micromechanical bulk acoustic mode oscillator," *Sens. Actuators, A*, vol. 101, no. 1-2, pp. 1–9, Sep. 2002.

[91] B. Le Foulgoc, T. Bourouina, O. Le Traon, A. Bosseboeuf, F. Marty, C. Breluzeau, J. Grandchamp, and S. Masson, "Highly decoupled single-crystal silicon resonators: An approach for the intrinsic quality factor," *J. Micromech. Microeng.*, vol. 16, no. 6, pp. S45–S53, Jun. 2006.

[92] R. Holland and E. P. EerNisse, *Design of Resonant Piezoelectric Devices.* Cambridge,MA: The MIT Press, 1969.

[93] X. Liu, J. F. Vignola, H. J. Simpson, B. R. Lemon, B. H. Houston, and D. M. Photiadis, "A loss mechanism study of a very high Q silicon micromechanical oscillator," *J. App. Phys.*, vol. 97, no. 2, p. 023524, Jan. 2005.

[94] J. Kaitila, "Review of wave propagation in baw thin film devices - progress and prospects," in *IEEE International Ultrasonic Symposium*, New York, NY, 28-31 Oct. 2007, pp. 120–129.

[95] V. Kaajakari, T. Mattila, A. Oja, J. Kiihamäki, and H. Seppä, "Square-extensional mode single-crystal silicon micromechanical resonator for low phase noise oscillator applications," *IEEE Electron Device Lett.*, vol. 25, no. 4, pp. 173–175, Apr. 2004.

[96] W. Newell, "Miniaturization of tuning forks," *Science*, vol. 161, no. 3848, pp. 1320–1326, Sep. 1968.

[97] W. S. Griffin, H. H. Richardson, and S. Yamanami, "A study of fluid squeeze-film damping," *Trans. ASME J. Basic Eng.*, vol. 88, pp. 451–456, Jun. 1966.

[98] J. J. Blech, "On isothermal squeeze films," *J. Lubr. Technol.*, vol. 105, no. 4, pp. 615–620, Oct. 1983.

[99] T. Veijola, H. Kuisma, J. Lahdenperä, and T. Ryhänen, "Equivalent-circuit model of the squeezed gas film in a silicon accelerometer," *Sens. Actuators, A*, vol. 48, no. 3, pp. 239–248, May 1995.

[100] M. Bao, H. Yang, Y. Sun, and P. French, "Modified Reynolds' equation and analytical analysis of squeeze-film air damping of perforated structures," *J. Micromech. Microeng.*, vol. 13, no. 6, pp. 795–800, Nov. 2003.

[101] T. Veijola, "End effects of rare gas flow in short channels and in squeezed-film dampers," in *Proceedings of the 5th International Conference on Modeling and Simulation of Microsystems*, San Juan, PR, 2002, pp. 104–107.

[102] Y. H. Cho, A. P. Pisano, and R. T. Howe, "Viscous damping model for laterally oscillating microstructures," *J. Microelectromech. Syst.*, vol. 3, no. 2, pp. 81–87, Jun. 1994.

[103] X. Zhang, W. Tang, F. Inc, and C. Springs, "Viscous air damping in laterally driven microresonators," in *MEMS'94 Proceedings of the IEEE Workshop on Micro Electro Mechanical Systems*, Oiso, Japan, 25-28 Jan. 1994, pp. 199–204.

[104] T. Veijola and M. Turowski, "Compact damping models for laterally moving microstructures with gas-rarefaction effects," *J. Microelectromech. Syst.*, vol. 10, no. 2, pp. 263–273, Jun. 2001.

[105] J. Wang, Z. Ren, and C.-C. Nguyen, "1.156-GHz self-aligned vibrating micromechanical disk resonator," *IEEE Trans. Ultrason., Ferroelect., Freq. Contr.*, vol. 51, no. 12, pp. 1607–1628, Dec. 2004.

[106] T. Veijola, T. Tinttunen, H. Nieminen, V. Ermolov, and T. Ryhänen, "Gas damping model for a RF MEM switch and its dynamic characteristics," in *2002 IEEE International MTT-S Microwave Symposium*, Seattle, WA, 2-7 Jun. 2002, pp. 1213–1216.

[107] W. P. Eaton and J. H. Smith, "Micromachined pressure sensors: Review and recent developments," *Smart Mater. Struct.*, vol. 6, no. 5, pp. 521–539, Oct. 1997.

[108] M. Bao, *Analysis and Design Principles of MEMS Devices*. Amsterdam, Netherlands: Elsevier Science, 2005.

[109] "SCP1000 series (120 kPa) absolute pressure sensor," VTI Technologies, Myllynkivenkuja, P.O. Box 27, Vantaa, 01621, FINLAND, 2007. [Online]. Available: www.vti.fi

[110] H. E. Elgamel, "Closed-form expressions for the relationships between stress, diaphragm deflection, and resistance change with pressure in silicon piezoresistive pressure sensors," *Sens. Actuators, A*, vol. 50, no. 1-2, pp. 17–22, Aug. 1995.

[111] W. Trimmer and R. Jebens, "Actuators for micro robots," in *IEEE Int. Conf. Robotics and Automation*, Scottsdale, AZ, May 14-19 1989, pp. 1547 – 1552.

[112] D. Thompson, *On Growth and Form*. Cambridge, MA: Cambridge University Press, 1942.

[113] I. Hunter and S. Lafontaine, "A comparison of muscle with artificial actuators," in *Hilton-Head'92. Solid State Sensor, Actuator and Microsystems Workshop*, Hilton Head Island, SC, 22-25 Jun. 1992.

[114] H. Fujita and K. Gabriel, "New opportunities for microactuators," in *Transducers'91. The 6th International Conference on Solid-State Sensors and Actuators*, San Francisco, CA, 24-27 Jun. 1991, pp. 14–20.

[115] J. Comtois, M. Michalicek, and C. Barron, "Characterization of electrothermal actuators and arrays fabricated in a four-level, planarized surface-micromachined polycrystalline silicon process," in *Transducers'97. The 9th International Conference on Solid-State Sensors and Actuators*, Chicago, IL, Jun. 16-19 1997, pp. 769–772.

[116] T. Y. C., L. L. S., and R. Muller, "IC-processed electrostatic micromotors: Design, technology, and testing," in *MEMS'89 Proceedings of the IEEE Micro Electro Mechanical Systems. An Investigation of Micro Structures, Sensors, Actuators, Machines and Robots*, Salt Lake City, UT, 20-22 Feb. 1988, pp. 666–669.

[117] S. C. Jacobsen, R. H. Price, J. E. Wood, T. H. Rytting, and M. Rafaelof, "The wobble motor: An electrostatic, planetary-armature, microactuator," in *MEMS'01. Proceedings of the 14th IEEE International Conference on Micro Electro Mechanical Systems*, Salt Lake City, UT, 20-22 Feb. 1989, pp. 17–24.

[118] F. Paschen, "Ueber die zum funkenübergang in luft, wasserstoff und kohlensäure bei verschiedenen drucken erforderliche potentialdifferenz," *Ann. Phys.*, vol. 273, no. 5, pp. 69–75, 1889.

[119] E. Bazelyan and Y. P. Raizer, *Spark Discharge.* Boca Raton: CRC Press, 1998.

[120] H. Guckel, T. Christenson, K. Skrobis, T. Jung, J. Klein, K. Hartojo, and I. Widjaja, "A first functional current excited planar rotational magnetic micromotor," in *MEMS'93 Proceedings of the IEEE Micro Electro Mechanical Systems. An Investigation of Micro Structures, Sensors, Actuators, Machines and Systems*, Fort Lauderdale, FL, Feb. 7-10 1993, pp. 7–11.

[121] H. Guckel, "High-aspect-ratio micromachining via deep X-ray lithography," *Proc. IEEE*, vol. 86, no. 8, pp. 1586–1593, Aug. 1998.

[122] M. Zalalutdinov, K. L. Aubin, R. B. Reichenbach, A. T. Zehnder, B. Houston, J. M. Parpia, and H. G. Craighead, "Shell-type micromechanical actuator and resonator," *Appl. Phys. Lett.*, vol. 83, no. 18, pp. 3815–3817, Nov. 2003.

[123] A. Flynn, L. Tavrow, S. Bart, R. Brooks, D. Ehrlich, K. Udayakumar, and L. Cross, "Piezoelectric micromotors for microrobots," *J. Microelectromech. Syst.*, vol. 1, no. 1, pp. 44–51, Mar. 1992.

[124] A. Lal and R. M. White, "Silicon microfabricated horns for power ultrasonics," *Sens. Actuators, A*, vol. 54, no. 1, pp. 542–546, Jun. 1996.

[125] X. Chen and A. Lal, "Integrated pressure and flow sensor in silicon-based ultrasonic surgical actuator," in *IEEE International Ultrasonic Symposium*, Atlanta, USA, Oct. 7-10 2001, pp. 1373–1376.

[126] U. Krishnamoorthy, D. Lee, and O. Solgaard, "Self-aligned vertical electrostatic combdrives for micromirror actuation," *J. Microelectromech. Syst.*, vol. 12, no. 4, pp. 458–464, Aug. 2003.

[127] A. Selvakumar and K. Najafi, "Vertical comb array microactuators," *J. Microelectromech. Syst.*, vol. 12, no. 4, pp. 440–449, Aug. 2003.

[128] N. Szita, R. Sutter, J. Dual, and R. Buser, "A fast and low-volume pipettor with integrated sensors for high precision," in *MEMS'02. Proceedings of the 15th IEEE International Conference on Micro Electro Mechanical Systems*, Miyazaki, Japan, 23-27 Jan. 2000, pp. 409–413.

[129] A. Lal, "Silicon-based ultrasonic surgical actuators," in *The 20th Annual International Conference of the IEEE Engineering in Medicine and Biology Society*, Hong Kong, China, 29 Oct.-1 Nov. 1998, pp. 2785–2790.

[130] L. Que, J.-S. Park, and Y. Gianchandani, "Bent-beam electro-thermal actuators for high force applications," in *MEMS'99. Proceedings of the 12th IEEE International Conference on Micro Electro Mechanical Systems*, Orlando, FL, 17-21 Jan. 1999, pp. 31–36.

[131] W. Riethmüller and W. Benecke, "Thermally excited silicon microactuators," *IEEE Trans. Electron Devices*, vol. 35, no. 6, pp. 758–763, Jun. 1988.

[132] E. Kreyszig, *Advanced Engineering Mathematics*. New York, NY: Wiley, 1993.

[133] A. Geisberger, N. Sarkar, M. Ellis, and G. Skidmore, "Electrothermal properties and modeling of polysilicon microthermal actuators," *J. Microelectromech. Syst.*, vol. 12, no. 4, pp. 513–523, Aug. 2003.

[134] J. Qiu, J. H. Lang, and A. H. Slocum, "A centrally-clamped parallel-beam bistable MEMS mechanism," in *MEMS'01. Proceedings of the 14th IEEE International Conference on Micro Electro Mechanical Systems*, Interlaken, Switzerland, 21-25 Jan. 2001, pp. 353–356.

[135] H. Matoba, T. Ishikawa, C. J. Kim, and R. S. Muller, "A bistable snapping microactuator," in *MEMS'94 Proceedings of the IEEE Workshop on Micro Electro Mechanical Systems*, Oiso, Japan, 25-28 Jan. 1994, pp. 45–50.

[136] H. Kogelnik and T. Li, "Laser beams and resonators," *Appl. Opt.*, vol. 5, no. 10, pp. 1550–1567, Oct. 1966.

[137] R. S. Muller and K. Y. Lau, "Surface-micromachined microoptical elements and systems," *Proc. IEEE*, vol. 86, no. 8, pp. 1705–1720, Aug. 1998.

[138] M. H. Kiang, O. Solgaard, R. Muller, and K. Y. Lau, "Micromachined polysilicon microscanners for barcode readers," *IEEE Photon. Technol. Lett.*, vol. 8, no. 12, pp. 1707–1709, Dec. 1996.

[139] M. H. Kiang, O. Solgaard, R. S. Muller, and K. Y. Lau, "Surface-micromachined electrostatic-comb driven scanning micromirrors for barcode scanners," in *MEMS'96. Proceedings of the Workshop on Microelectromechanical Systems*, San Diego, CA, 11-15 Feb. 1996, pp. 192–197.

[140] R. Conant, J. Nee, K. Lau, and R. Muller, "A flat high-frequency scanning micromirror," in *Hilton-Head'00. Solid State Sensor, Actuator and Microsystems Workshop*, Hilton Head, SC, 4-8 Jun. 2000, pp. 6–9.

[141] M. H. Kiang, O. Solgaard, K. Y. Lau, and R. S. Muller, "Electrostatic combdrive-actuated micromirrors for laser-beamscanning and positioning," *J. Microelectromech. Syst.*, vol. 7, no. 1, pp. 27–37, Mar. 1998.

[142] "Intermec Technologies Corporation." [Online]. Available: http://www.intermec.com/

[143] "Microvision, Inc." [Online]. Available: http://www.microvision.com/barcode/mems_scanner.html

[144] R. C. Johnson, "DLP pioneer tells how TI did it with mirrors," EE Times, Jan. 2007. [Online]. Available: http://www.eetimes.com/showArticle.jhtml?articleID=196902930

[145] P. Van Kessel, L. Hornbeck, R. Meier, and M. Douglass, "A MEMS-based projection display," *Proc. IEEE*, vol. 86, no. 8, pp. 1687–1704, Aug. 1998.

[146] L. Hornbeck, "Current status of the digital micromirror device (DMD) for projection television applications," in *IEDM'93 Proceedings of the 1993 IEEE International Conference on Solid-State Circuits*, Washington, DC, 5-8 Dec. 1993, pp. 381–384.

[147] L. J. Hornbeck, "Digital light processing^{TM} for high-brightness, high-resolution applications," in *Proc. SPIE*, vol. 3013, San Jose, CA, 10-12 Feb 1997, pp. 27–40.

[148] A. Neukermans and R. Ramaswami, "MEMS technology for optical networking applications," *IEEE Commun. Mag.*, vol. 39, no. 1, pp. 62–69, Jan. 2001.

[149] P. Chu, S.-S. Lee, and S. Park, "MEMS: The path to large optical crossconnects," *IEEE Commun. Mag.*, vol. 40, no. 3, pp. 80–87, Mar. 2002.

[150] C. Marxer, P. Griss, and N. F. de Rooij, "A variable optical attenuator based on silicon micromechanics," *IEEE Photon. Technol. Lett.*, vol. 11, no. 2, pp. 233–235, Feb. 1999.

[151] K. Isamoto, K. Kato, A. Morosawa, C. Chong, H. Fujita, and H. Toshiyoshi, "A 5-V operated MEMS variable optical attenuator by SOI bulk micromachining," *IEEE J. Select. Topics Quantum Electron.*, vol. 10, no. 3, pp. 570–578, May/Jun. 2004.

[152] W. Noell, P. A. Clerc, L. Dellmann, B. Guldimann, H. P. Herzig, O. Manzardo, C. R. Marxer, K. J. Weible, R. Dandliker, and N. de Rooij, "Applications of SOI-based optical MEMS," *IEEE J. Select. Topics Quantum Electron.*, vol. 8, no. 1, pp. 148–154, Feb. 2002.

[153] "MEMS variable optical attenuators," 2005. [Online]. Available: www.ozoptics.com

[154] M. C. Wu, O. Solgaard, and J. E. Ford, "Optical MEMS for lightwave communication," *IEEE Photon. Technol. Lett.*, vol. 24, no. 12, pp. 4433–4454, Dec. 2006.

[155] W.-H. Juan and S. Pang, "High-aspect-ratio Si vertical micromirror arrays for optical switching," *J. Microelectromech. Syst.*, vol. 7, no. 2, pp. 207–213, Jun. 1998.

[156] L. Field, D. Burriesci, P. Robrish, and R. Ruby, "Micromachined 1x2 optical fiber switch," in *Transducers'05. The 13th International Conference on Solid-State Sensors, Actuators and Microsystems*, 25-29 Jun. 1995, pp. 344–347.

[157] D. S. Greywall, P. A. Busch, F. Pardo, D. W. Carr, G. Bogart, and H. T. Soh, "Crystalline silicon tilting mirrors for optical cross-connect switches," *J. Microelectromech. Syst.*, vol. 12, no. 5, pp. 708–712, Oct. 2003.

[158] "Vaisala CARBOCAP® sensor technology for stable carbon dioxide measurement," Vaisala Instruments Catalog, 2007. [Online]. Available: www.vaisala.com

[159] "IMOD technology overview (white paper)," 2007. [Online]. Available: http://www.qualcomm.com/technology/imod/index.html

[160] R. F. Pierret and K. Harutunian, *Semiconductor Device Fundamentals*. New York, NY: Addison-Wesley, 1994.

[161] A. Dec and K. Suyama, "Micromachined electro-mechanically tunable capacitors and their applications to RF IC's," *IEEE Trans. Microwave Theory Tech.*, vol. 46, no. 12, pp. 2587–2596, Dec. 1998.

[162] J. Yao, S. Park, and J. DeNatale, "High tuning ratio MEMS based tunable capacitors for RF communications applications," in *Hilton-Head'98. Solid State Sensor, Actuator and Microsystems Workshop*, Hilton Head, SC, 7-11 Jun. 1998, pp. 8–11.

[163] L. Dussopt and G. M. Rebeiz, "High-Q millimeter-wave MEMS varactors: Extended tuning range and discrete-position designs," in *2002 IEEE MTT-S International Microwave Symposium Digest*, Seattle, WA, 2-7 Jun. 2002.

[164] G. M. Rebeiz and J. B. Muldavin, "RF MEMS switches and switch circuits," *IEEE Microwave*, vol. 2, no. 4, pp. 59–71, Dec. 2001.

[165] J. Yao, "RF MEMS from a device perspective," *J. Micromech. Microeng.*, vol. 10, no. 4, pp. 9–38, Dec. 2000.

[166] C. T. C. Nguyen, L. P. B. Katehi, and G. M. Rebeiz, "Micromachined devices for wireless communications," *Proceedings of the IEEE*, vol. 86, no. 8, pp. 1756–1768, Aug. 1998.

[167] C. Goldsmith, J. Randall, S. Eshelman, T. Lin, D. Denniston, S. Chen, and B. Norvell, "Characteristics of micromachined switches at microwave frequencies," in *1996 IEEE MTT-S International Microwave Symposium Digest*, San Francisco, CA, 17-21 Jun. 1996.

[168] Z. J. Yao, S. Chen, S. Eshelman, D. Denniston, and C. Goldsmith, "Micromachined low-loss microwave switches," *J. Microelectromech. Syst.*, vol. 8, no. 2, pp. 129–134, Jun. 1999.

[169] C. Goldsmith, Z. Yao, S. Eshelman, and D. Denniston, "Performance of low-loss RF MEMS capacitive switches," *IEEE Microwave Guided Wave Lett.*, vol. 8, no. 8, pp. 269–271, Aug. 1998.

[170] G. Rebeiz, *RF MEMS: Theory, Design, and Technology.*   Hoboken, NJ: John Wiley & Sons, Inc, 2003.

[171] H. S. Newman, "RF MEMS switches and applications," in *40th Annual Reliability Physics Symposium*, Dallas, TX, 7-11 Apr. 2002, pp. 111–115.

[172] C. Goldsmith, J. Ehmke, A. Malczewski, B. Pillans, S. Eshelman, Z. Yao, J. Brank, and M. Eberly, "Lifetime characterization of capacitive RF MEMS switches," in *2001 IEEE MTT-S International Microwave Symposium Digest*, Phoenix, AZ, 20-25 May 2001, pp. 227–230.

[173] S. Majumder, J. Lampen, R. Morrison, and J. Maciel, "A packaged, high-lifetime ohmic MEMS RF switch," in *2003 IEEE MTT-S International Microwave Symposium Digest*, Philadelphia, PA, 8-13 Jun. 2003, pp. 1935–1938.

[174] K. Petersen, "Micromechanical membrane switches on silicon," *IBM J. Res. Develop.*, vol. 23, no. 4, pp. 376–385, Jul. 1979.

[175] V. Kaajakari, "Closed form expressions for RF MEMS switch actuation and release time," *Electr. Let.*, vol. 43, no. 3, pp. 149–150, Jan. 2009.

[176] "Radant MEMS, Inc." 255 Hudson Rd., Stow, MA 01557, USA. [Online]. Available: http://www.radantmems.com/

[177] V. Kaajakari, A. Alastalo, and T. Mattila, "Electrostatic transducers for micromechanical resonators: Free space and solid dielectric," *IEEE Trans. Ultrason., Ferroelect., Freq. Contr.*, vol. 53, no. 12, pp. 2484–2489, Dec. 2006.

[178] V. Kaajakari, T. Mattila, A. Oja, and H. Seppä, "Nonlinear limits for single-crystal silicon microresonators," *J. Microelectromech. Syst.*, vol. 13, no. 5, pp. 715–724, Oct. 2004.

[179] M. Agarwal, S. Chandorkar, R. Candler, B. Kim, M. Hopcroft, R. Melamud, C. Jha, T. Kenny, and B. Murmann, "Optimal drive condition for nonlinearity reduction in electrostatic MEMS resonators," *Appl. Phys. Lett.*, vol. 89, no. 21, p. 214105, Nov. 2006.

[180] Epson Toyocom, "About QMEMS (Quartz MEMS)," Datasheet, 421-8 Hino, Hino-shi, Tokyo 191-8501, Japan, 2007. [Online]. Available: http://www.epsontoyocom.co.jp/english/info/2006/qmems.html

[181] ——, "FC-135 kHz range crystal unit," Datasheet, 421-8 Hino, Hino-shi, Tokyo 191-8501, Japan, 2007. [Online]. Available: http://www.epsontoyocom.co.jp

[182] W. Newell and R. Wickstrom, "The tunistor: A mechanical resonator for microcircuits," *IEEE Trans. Electron Devices*, vol. 16, no. 9, pp. 781–787, Sep 1969.

[183] W. Newell, R. Wickstrom, and D. Page, "Tunistors, mechanical resonators for microcircuits," *IEEE Trans. Electron Devices*, vol. 15, no. 6, pp. 411–412, Jun 1968.

[184] C. Nguyen and R. Howe, "An integrated CMOS micromechanical resonator high-Q oscillator," *IEEE J. Solid-State Circuits*, vol. 34, no. 4, pp. 440–455, Apr. 1999.

[185] K. Cioffi and W. Hsu, "32KHz MEMS-based oscillator for low-power applications," in *Proceedings of the 2005 IEEE International Frequency Control Symposium and Exposition*, Vancouver, Canada, 3-5 Oct. 2005.

[186] P. Rantakari, V. Kaajakari, T. Mattila, J. Kiihamäki, A. Oja, I. Tittonen, and H. Seppä, "Low noise, low power micromechanical oscillator,"

in *Transducers'05. The 13th International Conference on Solid-State Sensors, Actuators and Microsystems*, Seoul, Korea, 5-9 Jun. 2005, pp. 2135–2138.

[187] SiTime, "SiT0100, 5.1 MHz MEMS resonator die," Datasheet, 2007. [Online]. Available: www.sitime.com

[188] K. Petersen, "Sitime: Going from 0 to 3 million units in less than 3 years," Short course, 2007.

[189] Epson Toyocom, "SG-150 series crystal oscillator," Datasheet, 2007. [Online]. Available: http://www.epsontoyocom.co.jp

[190] SiTime, "SiT1 fixed frequency oscillator," Datasheet, 2007. [Online]. Available: www.sitime.com

[191] S. Pourkamali and F. Ayazi, "Electrically coupled MEMS bandpass filters: Part I: With couplingelement," *Sens. Actuators, A*, vol. 122, no. 2, pp. 307–316, Aug. 2005.

[192] L. Lin, R. T. Howe, and A. P. Pisano, "Microelectromechanical filters for signal processing," *J. Microelectromech. Syst.*, vol. 7, no. 3, pp. 286–294, Sep. 1998.

[193] C. T.-C. Nguyen, "Frequency-selective mems for miniaturized low-power communication devices," *IEEE Trans. Microwave Theory Tech.*, vol. 47, no. 8, pp. 1486–1503, Aug. 1999.

[194] M. U. Demirci and C. T.-C. Nguyen, "Mechanically corner-coupled square microresonator array for reduced series motional resistance," *J. Microelectromech. Syst.*, vol. 15, no. 6, pp. 1419–1436, Dec. 2006.

[195] G. Piazza, P. Stephanou, J. Porter, M. Wijesundara, and A. Pisano, "Low motional resistance ring-shaped contour-mode aluminum nitride piezoelectric micromechanical resonators for UHF applications," in *MEMS'05. Proceedings of the 18th IEEE International Conference on Micro Electro Mechanical Systems*, Miami, FL, 30 Jan.-3 Feb. 2005, pp. 20–23.

[196] A. Persson, "How do we understand the Coriolis force," *Bull. Am. Meteorolog. Soc.*, vol. 79, no. 7, pp. 1373–1385, Jul. 1998.

[197] P. Greiff, B. Boxenhorn, T. King, and L. Niles, "Silicon monolithic micromechanical gyroscope," in *Transducers'91. The 6th International Conference on Solid-State Sensors and Actuators*, San Francisco, CA, 24-27 Jun. 1991, pp. 966–968.

[198] K. Tanaka, Y. Mochida, S. Sugimoto, K. Moriya, T. Hasegawa, K. At-suchi, and K. Ohwada, "A micromachined vibrating gyroscope," in *MEMS'95. Proceedings of the 8th Annual International Workshop on Micro Electro Mechanical Systems*, Amsterdam, NL, 29 Jan.-2 Feb. 1995, pp. 278–281.

[199] J. Bernstein, S. Cho, A. King, A. Kourepenis, P. Maciel, and M. Weinberg, "A micromachined comb-drive tuning fork rate gyroscope," in *MEMS'93 Proceedings of the IEEE Micro Electro Mechanical Systems. An Investigation of Micro Structures, Sensors, Actuators, Machines and Systems*, Fort Lauderdale, FL, 7-10 Feb 1993, pp. 143–148.

[200] M. Lutz, W. Golderer, J. Gerstenmeier, J. Marek, B. Maihofer, S. Mahler, H. Munzel, and U. Bischof, "A precision yaw rate sensor in silicon micromachining," in *Transducers'97. The 9th International Conference on Solid-State Sensors and Actuators*, Chicago, IL, 16-19 Jun. 1997, p. 847850.

[201] M. S. Weinberg and A. Kourepenis, "Error sources in in-plane silicon tuning-fork MEMS gyroscopes," *J. Microelectromech. Syst.*, vol. 15, no. 3, pp. 479–491, Jun. 2006.

[202] W. A. Clark, R. T. Howe, and R. Horowitz, "Surface micromachined Z-axis vibratory rate gyroscope," in *Solid-State Sensor and Actuator Workshop*, Hilton Head Island, SC, 2-6 Jun. 1996, pp. 283–287.

[203] J. A. Geen and D. W. Carow, "Micromachined gyros," US Patent 6,505,511 B1, Jan. 2003.

[204] J. Geen, S. Sherman, J. Chang, and S. Lewis, "Single-chip surface-micromachined integrated gyroscope with 50°/hour root Allan variance," in *ISSCC'02 Proceedings of the 2002 IEEE International Solid-State Circuits Conference*, San Francisco, CA, 3-7 Feb. 2002, pp. 346–539.

[205] M. Saukoski, L. Aaltonen, T. Salo, and K. Halonen, "Readout and control electronics for a microelectromechanical gyroscope," in *IMTC 2006. Proceedings of the IEEE Instrumentation and Measurement Technology Conference*, Sorrento, Italia, 24 - 27 Apr. 2006, pp. 1741–1746.

[206] J. A. Geen, S. J. Sherman, J. F. Chang, and S. R. Lewis, "Single-chip surface micromachined integrated gyroscope with 50°/h Allan deviation," *IEEE J. Solid-State Circuits*, vol. 37, no. 12, pp. 1860–1866, Dec. 2002.

[207] "±150°/s single chip yaw rate gyro with signal conditioning," Datasheet, Analog Devices, One Technology Way, P.O. Box 9106, Norwood, MA 02062, USA, 2004. [Online]. Available: www.analog.com

[208] R. D. Geddes and A. M. Madni, "A micromachined quartz angular rate sensor for automotive and advanced inertial applications," *Sensors*, pp. 26–34, Aug. 1999.

[209] A. M. Madni, L. E. Costlow, and S. J. Knowles, "Common design techniques for BEI GyroChip quartz rate sensors for both automotive and aerospace/defense markets," *IEEE Sensors J.*, vol. 3, no. 5, pp. 569–578, Oct. 2003.

[210] S. Fujishima, T. Nakamura, and K. Fujimoto, "Piezoelectric vibratory gyroscope using flexural vibration of a triangular bar," in *Proceedings of the 1991 IEEE International Frequency Control Symposium and Exposition*, Los Angeles, CA, 29-31 May 1991, pp. 261–265.

[211] "Piezoelectric vibrating gyroscopes (GYROSTAR)," Murata, 10-1, Higashikotari 1-chome, Nagaokakyo-shi, Kyoto 617-8555, Japan, 2006. [Online]. Available: www.murata.com

[212] J. Geen, "Very low cost gyroscopes," in *Proceedings of 2005 IEEE Sensors*, Irvine, CA, 30 Oct.-3 Nov. 2005, pp. 537–540.

[213] S. Terry, J. Jerman, and J. Angell, "A gas chromatographic air analyzer fabricated on a silicon wafer," *IEEE Trans. Electron Devices*, vol. 26, no. 12, pp. 1880–1886, Dec. 1979.

[214] K. Petersen, "Fabrication of an integrated, planar silicon ink-jet structure," *IEEE Trans. Electron Devices*, vol. 26, no. 12, pp. 1918–1920, Dec. 1979.

[215] L. Jiang, J. Mikkelsen, J. M. Koo, D. Huber, S. Yao, L. Zhang, P. Zhou, J. G. Maveety, R. Prasher, J. G. Santiago, T. W. Kenny, and K. E. Goodson, "Closed-loop electroosmotic microchannel cooling system for VLSI circuits," *IEEE Trans. Comp. Packag. Technol.*, vol. 25, no. 3, pp. 347–355, Sep. 2002.

[216] D. J. Laser and J. G. Santiago, "A review of micropumps," *J. Micromech. Microeng.*, vol. 14, no. 6, pp. 35–64, Jun. 2004.

[217] J. Voldman, M. L. Gray, and M. A. Schmidt, "Microfabrication in biology and medicine," *Annu. Rev. Biomed. Eng.*, vol. 1, no. 1, pp. 401–425, Aug. 1999.

[218] P. Gravesen, J. Branebjerg, and O. S. Jensen, "Microfluidics – a review," *J. Micromech. Microeng.*, vol. 3, no. 4, pp. 168–182, Dec. 1993.

[219] D. Duffy, J. McDonald, O. Schueller, and G. Whitesides, "Rapid proto-typing of microfluidic systems in poly (dimethylsiloxane)," *Anal. Chem.*, vol. 70, no. 23, pp. 4974–4984, Dec. 1998.

[220] D. Erickson and D. Li, "Integrated microfluidic devices," *Anal. Chim. Acta*, vol. 507, no. 1, pp. 11–26, Apr. 2004.

[221] A. Manz, N. Graber, and H. M. Widmer, "Miniaturized total chemical analysis systems: a novel concept for chemical sensing," *Sens. Actuators, B*, vol. 1, no. 1-6, pp. 244–248, Jan. 1990.

[222] D. Beebe, G. Mensing, and G. Walker, "Physics and applications of mi-crofluidic in biology," *Annu. Rev. Biomed. Eng.*, vol. 4, no. 1, pp. 261–286, Aug. 2002.

[223] G. Yarnold, "The motion of a mercury index in a capillary tube," *Proc. Phys. Soc.*, vol. 50, no. 4, pp. 540–552, Jul. 1938.

[224] N.-T. Nguyen and S. T. Wereley, *Fundamentals and Applications of Mi-crofluidics.* Norwood, MA: Artech House, 2006.

[225] H. Jerman, "Electrically-activated, micromachined diaphragm valves," in *Hilton-Head'90. Solid-State Sensor and Actuator Workshop*, Hilton Head Island, SC, 4-7 Jun. 1990, pp. 65–69.

[226] H. T. G. van Lintel, F. C. M. van de Pol, and S. Bouwstra, "A piezoelectric micropump based on micromachining of silicon," *Sens. Actuators*, vol. 15, no. 2, pp. 153–167, Oct. 1988.

[227] J. R. Webster, M. A. Burns, D. T. Burke, and C. H. Mastrangelo, "Mono-lithic capillary electrophoresis device with integrated fluorescence detec-tor," *Anal. Chem.*, vol. 73, no. 7, pp. 1622–1626, Apr. 2001.

[228] T. Squires and S. Quake, "Microfluidics: Fluid physics at the nanoliter scale," *Rev. Mod. Phy.*, vol. 77, no. 3, pp. 977–1026, Jul. 2005.

[229] R. M. Moroney, R. M. White, and R. T. Howe, "Ultrasonically induced microtransport," in *MEMS'01. Proceedings of the 14th IEEE Interna-tional Conference on Micro Electro Mechanical Systems*, Nara, Japan, 30 Jan. - 2 Feb. 1991, pp. 277–282.

[230] R. B. Fair, V. Srinivasan, H. Ren, P. Paik, V. K. Pamula, and M. G. Pollack, "Electrowetting-based on-chip sample processing for integrated microfluidics," in *IEDM'03 Proceedings of the 2003 IEEE International Conference on Solid-State Circuits*, Washington, DC, 8-10 Dec. 2003, pp. 32.5.1–32.5.4.

[231] M. G. Pollack, R. B. Fair, and A. D. Shenderov, "Electrowetting-based actuation of liquid droplets for microfluidic applications," *Appl. Phys. Lett.*, vol. 77, no. 11, pp. 1725–1726, sep 2000.

[232] K. Petersen, W. McMillan, G. Kovacs, A. Northrup, L. Christel, and F. Pourahmadi, "The promise of miniaturized clinical diagnostic systems," *IVD Technol.*, vol. 4, no. 4, pp. 43–49, Jul. 1998.

[233] D. de Vries, P. Semicond, and F. Crolles, "Investigation of gross die per wafer formulas," *IEEE Trans. Semiconduct. Manufact.*, vol. 18, no. 1, pp. 136–139, Feb. 2005.

[234] A. Franke, J. Heck, T. King, and R. Howe, "Polycrystalline silicon-germanium films for integrated microsystems," *J. Microelectromech. Syst.*, vol. 12, no. 2, pp. 160–171, Apr. 2003.

[235] H. Xie and G. Fedder, "A CMOS-MEMS lateral-axis gyroscope," in *MEMS'01. Proceedings of the 14th IEEE International Conference on Micro Electro Mechanical Systems*, Interlaken, Switzerland, 21-25 Jan. 2001, pp. 162–165.

[236] "VTI Technologies Oy's consolidated financial statements for 2007," VTI Technologies, Myllynkivenkuja, P.O. Box 27, Vantaa, 01621, FINLAND, May 2008. [Online]. Available: http://www.vti.fi/en/news-events/press-releases/view/1210145036.html

[237] L. D. Landau and E. M. Lifshitz, *Mechanics*. Oxford: Butterworth-Heinemann, 1999.

[238] J. J. Wortman and R. A. Evans, "Young's modulus, shear modulus, and Poisson's ratio in silicon and germanium," *J. App. Phys.*, vol. 36, no. 1, pp. 153–156, Jan. 1965.

[239] K. Tsubouchi, K. Sugai, and N. Mikoshiba, "AlN material constants evaluation and SAW properties on AlN/Al_2O_3 and AlN/Si," in *IEEE International Ultrasonic Symposium*, Chicago, IL, 14-16 Oct. 1981, pp. 375–380.

Index

CPSIA information can be obtained
at www.ICGtesting.com
Printed in the USA
LVHW060825280122
709465LV00017B/44

9 780982 299104